The American Forestry Series

HENRY J. VAUX, *Consulting Editor*

ALLEN An Introduction to American Forestry
BAKER Principles of Silviculture
BAKER The Theory and Practice of Silviculture
BOYCE Forest Pathology
BROWN, PANSHIN, AND FORSAITH Textbook of Wood Technology: *Volume I,* Structure, Identification, Defects, and Uses of the Commercial Woods of the United States; *Volume II,* The Physical, Mechanical, and Chemical Properties of the Commercial Woods of the United States
BRUCE AND SCHUMACHER Forest Mensuration
CHAPMAN AND MEYER Forest Mensuration
CHAPMAN AND MEYER Forest Valuation
CLAWSON The Western Range Livestock Industry
DOANE, VAN DYKE, CHAMBERLIN, AND BURKE Forest Insects
GRAHAM Forest Entomology
GUISE The Management of Farm Woodlands
HARLOW AND HARRAR Textbook of Dendrology
KITTREDGE Forest Influences
MARQUIS Economics of Private Forestry
MATTHEWS Cost Control in the Logging Industry
MATTHEWS Management of American Forests
PANSHIN, HARRAR, BAKER, AND PROCTOR Forest Products
PRESTON Farm Wood Crops
SHIRLEY Forestry and Its Career Opportunities
STODDART AND SMITH Range Management
TRIPPENSEE Wildlife Management
WACKERMAN Harvesting Timber Crops

Walter Mulford was Consulting Editor of this series from its inception in 1931 until January 1, 1952.

Forestry means the growing of timber crops and their use for human welfare. (Edward Hines Lumber Company. A.F.P.I. photo.)

Forestry also means managing land to protect soil, water, scenic grandeur, and wildlife for man's benefit and pleasure. Butte Lake, Lassen Volcanic National Park. (National Park Service photo.)

The American Forestry Series
HENRY J. VAUX, *Consulting Editor*

FORESTRY AND ITS CAREER OPPORTUNITIES

Forestry and Its Career Opportunities

HARDY L. SHIRLEY
Dean and Professor of Forestry
State University of New York
College of Forestry
at Syracuse University

FIRST EDITION

McGRAW-HILL BOOK COMPANY, INC.
1952 New York Toronto London

FORESTRY AND ITS CAREER OPPORTUNITIES

Library of Congress Catalog Card Number: 52-6001

Preface

THE SECOND WORLD WAR marked the end of an epoch for citizens of the United States. No longer could the individual American stand four-square before humanity and declare himself to be unconcerned with the vicissitudes that beset the rest of mankind. No longer could he "look the whole world in the face because he owes not any man." He realized that for good or ill his own fate is inextricably interwoven with the fate of humanity throughout the world, that the bountiful resources his own country possesses and the prodigal use he has made of these in the past will not save him in the future. No open frontier lies to his west and no unexploited or unoccupied section of the world remains that he may colonize or exploit. Henceforth his economy and his welfare must depend upon the skill with which he develops and utilizes those resources which his own land can permanently supply. He may trade with the rest of mankind, but this trade must be on a mutually helpful basis.

The purpose of this book is to present a broad picture of forestry in relation to its historic development and its place in national and world economy. Technical forestry is treated lightly, as good books covering the several phases individually are available. This book is written to help beginning students to understand the many facets of forestry and to give them background for deciding if forestry is to be their life career. It should also prove helpful to vocational-guidance counselors and the general reader interested in the scope and magnitude of forestry.

The book has been developed from lecture notes and other materials used in a course in general forestry for freshmen at the State University of New York, College of Forestry.

The concept of forestry as developed through the programs of instruction offered at Syracuse is broad. It embraces the management of forest lands and the use of their products and services by mankind. No apology is made for such broad treatment in this book. Rather, the effort is made to explore the several special fields of forestry and their interrelationship.

Many people have made substantial contributions to this book. Most helpful have been the instructors who assisted the author in general forestry. These include John W. Barrett, William H. Bennett, Alfred H. Bishop, David H. Caldwell, James P. Casey, S. Earll Church, Harold A. Core, C. Eugene Farnsworth, Harold B. Gatslick, Robert S. Hitchings, Jacob L. Kovner, Jack L. Krall, Charles C. Larson, Lawrence Leney, John L. Morrison, Harry S. Mosebrook, Lucian P. Plumley, Stanley L. Pringle, Dean B. Richards, Edward G. Roberts, John C. Sammi, Nelson T. Samson, Gerald H. Smith, Bruce T. Stanton, LeRoy C. Stegeman, David P. Thomas, Robert B. Williams, and Aubrey E. Wylie.

Other faculty members who furnished material and reviewed portions or all of the manuscript include Maurice M. Alexander, Harry P. Brown, Nelson C. Brown, Paul F. Graves, Svend O. Heiberg, Joseph S. Illick, Ralph T. King, Edward H. McCarthy, James M. Owens, and Robert A. Zabel.

Sections of the manuscript have also been reviewed by these members of the U.S. Forest Service: C. Edward Behre, H. Dean Cochran, Lynne M. Correll, Crosby Hoar, H. R. Josephson, William P. Kramer, W. H. Larrimer, J. W. McNair, Fred H. Miller, James Rettie, and Lloyd W. Swift.

U.S. Department of the Interior reviewers include L. D. Arnold, L. F. Cook, Arthur A. Davis, Richard C. Griffith, W. H. Horning, Floyd D. Larson, and John F. Shanklin.

Other reviewers who have furnished helpful criticisms include Roy C. Brundage, Purdue University; Henry Clepper, Society of American Foresters; Gordon M. Day, Rutgers University; Otis F. Hall, University of Minnesota; R. D. Hostetter, Society of American Foresters; Frank Kaufert, University of Minnesota; E. R. Martell, Purdue University; James C. McClellan, American Forest Products Industries.

Valuable suggestions and material were supplied by many reviewers, especially those in Federal agencies. Professor Douglas G. Haring of Syracuse University reviewed Chap. 2 and furnished material on primitive man. Instructors at the College of Forestry helping with specific chapters include John W. Barrett, Chap. 7; Gerald H. Smith, Chaps. 8 and 9; David P. Thomas, Chap. 9; James P. Casey, S. Earll Church, Edwin C. Jahn, C. Earl Libby, Chap. 10; George J. Albrecht, Robert B. Williams, Chap. 13; C. Eugene Farnsworth, Charles C. Larson, Chap. 25.

The book in preliminary form has been used two years in the classrooms at the State University of New York, College of Forestry. Following extensive review by those mentioned above, the entire manuscript was rewritten. The final draft has been critically reviewed by Dr. Robert A. Zabel of the College of Forestry faculty.

To all of these men the author expresses his deepest gratitude. Final responsibility for material included, presentation, and opinions expressed rests with the author.

He is grateful for use of photographs furnished by the U.S. Forest Service, the Bureau of Land Management, the U.S. Park Service, the American Forest Products Industries, Timber Engineering Company, and *American Forests.* Unacknowledged photographs were supplied by the State University, College of Forestry, or the author.

The author expresses his high appreciation to the college staff. The librarians helped in locating material and checking references. These and other staff members have worked beyond the call of normal duty to expedite this publication.

To Genevieve Reidy, his efficient secretary, he owes unbounded appreciation for intelligent help with the countless tasks involved in collecting material and preparing the manuscript and index.

Hardy L. Shirley

Syracuse, N.Y.
July, 1952

Contents

Preface		vii
1	The Meaning of Forestry	1
2	Forestry throughout the Ages	8
3	The Development of Forestry in the United States	17
4	Forest Regions	44
5	Protecting Forests	72
6	Forest Care and Use	96
7	Harvesting Timber Crops	120
8	Lumber Manufacture and Sale	133
9	Wood, Its Nature, and Uses	154
10	Wood Chemistry, Paper, and Plastics	177
11	Secondary Forest Products	195
12	Wildlife and Range Management	210
13	Forests and Recreation	235
14	Protecting Soil and Watersheds	249
15	The Business of Forestry	260
16	Social Benefits of Forestry	273
17	World and National Forest Resources	287
18	Recent Programs and Progress	310
19	Proposed Programs of Forestry	329
20	The Profession of Forestry	341
21	Employers of Foresters	354
22	Forest Employment by Regions and Fields	388
23	Education in Forestry	412
24	Research in Forestry	427
25	Administration in Forestry	440
26	Trends and Outlook in Forestry	456
Index		469

CHAPTER 1. *The Meaning of Forestry*

FEW PROFESSIONS in the United States have grown as rapidly during the past half century as forestry. This growth has been both in numbers and in scope of activities. Starting in 1898 with the establishment of the first schools of forestry, the profession grew from less than 10 to more than 13,000 members by 1950. In 1900 seven men organized the Society of American Foresters as the professional body to represent all foresters. By 1950 this society had 7,000 members.

Following its European tradition, the profession at first dealt with the technical phases of growing crops of timber and protecting steep mountain slopes from erosion. Transplanted to America, it rapidly outgrew this concept to include the management of all biologic resources of forest land. Gradually it expanded even further to embrace the management and use of forest land for the many products and services the forest can permanently provide.

The scope of forestry in the United States in 1951 is indicated by the specialized curriculums offered in some of the 34 colleges of forestry. The curriculums include professional training in forest-land management, range management, landscape and recreation management, wildlife management, logging engineering, forest-products manufacture and use, retail merchandising of lumber and the light-construction industry, wood technology, furniture manufacture, pulp and paper manufacture, wood chemistry, and wood-based plastics. The Society of American Foresters through its members has sponsored The Society of Wildlife Management, The Society of Range Management, and The Forest Products Research Society.

Opinions differ sharply on how many of these activities rightly belong in professional forestry. No sharp dividing line exists. The forest landowner may profitably lease summer cottages and sell water and hunting and fishing rights. He may do his own logging, operate a sawmill, and sell lumber at wholesale and retail. Large companies such as the Weyerhaeuser Company do most of the above and also manufacture veneer, plywood, pulp, paper, various sawmill by-products, prefabricated houses, and a host of other products dependent upon trees and their technical properties.

At mid-century in America it is no longer appropriate to consider the forest apart from the industries that look to it for wood, the communities that use its water, or the people who harvest and process its crops. We can expect, therefore, no sharp delineation of the profession of forestry. Like other professions it is likely to drop some activities and to accept new ones as future necessity dictates.

A sharper picture of forestry can be gained by defining some of the terms and concepts foresters use.

Forest Land

To the forester, forest land includes not only land that is covered with trees but land which finds its highest use and service to man in supporting woody vegetation. This concept embraces not only what the land supports today but also what it may support tomorrow if put to its best use.

Land may be considered to be forest land only if climate, topography, and soil are favorable to tree growth. Land with climates unfavorable to trees may be classed as tundra, prairie, desert scrub, savanna, or park land but not as forest land. The more favorable the climate for vigorous tree growth, the more valuable the land for forest purposes, provided soil and other factors remain favorable.

Climate includes such factors as sunlight, rainfall, air movement and temperature, all of which affect the rate of tree growth.

Topography determines the ease with which soil can be formed and held in place. Mountain peaks above timber line, watercourses, lakes, open swamps, and bare rock faces are not forest land, though they may be included within a large forest area. Rugged topography and steepness of slope may preclude harvesting commercial timber from mountains, but such land if tree-clad is not excluded for this reason from being classed as forest land. Watercourses and lakes are a result of climate and topography and are influenced greatly by the forest, which helps to regulate ground-water supply. Streams may provide transportation systems for forest products, and they oftentimes attract recreationists. The scenic and recreational attractions of forests are related to the land, and from the forester's point of view are an important feature of the forest as a whole.

Soil affects the growth rate, composition, and to a large extent the value of the forest. The forester is concerned with the origin of the soil as this affects its depth and texture. He considers the natural drainage and the presence or absence of impervious layers near the surface. He is interested in the tilth of the soil, which determines how rapidly it will soak up water and how easily it may be penetrated by roots. He is concerned with its humus content and acidity, which determine whether or not it is subject to leaching and possible hardpan formation, and with the mineral nutrient content that is essential for tree growth.

Tree

A tree is defined as "a woody plant having one well-defined stem and a more or less definitely formed crown, usually attaining a height of at least 8 feet."[1*] A tree should be distinguished from a shrub, which is defined as "a woody perennial plant differing from a perennial herb by its persistent and woody stems, and from a tree by its low stature and habit of branching from the base."[1]

Trees differ in kind, size, growth rate, shape, quality of wood, growth requirements, and in susceptibility to fire, insects, disease, and storm damage. The United States has 918 species of trees, 174 varieties, and 89 hybrids. Of these more than 160 have commercial value.

Forest

To a hunter, the forest is a source of game and the keen enjoyment of pursuing it. To a logger, it is a source of timber. To an ecologist, it is nature's grandest biologic community (Fig. 1). To a water user, it is a regulator of stream flow. To an artist, it is a source of creative inspiration.

To a forester a forest means what it means to hunter, logger, ecologist, water user, and artist, and very much more. Simply stated, the forester's concept includes the trees, everything that affects their growth and value, and all that forests yield of any worth to man (Frontispiece *a* and *b*). A forest is defined as "a plant association predominantly of trees and other woody vegetation occupying an extensive area of land."[1]

The Community Concept of the Forest

The forest visitor sees particular trees of outstanding interest and form. He may notice individual animals, individual wildflowers, individual edible mushrooms, and other individual examples of life that occur in the forest. To him these trees, flowers, mushrooms, and animals seem to be individuals very much as he is an individual. He may think of them as each living its independent life as he in turn lives his independent life.

A man is an individual, but he is also a part of some community. The community in turn is a part of a larger economic and social unit in which many individuals play a part and in which all are more or less inter-

* Superior figures are used to refer to items listed under Literature Cited at the end of each chapter.

dependent. So it is in the forest. The trees are to a certain extent mutually dependent upon one another (Fig. 1). But they also compete for soil, water, and light. Each tree depends upon its neighbor for shelter from wind, for protecting the soil around its roots, and for providing litter and shade to make the soil suitable as a home for small plants and animals that in turn perform useful functions in the life of the forest as a whole (Fig. 2). Seed-eating birds and mammals transport seed. Red squirrels hoard pine cones and acorns.

After performing their function on the tree, hardwood leaves drop to the ground where they become food for earthworms, which chew up the leafy material and pass it through their bodies. The castings, rich in humus, are further worked over by microorganisms that break down the

FIG. 1. Virgin Douglas-fir, Rogue River National Forest, Ore. (*U.S. Forest Service photo.*)

FIG. 2. The forest is a community of trees, shrubs, and herbs, and the animal life these support. A large beaver house on Deer Pond, Huntington Forest, N.Y.

cellulose and lignin fractions, releasing plant nutrients from the humus. Nitrogen, calcium, phosphorus, potassium, iron, magnesium, and other plant nutrients are thus made available for reuse by the trees to form leaves, twigs, stems, and roots.

The Dynamic Forest

Just as human communities change from year to year, so also do forests. Ordinarily this change is slow. Trees crowd each other, and one by one the weaker die. Shrubs and other vegetation wax or wane as the tree canopy opens or closes. Trees may slowly invade an abandoned eastern pasture. First, hawthorn, redcedar, or gray birch gain a foothold. These may later be followed by ash, red maple, and cherry, which form a denser canopy and eventually crowd out the pioneer invaders. Gradually these in turn give way to sugar maple, beech, yellow birch, and hemlock that form the climax forest.

Even the climax forest is far from static. Examined closely, it is often found to consist of groups of trees more or less uniform in age. The history of such forests reveals that most of the trees owe their origin to forest catastrophes of the past, such as severe droughts, fires, windstorms, insect invasions, or tree diseases. Examples of such natural catastrophes are the New England hurricane of 1938, the New York windstorm of

1950, that felled almost 2 million cords of wood, bark beetle outbreaks in the western states, and the spruce budworm outbreaks in Canada and the northern United States.

Forest Products

Forest products may be defined as those useful commodities that may be removed from the forest for the benefit of man. Of greatest financial importance are such primary wood products as saw logs, fuel wood, pulpwood, veneer logs, fence posts, and other items that may be cut from the stems of trees. Important also are gums for naval stores and other products, bark for tannin, maple sugar and syrup, Christmas trees and greens, various oils and waxes, nuts, fruits, charcoal, wood chemicals, pharmaceuticals, and other vegetable products. Forage is a product of some forests that is harvested by domestic animals and game. Game also is a product of the forest. Together forest products directly and indirectly support a large segment of the American economy.

Forest Services

The forest provides a number of services to man in addition to these products. It builds new soil and holds old soil in place. The tree canopy, leaf litter, and forest humus break the force of falling rain and permit it readily to enter the porous forest soil and recharge the underground storage reservoirs. From these it is gradually discharged, cool and clear, suitable for irrigation, industrial, and domestic use. The forest ameliorates local climate. It protects the soil from the heating and drying of midday sun and from rapid cooling by earth radiation at night. Wind velocity is greatly checked; consequently evaporation is slackened. Aside from producing game and fur, the forest provides a home for many useful nongame birds and animals that destroy insects and add to man's pleasure.

Finally forests provide recreation for man. Millions of vacationers annually seek the forests to camp, picnic, hike, take pictures, fish, hunt, and especially to enjoy the relaxation and peace of mind this natural setting invokes.

Definition of Forestry

Forestry is the management of forest lands for the many products and services the forest can permanently supply, and the sustained use of these products and services for the benefit of mankind. Or more briefly, forestry is the management and use of forests and their products for permanent human benefit.

Two points in the above definition need emphasis: (1) Forestry involves multiple use, embracing all renewable products and services of forest land; (2) forestry involves planned, sustained use at a uniformly high level. As for timber use, forestry involves the regulation of the forest for the proper distribution of age classes, which makes possible harvests of roughly constant timber volume each year; the prompt provision of new trees to replace those cut; and the protection of the forest so that productivity may not be lowered. Harvesting secondary products is also a function of forestry. This should be done on a sustained-yield basis and include all products that it is economically feasible to use, provided such use does not destroy other values that outweigh those realized. As for game and wildlife, forestry involves maintaining a vigorous population and harvesting surplus animals by hunting and trapping. In the field of recreation it means regulating other uses so as to minimize their interference with public enjoyment of the forest. Basic to all these, forestry means protecting the soil that supports all other uses.

But forestry does not end in the forest. It includes harvesting timber and other forest crops, transporting these to primary processing plants, and the manufacture of lumber, veneer, and other products. Such products often undergo secondary processing such as seasoning, preserving, planing, laminating, gluing, and even fabrication into household furniture. No sharp break in the chain of care and use exists from the man who protects the young seedling to the man who sets a crosstie, builds a house, or finishes a television cabinet. Throughout the chain of use, knowledge of trees and wood is essential to efficient use, and this knowledge is the forester's contribution.

LITERATURE CITED

1. Hawley, Ralph C., chairman. 1944. "Forestry Terminology," Society of American Foresters, Washington, D.C. 84 pp.

CHAPTER 2. *Forestry throughout the Ages*

THE ATTITUDES of man toward the forest and the laws he has established to govern its use have their roots deep in man's biologic needs and social make-up. Rights of use and land-ownership patterns are conditioned by practices dating back beyond the dawn of recorded history.

Primitive Man and the Forest

Prehistoric man's relation to the forest can only be surmised from the relics he has left and from the use primitive tribes today make of forests. The following is an attempt to weave certain known facts into a plausible account of this relationship.

The earliest known men were ground dwellers rather than tree dwellers. But they did use wood. Both the Peking man and Java man used wood for fire 750,000 or more years ago.

In the forest primitive man found shade from the heat of the midday sun. He found it warmer to sleep in the forest at night than under the open sky. Beneath leafy crowns, under leaning trunks, and in hollow trees, he found shelter against rain. Fallen trunks provided a reflector for his fire and a shelter against wind. He learned to use the berries, nuts, roots, the fruits of the forest, and the forest-dwelling animals as food to sustain his life and as medicines to soothe him when ill. He learned to use wood directly for his tools, weapons, and traps. Trees also provided means of escape from aroused animals for which man with his rude weapons was no match.

The earliest known organized communities were found along watercourses that flowed through arid lands in Egypt, Mesopotamia, and India. Here sod huts were erected and cultivation of heavy seeded grasses, the cereals, began. Scattered trees and shrubs probably occurred along the river banks and furnished fuel, handles for tools, and timber for construction. As these were used up, man probably sent expeditions farther and farther upstream to obtain wood, which could be floated down to his villages.

Cereals were the basic element which permitted organized society to develop. They provided a concentrated food, easily cultivated and capable of storage. The river borders that supported grasses attracted birds and animals useful for food and man later learned to domesticate animals as well as plants. Thus a civilization arose.

Early separation of the forest dweller and plains dweller occurred. The former lived a more primitive life but developed considerable skill in use of his forest environment. He, too, frequented the stream borders. It was only natural that watercourses became man's arteries of travel. At first he kept to the banks, but as he learned first to use logs, later rafts, and finally dugout canoes, he used the stream to float himself and his belongings from one place to another. As his skills developed in hunting, in use of tools, and particularly in use of fire, he dared to penetrate into the forest areas and there to make crude openings by girdling and burning trees. He also took advantage of natural openings caused by wind throw, insect attack, or other destructive agents. Here he planted small fields and so extended his dominion over forest as well as plains.

Various types of litter were made on which man could carry food, animals he had killed, clothing, and other human beings who were unable to walk or who had risen to positions of authority and could command others to provide their transportation. When animals had been domesticated, litters were devised that could be drawn by ox or donkey. Short logs were used as rollers under heavy objects man wished to transport. Some brilliant inventor of prehistoric times conceived of combining the log and litter to form a type of cart. By chipping away wood in the center of the log, he made his cart roller into two wheels and an axle. The true wheel, consisting of hub, spokes, fellies, and rim, was developed in India.

As the river villages prospered through use of grain and domestic animals, their need for wood increased. They desired, in addition to fuel, wood for workshops, trading centers, temples of worship, and council chambers. Woodcutters were sent into the hills to procure timbers needed by ancient civilizations. We find this well described in the biblical account of Solomon's temple. Organized societies found ever-increasing uses for timber. Larger buildings were erected; more homes constructed. Boats became numerous and larger in size. Eventually they were made large enough to cross the oceans.

Man came to see in the forest a source of material for his welfare and livelihood. As timber became scarce, efforts were finally made to restrict use of forests.

Stages in Use of Forests

Primitive man was largely content to accept his food and shelter from the forest with little effort to modify his environment to suit his needs.

FIG. 3. Virgin old-growth timber with scattered openings. Primitive man was no match for such a forest. This forest now awaits access roads before harvesting. (*U.S. Department of the Interior, Bureau of Land Management photo.*)

The forest was, in fact, too strong for him to conquer (Fig. 3). Only by use of fire could he weaken the forest's hold on the land, and when fires burned vast areas, he was obliged to migrate.

With the beginning of agriculture man began a struggle against the forest to clear land for his crops and domestic animals. As one cleared area became infertile, he abandoned it for another. The abandoned fields soon were reclaimed by woody growth. The cycle was often repeated, giving rise to a system of shifting agriculture that may be found in the tropics today. It was in common use by North American Indians and other primitive peoples. Some primitive people have, in fact, come to appreciate that their lands are too infertile for continuous cultivation and look to shifting agriculture as the only method of maintaining soil fertility.

As skill in husbandry increased, men sought to avoid the heavy labor involved in clearing new land. By applying natural or artificial fertilizers and by various cropping techniques, they attempted to maintain the land in continuous cultivation. Surplus products were exchanged in villages for other products.

Permanent occupancy and cultivation of the land often led to development of property rights both in the villages and on the land. Forest land was at first held as communal property where any man might harvest such timber products as he required. Villages provided a permanent market for fuel wood. When fuel wood became scarce, property rights were extended to the forest. These took two forms: small individual holdings and community forests in which the village authorities or ruler exercised control over the use of the land.

The growth of communities into cities and the development of agriculture increased the use of forest products for construction so that even higher values were assigned to the forest. Lumber and other timber operations solidified the rights of ownership, and exploitation of the forest was carried out with no regard for future productivity. Unrestricted cutting led to local wood shortages. Gradually such shortages spread. Timber shortages in China probably occurred as early as 5,000 years ago; in Egypt, 4,000 years ago. The Greeks and Romans felt timber shortages more than 2,000 years ago, and the central Europeans 1,000 years ago.[2]

Timber shortages caused farsighted rulers and landowners to regulate cutting in efforts to perpetuate the timber supply. Such regulations were made by the Chinese priests, by the Roman emperors, and by the Egyptians before the time of Christ. They were carried out in France, Germany, and Great Britain before the time of the Crusades. The Chinese philosopher Mencius, who lived from 372 to 289 B.C., stated the case for conservation eloquently in his advice to King Hwuy of Leang.[1]*

> If the seasons of husbandry be not interfered with, the grain will be more than can be eaten. If close nets are not allowed to enter the pools and ponds, the fishes and turtles will be more than can be consumed. If the axes and bills enter the hills and forests only at the proper time, the wood will be more than can be used. When the grain and fish and turtles are more than can be eaten, and there is more wood than can be used, this enables the people to nourish their living and bury their dead, without any feeling against any. This condition, in which the people nourish their living and bury their dead without any feeling against any, is the first step of royal government.

The stages mentioned above have not always occurred in sequence; nor have all stages appeared in every forested area. They are well exemplified in the history of the Cedars of Lebanon as it is reconstructed by Lowdermilk.[4, 5, 6] †

> King Solomon is recorded to have made a bargain with Hiram, King of Tyre, to exchange wheat and olive oil out of his Kingdom, including Palestine, for

* Quoted from Robert O. Ballou, ed., "The Bible of the World," The Viking Press, Inc.

† Quotation courtesy of *American Forests.*

timber of fir and aromatic Cedars of Lebanon for the construction of the Temple at Jerusalem. According to the record Solomon furnished 80,000 lumberjacks as hewers of wood for the wood operation in the forested mountains and 70,000 burden bearers to skid out the timbers. The forests that covered the lofty shoulders of the Lebanon Mountains must have been extensive to keep such a force of 150,000 workers busy.

The forests of Lebanon were incorporated into a state forest under the reign of Emperor Hadrian, and efforts were apparently made to stop indiscriminate cutting. Throughout Roman times the forests were used to build great temples and ships that sailed in the Mediterranean. After the Roman Empire fell, the lands were invaded by Arabs, who neglected the conservation works of the Romans and let them fall into disrepair. The forests dwindled and were further exploited during the Turkish period that followed. Today the forests are all but extinct. Only four groves remain. Grazing by goats and sheep has completed the destruction. "Since the Arab invasion, thirteen centuries ago, the forests of Lebanon have been exploited and neglected until the slopes have been rendered barren, gray, and rocky, making impossible the restoration of forests equal to the original stands." [6] Lowdermilk concluded from his extensive travels in the Mediterranean area, Far East, and the Caribbean: "I find that where man has lived the longest in organized societies, there the land is in worst condition." [5] Forests are destroyed, and peoples are reduced to the lowest states of general well-being. The destruction of the Cedars of Lebanon that took place over a period of some 2,000 years has been duplicated in forests of the Andes in a period of less than two generations. Here the Indians had developed terracing to a high degree. But the Spaniards found such agriculture unprofitable. Gradually one piece of land after another was abandoned and more and more virgin forests cleared for cultivation. Here the steep soils unprotected by terraces eroded rapidly under the heavy tropical rainfall so that serious gullying and complete soil exhaustion has been brought about at a rapid rate (Fig. 4).[7]

Early Forestry Measures in Asia, Europe, and the Americas

The wholesale destruction of forests in China has probably been due primarily to two influences: first, the dividing of the land into small-sized parcels owned by individuals, and second, the general press of the population on the land. Freehold units in China now have reached about $2\frac{1}{2}$ to 3 acres in size. These must be intensively cultivated and irrigated to support the Chinese families. Whole mountainsides, particularly in the Yunnan Province, were cleared for many thousand feet above the valley floor and terraced for agriculture. Eventually drought caused

FIG. 4. Concentrated sheep trailing has denuded this slope. (*U.S. Department of the Interior, Bureau of Land Management photo.*)

widespread starvation of the inhabitants, and today the terraces are neglected, the hillsides badly eroded, and stunted pines occupy the uncultivated soil.[2]

Forests fared somewhat better in India than in China. The Buddhist religion, as well as the activities of Indian monarchs and Hindu priests, encouraged tree-planting programs.[2] The unappropriated forests were held by the Indian princes, who harvested such rare woods as rosewood, ebony, teak, and sandalwood for export to other countries.

Forestry entered the Mediterranean region before the time of Christ. The Egyptians imported a number of trees for forest planting and had a chief forester as early as 3500 B.C. Reference is made to keepers of the king's forest and to actual foresters in Assyria, Palestine, and Asia Minor. A number of sacred groves were maintained in Lebanon and elsewhere. The Romans took a lively interest in trees, importing tree seedlings from the Indias, Africas, Spain, and Germany.[3] Greeks and Romans alike maintained sacred groves of trees comparable to those in Carthage, Lebanon, and elsewhere.

The Aztec king Nezahualcoyotl, who reigned in the fifteenth century,[2] mitigated the severe penalties imposed on citizens who trespassed on the chieftain's forest. He allowed the tribes to gather wood from the ground, but harvesting of standing timber was punished with death.

By the eighth century kings of England forbade trespass on royal property under fine of 60 shillings and thereby established the banned forests. By the tenth century English kings extended the royal right to chase, over fields, pastures, waters, villages, and woods. Ordinances were passed by the princes and barons governing the amount of wood that might be cut for family use and the uses to which such wood might be put, including charcoal making and burning for potash. Reforestation was practiced by Henry VII of England and by the medieval city of Nuremberg in Germany.

During the Middle Ages clearing of land for agriculture had gone so far that decrees in both Austria and Germany forbade further land clearing. The purpose, in part, was to protect hunting privileges of the nobility, but it was also to ensure a fuel-wood supply for local industries and home use.

The first laws to prohibit the clearing of forests appeared in 1165. These were followed by ordinances in 1177, 1226, 1291, and 1237. A great number of laws were passed thereafter. The early regulation laws were simple and designed primarily to prohibit clearing of forests. Later regulations forbade pasturing newly cut areas. Some regulations even went so far as to direct that buildings be erected of stone to conserve wood.

Early community forests in France go back to the time of the Gauls in A.D. 500 to 510. Clovis in the sixth century took over royal and state forests. Eventually church property became royal property. After 1791 royal property was declared national land which, with the addition of church property, brought the total up to 4.3 million acres. Forest officers were appointed as early as 1291, and by 1669 the first regulations governing timber cutting on public and private lands were established.[3]

In Switzerland, community forest ownership has dominated forest property throughout its history. The first forest ordinance regulating use dates around 1304, and the first working plan was made for the Sihlwald near Zurich which with revisions has remained in effect for more than 600 years. Special regulations governing cutting of the steep Alpine region were introduced early and were rigidly enforced.

In Austria the city forests were placed under a form of management as early as A.D. 1100. Various regulations governing the felling of timber were adopted from time to time. A comprehensive forest ordinance adopted in the reign of Empress Maria Theresa in 1754 revised many of the restrictions on private landowners and on the management of communal forests.

Regulations of timber-cutting operations in Scandinavia date back to 1638. A forest-planting law was established in 1647. In Norway, attempts were made to stop forest devastation in 1500, and a forest survey was authorized in 1600 by Christian IV. By 1725 a forest and sawmill commission was established, and export restrictions were imposed in 1795. Danish forest laws date back to 1557, and the state forests were placed under organized forest management as early as 1781.

European Forestry up to 1900

The rise of revolutionary movements and the downfall of feudalism in Europe caused a pronounced change in the attitude toward forest properties. Feudal holdings were seized in many countries and made a part of the public domain. Church property was confiscated. At first, some states sought to liquidate these lands by sale to increase state revenue. Germany adopted such a course but early in the nineteenth century appointed Hartig as chief of the forest administrative branch of the treasury department. Hartig curtailed further dismemberment of state land and instituted forest management. Since the Swiss had most of their lands in communal ownership from the beginning, little change occurred there. France had a difficult time in stopping the overcutting and abuse that followed the French Revolution. However, early in the nineteenth century regulations were established governing forest use.

Concepts of forest management developed rapidly in Germany, France, and Scandinavia during the nineteenth century. Forest properties were subdivided into compartments, and working plans established governing cutting rates and cutting areas in efforts to obtain the highest return from the soil. In France, special attention was paid to protecting watersheds against landslides and erosion, and to stabilizing sand dunes.

Regulation of cutting on both public and private land was introduced by most European states during the late nineteenth century, if not before. State forests were placed under intensive management and efforts made to extinguish long-established rights of users, such as collecting litter and grazing livestock, that reduced the yield of the forest. Important strides were made in private forestry during this period, in part because of state regulation and in part because forest properties became sound investments for their owners. Forests constituted the base of family wealth for a number of aristocratic owners. Entailed estate was established whereby the owner contracted to operate on a sustained-yield basis and to bind his heirs to do likewise in return for certain technical aid and tax inducements offered by the state.

Formal education in forestry began about 1825 when private forestry schools were established. These were an outgrowth of the old master schools that had existed before that time. In 1816 the Cotta Master School grew into the forestry college at Tharandt, which became one of the leading forestry schools in Germany. The French school of forestry was established at Nancy in 1825. It has been in continuous existence since that time. Other colleges of forestry were established during succeeding years of the nineteenth century.

Forest research began almost as early as forest education and has been carried forward by European schools and forest officers since that time.

Europe maintained its leadership in forestry up until the outbreak of the Second World War. By this time the United States had developed forest technology on a high level. The war caused a considerable setback to forestry in Europe, and occupied lands after the war were subjected to forest regulations established by the occupying forces. The United States now stands as a world leader in forestry, particularly in forest utilization.

LITERATURE CITED

1. Ballou, Robert O., ed. 1939. "The Bible of the World," pp. 429–430, The Viking Press, Inc., New York.
2. Caldwell, David H. 1949. Unpublished manuscript, New York State College of Forestry, Syracuse, N.Y.
3. Fernow, Bernhard E. 1911. "A Brief History of Forestry in Europe, the United States and Other Countries," rev. ed., University Press, Toronto, and *Forestry Quarterly,* Cambridge, Mass. 506 pp.
4. Lowdermilk, Walter C. 1941. "Conquest of the Land," U.S. Department of Agriculture, Soil Conservation Service, Washington, D.C. 34 pp. mim.
5. Lowdermilk, Walter C. 1946. Soil Science in Wildland Management, *Journal of Forestry,* 44:788–798.
6. Lowdermilk, Walter C. 1941. The Cedars of Lebanon—Then and Now, *American Forests,* 47(1):16–20, 34.
7. Meyer, H. Arthur. 1945. "Conuqueros"–Forest Destroyers of the Andes, *American Forests,* 51(4):178–179, 200.

CHAPTER 3. *The Development of Forestry in the United States*

THE STORY of forestry in America is inseparable from the story of the early pioneers and their indomitable struggle to make a home in the wilderness they found in the new world. It is also a part of the story of those great spirits that rose up among the pioneers to create a new nation on American soil. On the one hand it reflects the actions of a vigorous people surging eagerly forward to exploit for personal profit the rich resources of the new land. On the other it deals with cooperative action to build churches, schools, villages, and states, and with community-imposed restraints to equalize opportunity for achievement.

Two great struggles characterize America: the struggle to occupy the land and use its resources and the struggle in the hearts of men to free the creative spirit of individualism while guiding it in the common interest. The first struggle was won with the passing of the frontier; the second will continue as long as we remain a free people.

Forestry in Colonial Times

The early American colonist found on the eastern coast of our country an almost unbroken forest that stretched west to the prairie region. Extensive forests also existed in the Rocky Mountains and in the West Coast region. Hardwood forests occupied some 403 million acres; coniferous forests, 419 million acres. Woodland in the prairie region and semidesert regions of the West occupied 93 million acres, making a grand total of 915 million acres of forest and woodland. This was 48 per cent of the total land area of the United States.

It is small wonder that the early colonist considered the forest his enemy. The forests were dense, the trees large in size, and many of them of hard and durable wood that he could dispose of only by burning. The virgin forests contained windfalls and tops of trees broken by storm. They occupied the land so completely that only by heroic effort could the pioneer wrest from the forest wilderness a few acres of land that

FIG. 5. Replica of old French Fort erected in colonial days near present site of Syracuse, N.Y.

he might cultivate. Fire was his natural ally in the land-clearing process. The forest, to be sure, provided the settler with fuel, game, nuts, useful bark, and poles for erecting cabins and forts (Fig. 5). But timber to the early American colonist had largely a negative value. It was something to be got rid of so that he could free the land to grow crops and to pasture his livestock.

Yet even in colonial days farseeing citizens advocated forest-conservation measures. Transportation was crudely developed. What timber was cut had either to be floated down the streams or laboriously hauled by horses or oxen on crude wagons to the villages where it was used. Iron foundries and glass factories of colonial days depended upon wood for fuel to smelt the ore and melt the glass. Frequently they ran short of the local supplies. Cities such as Philadelphia and Providence were short of fuel even before the Revolution. Benjamin Franklin in his article describing the Franklin stove stated in 1774: [3*] "Wood, our common fuel, which within these hundred years might be had at any man's door, must now be fetched near one hundred miles to some towns, and makes a very considerable article in the expense of families." After describing his stove and its economical use of wood, he concludes:

* Quoted by permission of Gilbert Chinard.

By the help of this saving invention our wood may grow as fast as we consume it, and our posterity may warm themselves at a moderate rate, without being obliged to fetch their fuel over the Atlantic; as, if pit-coal should not be here discovered (which is an uncertainty) they must necessarily do.

Coal did not become generally available until the mid-nineteenth century.

Perhaps the first attempt at forest conservation was a provision written into land-purchase contracts by William Penn in 1681: ". . . that in clearing the ground care be taken to leave one acre of trees for each five acres cleared, and especially to preserve oak and mulberry for silk and shipping." The charter creating the province of Massachusetts Bay in 1691 reserved the best trees, particularly white pine for shipmasts for the Royal Navy. Such trees were to be marked with a broad arrow. This restriction was extended to New York and New Jersey later on. Because the officers of the crown were slow to mark the trees that might not be cut, and because the colonists had left Europe to escape from numerous regulations, the broad-arrow policy was difficult to enforce. Trespass was common even though severely punished when apprehended. These punishments only encouraged the colonists to go even further in defying the King's authority.[10] During the Revolutionary War patriots in Maine, Massachusetts, and New Hampshire did their utmost to prevent the British Navy from getting control over mast and spar timbers from American forests. Their efforts in the end had a considerable effect in crippling that navy.

During the colonial period no foundation was laid for good forest practice in America; nor did the situation change for the better with the coming of independence. Independence removed the restrictions of the mother country and led to more reckless use and abuse of the forest. By this time sawmills had been established and the forest was being exploited for its own worth. An active market for the products of the forest had grown up. The forest was no longer merely an encumbrance of the soil but a source of vitally needed raw material for the rapidly growing cities and villages in the new country.

Commenting on his observations in the Confederation between 1783 and 1784 Johann David Schoepf observed: [13]

In America there is no sovereign right over forests and game, no forest service. Whoever holds new land, in whatever way, controls it as his exclusive possession, with everything on it, above it, and under it. It will not easily come about therefore that, as a strict statutory matter, farmers and landowners will be taught how to manage their forests so as to leave for their grandchildren a bit of wood over which to hang the tea-kettle. Experience and necessity must here take the place of magisterial provision.

His words were singularly prophetic.

Timbers from the forests of North America played a very important role in the rapid development of the country. Much timber was cut by ax and fashioned with whipsaws, broad axes, adzes, and froes to supply squared logs, beams, framing members and shakes for construction. With the coming of sawmills, lumber became available and greatly facilitated the building up of our country. Almost 90 per cent of the one-family dwellings in the United States are of wood-frame construction. Heavy construction was also of wood in the earlier days. Masonry was used for exterior walls and general supports, wood for partitions. Railways used enormous amounts of wood for crossties, bridge timbers, stations, fuel, and other purposes. About 100 ties must be replaced annually on each mile of main track. Formerly, large amounts were used for car construction. Wood is still used in cars, though less extensively than steel. Mines have always been important users of wood in the form of timbers to support the roof, lagging for side walls, mine ties, and other products that are essential for safe mining operation and efficient handling of coal and ores. Farmers use wood in farm buildings and for fuel, fence posts, and corrals.

Federal Forestry Measures

The Public Domain and Its Disposal. An important source of conflict among the colonies even before they had won their independence was the ownership of the lands extending to the west of their individual boundaries. Virginia claimed all land west to include Kentucky and even parts of what is now Indiana.[11] In fact, her original charter embraced "Land, throughout from Sea to Sea, West and Northwest." Massachusetts claimed land beyond the Delaware. Connecticut and Pennsylvania disputed control over the rich farming land in Wyoming Valley, Pa., where Wilkes-Barre now is situated. Actual armed conflict broke out in this case and several skirmishes of a minor nature were fought. To remove this cause of disunity, the Continental Congress in 1780 passed a resolution urging the states to cede their western lands to the Union. From 1781 to 1802 the state lands ceded to the Federal government totaled some 260 million acres.[8]

Through the various land-purchase acts, including the Louisiana Purchase, the Florida Purchase, the various purchases and cessions of land from Mexico, and the Alaska Purchase, the Federal government acquired a vast public domain. It included three-fourths of continental United States and all of Alaska, a total of 1,800 million acres.[18] The public domain became both a source of wealth and a source of embarrassment to the Federal government. It was sold for public revenue, granted to soldiers in lieu of cash payments, granted to companies to promote the building of railways, and to states to further education. It became a source of embarrassment, because our young nation had no trained land administrators nor any police force to prevent trespass. Squatting on the public lands was forbidden by an act of 1807. The early policy for this vast estate was to preserve it for public revenue and national development.

This attitude on the part of the Congress and Federal officers in Washington was in striking conflict with the desires and actions of the frontiersmen. They occupied land wherever it was not occupied by another and used it for their own purposes. A strong demand arose to open the public lands for private settlement. The Preemption Act, passed in 1841, gave homesteaders the right to squat on land and later to purchase it. During the administration of Abraham Lincoln, the Homestead Act was passed. This opened the public domain for entry by homesteaders who desired to settle upon it and authorized the granting of 160 acres free to settlers who lived upon the land and developed it. The law was later liberalized by a series of additional laws. The Timber Culture Act permitted the government to grant 160 acres of land to anyone who would plant trees on 40 acres and keep them in healthy condition over a period of 10 years. The Timber and Stone Act, the Desert Land Act, and the Carey Act all liberalized the conditions under which settlers were able to acquire title to public land through homesteading.[18]

The Swampland Grant Act permitted the Federal government to grant lands chiefly covered by swamp to the states in which they were located to be used for state purposes.

Railway land grants in alternate sections on either side of the railway resulted in a checkerboard pattern of ownership embarrassing to forest-land management from that day to this. The Forest Lieu Act of 1897 [10] permitted railway companies and other land grantees to select lands outside their original grants in lieu of lands within that had already been appropriated. Some of our nation's best timberlands passed from public ownership under this Act. Administration of all these acts was weak; hence fraud was widespread. Little attempt was made to determine whether homesteaders lived up to the regulations or whether the land they homesteaded was suitable for agricultural use.

STEMMING THE LAND BOOM. The disposal of the public domain created the greatest land boom in history. Its momentum was tremendous. Early attempts toward restrained use got scarcely passing attention. Still, these are worthy of recall because they encouraged those who did successfully stem the tide.

François-André Michaux, who in 1819 published the first "Sylva of North America," wrote in his introduction: [3]

I have also endeavoured to impress on the American farmers the importance and pecuniary advantages which would result to them and to their families from the preservation of different species of timber of which they ought to ensure the growth, and on the contrary I have noted those which ought to be destroyed; for I am of opinion that a bad tree ought not to be suffered to exist where a better could grow.

Private societies in Philadelphia and Massachusetts offered premiums for tree growing. As early as 1795 a New York agricultural society recommended that inferior agricultural land be devoted to trees, a recommendation that was not carried out on a large scale until 1929.

A definite Federal effort was made to acquire land supporting live oak for ship timbers. At first $200,000 was set aside to purchase such land.[15] This was followed by granting the President authority to reserve public lands bearing live oak and cedar in Florida, Alabama, and Louisiana and to conduct experiments in planting and cultivating live oak. In 1831 Congress passed a law prohibiting the cutting of live oak from naval reservations or any other lands belonging to the United States. The law was never enforced.

Carl Schurz, Secretary of the Interior, 1877 to 1881, strongly advocated proper care of the public domain and attempted to enforce the Law of 1831 preventing trespass on public land. This he was stopped from doing by Congress in 1880. The Reverend Frederick Starr in 1865 predicted a timber famine within 30 years and advocated immediate research to determine how to manage forests and how to establish plantations. Secretary of Navy Woodberry during Andrew Jackson's administration tried to determine the amount of oak ship timbers that would be needed by the Navy and estimated the amount of land that would be needed to provide such timbers on a sustained-yield basis.[1] New York in 1872 set up a special commission to study the preservation of Adirondack forest for its influence on the Hudson River and the Erie Canal. In 1872 Jay Sterling Morton, Governor of Nebraska, inaugurated the first Arbor Day, and in 1873 the Timber Culture Act which provided 160-acre claims to settlers who would plant first 40 and later 10 acres to trees was passed. This law had considerable influence on the planting of groves and shelter belts in the prairie regions during the early days. It was the forerunner of the vast shelter-belt program of the 1930's.

By the late 1890's the end of free land was definitely in sight. What remained was, for the most part, poorly suited for agricultural use. Loose administration aroused the nation to the need for saving at least some land for permanent public use. Of even more concern was the rapid inroad being made on the timber resources.

Cutting of timber for lumber proceeded at a moderate pace during the colonial period and the early life of the republic. Sawmills were primitive in design and could be operated successfully only where water

FIG. 6. Even the modern concrete, steel, and brick structure requires large amounts of lumber for scaffolding, forms, staging, fencing, and other temporary use.

power was available. By 1811 the first steam sawmill was introduced, and by 1840 lumbering got under way at a very rapid rate. Maine led all other states in lumber production up to 1850, but in that year New York State took the lead, in 1860, Pennsylvania, and by 1870, peak production had shifted to the Lake States. The Northeast, which produced 55 per cent of all lumber in 1850, produced only 20 per cent in 1890. Local manufacturers in the East complained about the difficulties of getting good supplies of ash, hickory, white oak, walnut, and high-grade white pine.[16] The timber famine of which the Reverend Starr had so dramatically warned the country in 1865 seemed rapidly approaching. It is hard for us today to think of a timber famine in our country as far back as 1890. The people at that time were unaccustomed to using beech, the lower grades of pine, and many other species that to them were unsuited for lumber. We use these today because better quality woods are scarce and improved manufacture and seasoning make low-quality woods usable (Fig. 6).

Beginnings of Federal Forestry. The real spark that started the Federal government upon a permanent program of forestry grew out of a memorial by the American Association for the Advancement of Science addressed to the Congress in 1874. This memorial was prepared as a result of a talk by Franklin B. Hough of Lowville, N.Y. He stated:

> The preservation and growth of timber is a subject of great practical importance to the people of the United States, and is becoming every year of more and more consequence, from the increasing demand for its use; and while this rapid exhaustion is taking place, there is no effectual provision against waste or for the renewal of supply. . . . Besides the economical value of timber for construction, fuel, and the arts . . . questions of climate . . . the drying up of rivulets . . . and the growing tendency to floods and drought . . . since the cutting off of our forests are subjects of common observation. . . .[15]

Congress at the association's urging provided in 1876 for a survey of the existing forest resources and the demands that were being placed upon them by consuming industries. Inquiry also was to be made into the influence of forest on climate, and of the best methods of forest practice used in Europe. Dr. Hough was employed by the Department of Agriculture at the salary of $2,000 a year. He prepared three voluminous reports that contained the best information then available on America's forest resources and forest industries. Dr. Hough was succeeded by N. H. Egleston in 1883 who prepared a fourth report. By 1881 Congress had established a Division of Forestry under the Commissioner of Agriculture.

Two new organizations to promote forestry came to the front at this time, the American Forestry Association and the American Forestry Congress. They merged in 1882 to sponsor an American Forestry Congress. This gala affair [2] was held in Cincinnati. Hough, Egleston, and B. E. Fernow, a forester with technical training in Germany who came to the United States when still a young man, took an active part in promoting the congress. Governors of many states participated. Baron von Steuben, a Prussian forester and descendant of General von Steuben of Revolutionary War fame, was guest of honor. The meeting was climaxed by a huge parade and tree-planting ceremony. In a later congress held in Denver in 1886, the new association urged that public lands necessary for the preservation of water supply be granted to the several states in which they lay and kept by these states in perpetuity to protect the headwaters of all rivers. It also advocated national and state forest-fire control.

PUBLIC FOREST RESERVES. Carl Schurz during his term as Secretary of the Interior repeatedly urged reserving public-domain timberland for long-term management. Numerous bills were introduced from 1876 on but were defeated by special-interest groups. A rider was attached to the Act of March 3, 1891, "for the repeal of the Timber and Stone Act and for other purposes," authorizing the President to set apart from time to time reserves of land bearing forests, whether commercial or not, from the public domain. Gifford Pinchot called this Act the most important legislation in the history of American forestry.[12]

An active program of reservation was undertaken by Presidents Harrison and Cleveland. These forest reserves remained in the Department of the Interior, which had no trained personnel to administer them. Bernhard E. Fernow, who succeeded Egleston in 1886 as Chief of the Division of Forestry in the Department of Agriculture, was a vigorous man who worked hard to promote forestry in the United States. Repeatedly he advocated transfer of the forest reserves from the Department of the Interior to the Department of Agriculture and their administration under the Division of Forestry.

At the request of President Cleveland the National Academy of Sciences set up a special forestry commission to report on the forest reserves and to recommend measures for their proper use. Charles Sprague Sargent of Harvard University was chairman. The commission recommended increasing the reserve by 21 million acres but made no recommendations for managing them, an omission which Pinchot felt was a serious one on the commission's part. A storm of protest arose throughout western states over what the local people called locking up the public lands.[12]

This protest led Congress to suspend all but two of the reserves created by Cleveland. At the same time it provided for managing existing and future reserves and stated the conditions under which future forest reservations might be established. This Act of June 4, 1897, stated: "No public forest reservation shall be established except to improve and protect the forest, secure favorable conditions of water flow," and "furnish a continuous supply of timber for the use and necessities of citizens of the United States."[15] This Act received the support of the western people and laid the groundwork for a rational policy of managing and protecting the forest reserves.

The Secretary of the Interior as directed in 1897 attempted to protect and to administer the forest reserves. A field force of superintendents, rangers, and others was established, and Filibert Roth was placed in charge. Roth stayed only one year, however, and the Interior Department's efforts to build up satisfactory administration languished.

THE ROOSEVELT-PINCHOT ERA. The attitude of the people toward forestry in the nineties was shaped by many outstanding personalities among whom Fernow, Pinchot, Hough, Schurz, and Governor Morton were conspicuous. The country was ready for a huge conservation program and awaited only proper leadership in the White House.

This came with dramatic suddenness when McKinley was assassinated on September 14, 1901. Theodore Roosevelt, the new President, had become imbued with the principles of conservation through his interests in wildlife and through a friendship he had formed with Gifford Pinchot, who became chief of the Division of Forestry in 1898. The Division of Forestry became the Bureau of Forestry in 1901. In his first message to Congress, President Roosevelt, at the urging of Pinchot, Buell, Maxwell, and McGee, asserted: *

> The fundamental idea of forestry is the perpetuation of forests by use. Forest protection is not an end in itself. It is a means to increase and sustain the resources of our country and the industries which depend upon them. The preservation of our forest is an imperative business necessity. We have come to see clearly that whatever destroys the forest except to make way for agriculture threatens our well-being.
>
> The practical usefulness of the National Forest Reserves to the mining, grazing, irrigation, and other interests of the regions in which the reserves lie has led to a widespread demand by the people of the West for their protection and extension.

The President went on to advocate that administration of the reserves be united in the Bureau of Forestry.

Final action to carry out the President's recommendation did not take place until four years later. Meanwhile, Gifford Pinchot was building up the Bureau of Forestry. By offering services to large private landowners and others interested in inaugurating forestry programs, the bureau was acquainting influential men with the meaning of forestry. Pinchot's energy and influence caused many people, the President included, to become passionately interested in conservation of our nation's natural resources.

Year after year bills were introduced into Congress to transfer the reserves to the Department of Agriculture. Additional public sentiment for such transfer was created through a special American Forestry Congress that met in Washington in 1905. The Department of Agriculture, the Geological Survey, the Reclamation Service, the General Land Office, the President of the National Lumbermen's Association, the President of the National Livestock and National Wool Growers Association, the Union Pacific and the Great Northern Railroads, and the Weyerhaeuser Lumber Companies all joined hands in recommending the transfer of the forest reserves to the Department of Agriculture. The Transfer Act was passed February 1, 1905, and signed by President Roosevelt on that day. A month later the Department of Agriculture was given full legal authority to regulate the forest reserves.

REGULATING THE USE OF NATIONAL FORESTS. The day after the forest reserves were transferred to the Bureau of Forestry, the name of which

* Quoted from Gifford Pinchot, "Breaking New Ground," Harcourt, Brace and Company, Inc.

was at that time changed to the Forest Service, the Secretary of Agriculture, James Wilson, sent a letter to the Chief of the Forest Service outlining the administrative policies to guide the management and use of the forest reserves. This letter has become a classic in Federal forest policy.*

In the administration of the forest reserves it must be clearly borne in mind that all land is to be devoted to its most productive use for the permanent good of the whole people, and not for the temporary benefit of individuals or companies. All the resources of forest reserves are for use, and this use must be brought about in a thoroughly prompt and businesslike manner, under such restrictions only as will insure the permanence of these resources. The vital importance of forest reserves to the great industries of the western states will be largely increased in the near future by the continued steady advance in settlement and development. The permanence of the resources of the reserves is therefore indispensable to continued prosperity, and the policy of this department for their protection and use will invariably be guided by this fact, always bearing in mind that the conservative use of these resources in no way conflicts with their permanent value.

You will see to it that the water, wood, and forage of the reserves are conserved and wisely used for the benefit of the home builder first of all, upon whom depends the best permanent use of lands and resources alike. The continued prosperity of the agricultural, lumbering, mining, and livestock interests is directly dependent upon a permanent and accessible supply of water, wood, and forage, as well as upon the present and future use of their resources under businesslike regulations, enforced with promptness, effectiveness, and common sense. In the management of each reserve local questions will be decided upon local grounds; the dominant industry will be considered first, but with as little restriction to minor industries as may be possible; sudden changes in industrial conditions will be avoided by gradual adjustment after due notice; and where conflicting interests must be reconciled, the question will always be decided from the standpoint of the greatest good of the greatest number in the long run.[12]

This policy has guided the management of the national forests from that day to the present. The phrase "greatest good of the greatest number in the long run" has become the motto of the Forest Service.

BUILDING THE FOREST SERVICE. The forest reserves placed a heavy administrative responsibility on the small organization of foresters. Few precedents were available to guide them. Pinchot threw his dynamic energy into welding a strong Forest Service and to building up a program of national forest administration.

The task was not easy. Cattle and sheep ranchers had been grazing

* Quoted from Gifford Pinchot, "Breaking New Ground," Harcourt, Brace and Company, Inc.

their animals on the public domain without Federal permit and with only the rules they were personally able to impose. Naturally they resented the Federal intervention. A good deal of diplomacy and sheer courage were required of the early forest officers. The Service sought from the beginning to employ men familiar with local conditions and men who understood, and, if possible, commanded the respect of, the sheepmen and cattlemen with whom they must deal. By fairness, scrupulous honesty, and devotion to the ideals set forth in Secretary Wilson's letter, the Forest Service established itself in the western country.

While Theodore Roosevelt was President, a total of 148 million acres were transferred from the public domain to the national forests. Opposition developed in the West and soon made itself felt in Washington. Eventually Congress withdrew from the President the right to set aside new national forests from the public domain. Before signing the bill, President Roosevelt set aside a large additional area in national forest, enough he thought at the time to establish in Federal ownership all important public-domain forest land.

A White House Conference of Governors was called by Roosevelt in 1908. It stimulated interest of the states in forestry and set up the National Conservation Commission authorized to consider the four great classes of resources: water, forest, land, and minerals.

In 1909 President Roosevelt called a North American Conservation Congress. Representatives from Canada, Mexico, and the United States assembled. A declaration of principles was set forth by the congress in which they recognized that natural resources are not confined by national boundaries, that no nation acting alone can properly conserve them, and that all resources should be made available for the use and welfare of man.

Federal-State Cooperative Programs. Gifford Pinchot, the prime mover behind Theodore Roosevelt's conservation policies, was removed from the U.S. Forest Service in 1910 by President Taft because of a controversy between him and the Secretary of the Interior, Richard Ballinger. He was succeeded by Henry Graves who during his 10 years as chief of the U.S. Forest Service consolidated the gains won by Pinchot and Theodore Roosevelt. But Graves had a broader and more important function to perform. The states had begun forestry programs, but no Federal leadership to coordinate and support their efforts was available. The Weeks Law passed in 1911 extended the forestry effort to private lands and to all state lands. Under the terms of the Weeks Law the Federal government was authorized to contribute to the states on a matching basis funds for the protection of forest lands against fire. By 1919, 22 states were cooperating with 110 million acres under organized protection. By 1950, 43 states cooperated, protecting 357 million acres.[9, 17]

The Weeks Law, and particularly the administrative processes that grew up under it and its successor, the Clarke-McNary Law, established a new basic principle in Federal forest administration. Initially, the Federal government actually paid the salaries of forest guards who worked under the supervision of the states. Later on, complete administrative authority was lodged with the states. The Federal government gave technical advice but made only general audits and inspections to make certain the Federal funds were properly used.[9]

National Forests in the East. The Weeks Law also authorized the Federal government to acquire private land for national forests where these were necessary for the protection of navigable streams. This provided the basis for the national forests of the eastern states. The authorization was later changed to include also lands needed for growing timber. The White Mountain forest in New Hampshire, the Pisgah in North Carolina, the Allegheny in Pennsylvania were acquired. Later the system was extended to include 43 national forests and national-forest purchase units with lands in 23 states east of the Rocky Mountains.

Forestry in the First World War. The First World War interrupted the progress of United States forestry but presented new opportunities for service to foresters. The 10th and 20th Engineer Regiments were formed to get out timber needed by our fighting troops in France.[19] Graves, Greeley, Guthrie, Peck, Kelley, and many others destined to play important roles in shaping American forestry went to France as officers. Here they became acquainted with French forestry and established a fine reputation for effective service to their country. Meanwhile, Zon and Sparhawk were preparing their classic book, "Forest Resources of the World," a significant contribution to the field of world forestry. The Forest Products Laboratory, established in 1910, had a real chance during the war to prove its mettle by finding substitutes for scarce woods, by developing standards for use, and by many other valuable research contributions.

In 1920 Henry Graves returned to Yale as dean of forestry, a post he held with distinction until his retirement in 1939. From 1945 to 1947 he served as chairman of the Interim Committee on Forestry of the Food and Agriculture Organization, which drew up the proposals leading to establishment of the Forestry Division of the United Nations.

Between War and Depression. William B. Greeley succeeded Graves as Chief Forester in 1920.

Scarcity of lumber and increasing prices during the war sharpened the nation's appreciation of the role forests play in our national defense. A special study of forest resources, made under a Senate Resolution sponsored by Senator Capper, emphasized the shrinking timber supplies of our country and pointed up the general lack of forestry measures on private timber operations. Gifford Pinchot headed a committee of foresters advocating public regulation of private timber cutting, but the Capper and Snell bills, providing for such regulation, failed of passage.

For the country as a whole the early 1920's was a period of disillusionment following the war. This gave way later to reckless optimism that nursed the boom of 1926 to 1929 and vanished in the severe business crash in 1929 to 1932.

The hectic twenties, nevertheless, saw substantial advance in public forestry. Sensing that public sentiment at this time would oppose new Federal regulations, and having faith that much could be accomplished by milder measures, Greeley set his course. A special committee of industrialists and forestry leaders known as the National Forestry Program Committee was formed [7] with Royal S. Kellogg, executive secretary of the Newsprint Service Bureau, as chairman. Other members included E. T. Allen, Phillip Ayres, Hugh P. Baker, Ovid Butler, Wilson Compton, John Foley, William L. Hall, Charles Lathrop Pack, and J. Randall Williams. Strong ties were made with the United States Chamber of Commerce and other organizations. The decision was made to concentrate effort on noncontroversial legislation, and four important basic laws aiding forestry resulted. The General Exchange Act of March 20, 1922, permitted exchange of private lands within national forest boundaries for other public lands of equal value or for national-forest stumpage in the same state. This Act has been very useful in blocking up national forests. The Clarke-McNary law increased aid to states for forest protection and provided funds for forest extension and for supplying trees to farmers for planting. The McNary-Woodruff law of 1928 authorized appropriations of 40 million dollars over a 10-year period for purchase of national forests. The McSweeney-McNary law passed in 1928 provided organic legislation for research. It authorized 12 regional forest experiment stations and the Forest Products Laboratory for carrying on research in forestry and outlined a 10-year program of financial support for these organizations.[7] This greatly strengthened the Federal forest-research program.

Under Greeley's administration the organization and procedures of the Forest Service were further strengthened. He resigned in 1928 to become executive secretary of the West Coast Lumbermen's Association, a post he filled with high efficiency. After retiring, he served effectively as Chairman of the Board of the American Forest Products Industries, Inc.

Robert Y. Stuart became Chief Forester in 1928. His term marked the rapid expansion of new activities authorized under the legislation of the twenties. The national survey of forest resources was launched and vigorously carried forward.[4] The voluminous study of forest resources prepared in response to a Senate Resolution and known as the Copeland Report contained information on every phase of forestry and provided background material for subsequent forest policy. Forestry was making gratifying headway until the depression set in in earnest in 1931.

The depression rocked the country even more rudely than did the First World War. The stock market crashed, industrial production dropped precipitously, and millions of workers lost employment. The absence of consumer buying power caused further price cuts, layoffs, and business failures. Banks felt the strain, they soon began to fail at an alarming rate, and the complete bank closure came in March 5, 1933. Faith in the country's financial strength and in the capitalistic system wavered, and foresters prepared for a drastic reduction in activities. But the exact opposite occurred.

The Enlargement of Federal Forestry through Emergency Programs. The cooperative approach inaugurated by Greeley seemed to have spent its force by 1930. In the depression private industry, local and state governments, all appeared feeble. Businessmen, unemployed veterans, governors, and mayors thronged to Washington to beseech aid of the Federal government. In 1932 substantial appropriations were requested by President Hoover and granted by Congress to expand public works to create employment. Forestry activities shared in these funds. In 1933 President Franklin D. Roosevelt proposed a vast program of public works embracing forestry, soil conservation, river-valley improvements, flood control, public parks, as well as new schools, post offices, highways, airports, hospitals, sewage systems, and a host of other public facilities. A number of emergency agencies were established that performed useful forestry work, among them the Federal Emergency Relief Administration, the Civil Works Administration, Public Works Administration, and Works Progress Administration.

CIVILIAN CONSERVATION CORPS. The most imaginative and useful to forestry was the Civilian Conservation Corps, which enrolled needy young people out of work. Camps were established on national and state forests, in soil-conservation districts, and in national parks. The men engaged in tree planting, care of plantations, timber-stand improvement, construction of forest roads and trails, lookout towers, and various other good forestry measures on public land (Fig. 7). At its peak 2,600 camps, each organized to handle 200 men, were operating, and half of the camps were engaged in forestry projects. The Civilian Conservation Corps continued until 1941 when the project was closed because of war. The corps served to acquaint a good many people with forestry as a major government activity. It did a great deal to improve national and state forest lands and also provided young men a wholesome life and useful training for private jobs.

FIG. 7. Hand tree planting by the Civilian Conservation Corps. Lolo National Forest. (*U.S. Forest Service photo.*)

THE LUMBER CODE. The National Industrial Recovery Act of 1933 set up fair trade practices among industries. Article 10 in the lumber code under this Act specified certain forest practices to be followed by industry, which involved leaving the forest in good condition to reproduce another crop. Where practical, partial cutting rather than clear cutting was to be used. The Act was declared unconstitutional in 1935. Lumber associations, however, urged their members to continue to comply with the code so far as forest practices were concerned, and many did so. Lumbermen thus got a good understanding of what foresters meant by selective cutting and of the advantages that might accrue from this and other good forestry measures.

Stuart, who died in office in 1933, at the beginning of these programs, was succeeded as Chief Forester by Ferdinand Silcox. Silcox brought to the Forest Service a new interest in and sympathy for the people who lived in the forest and gained their livelihood from harvest and sale of forest products. He saw in the forest the economic base on which the livelihood of the nearby communities largely depended, and he sought to have the forests so managed that they served the welfare of those who depended directly on them to the best possible advantage. Great expansion in Federal activities occurred during this period. National forest purchasing increased at a rapid rate (Fig. 8). Lumbermen were begging the government to take up their cutover lands. New purchase units were established in Illinois, Indiana, Missouri, Ohio, Kentucky, Alabama, Florida, Mississippi, and other states.

The vast prairie-plains shelter-belt program was launched. This led to planting 18,510 miles of windbreaks. In 1942 this task was transferred to the Soil Conservation Service, which had planted 8,363 additional miles by 1949.[17] Flood control surveys were also started.

Silcox aided by Earle Clapp and others formulated a new policy for Federal forestry calling for increased cooperation with private and state forestry, expansion of public forests, and regulation of cutting practices on private forest lands. Proposals for range improvement also followed a substantial report entitled "The Western Range" that appeared in 1936. This report on the range resources was comparable in tenor and scope with the Copeland Report on forestry.

Earle Clapp succeeded Silcox in 1939 as Acting Chief of the Forest Service. A joint congressional committee was appointed in 1938 to study the forest situation, hearings were held, and a report was issued in 1941. This recommended Federal cooperation with the states in regulating forest practice on private lands. A number of bills introduced in Con-

Fig. 8. A fire-devastated area along Moose River, Wis., is being reclaimed by planted jack pine. Chequamegon National Forest. (*U.S. Forest Service photo.*)

gress between 1939 and 1951 provided for public regulation of private forestry. Some would authorize Federal cooperation with states, others straight Federal regulation. Considerable sentiment was behind the drive for regulation, but opposition by industry, farm groups, and others prevented passage of Federal legislation.

Attempts at wide extension of Federal forest holdings at this time were also defeated. Even the joint congressional committee recommended no increase in public forests. The trend toward rapid increase in Federal forestry was being checked.

This period saw substantial gains in conservation in the Department of the Interior. Civilian Conservation Corps and other work programs carried out important development projects on national parks and monuments. The Soil Erosion Service was started in 1933 to work with farmers. Later transferred to the Department of Agriculture and renamed the Soil Conservation Service, it sponsored tree planting and other forest-improvement measures on farms. The Taylor Grazing Act of 1934 and subsequent acts placed 147 million acres of the remaining public domain in districts to be administered for the forage resource. Legislation in 1937 provided for management under a sustained-yield plan of the Oregon and California Railroad revested lands, and regulations governing the management of Indian forest lands were revised and greatly improved. In 1939 the first appropriation was made for control of fires on the public-domain lands of Alaska.

Forestry in the Second World War. Lyle Watts became Chief of the Forest Service just when defense activities began to dominate Federal programs. Lumber, which was at first proposed as a substitute for steel, soon became critically short. Foresters at home and in the armed services engaged in logging and lumber manufacture, and the Forest Service organized the Timber Production War Project. In cooperation with state forestry organizations it sought to stimulate lumber production by helping mill operators to obtain needed supplies, equipment, and labor. Monthly surveys of lumber production were made to guide the War Production Board in allocating supplies. Special studies of lumber prices were made. Guayule, a rubber-bearing shrub, and Koksaghys, a latex-producing dandelion, were cultivated to augment raw-rubber supplies (Fig. 9). At the Forest Products Laboratory new wood products were developed by research. Special schools dealt with the packaging of materials for overseas shipment and use of wood in aircraft. Specifications for wood products for military use were prepared. From Alaskan forests Sitka spruce of aircraft quality was cut. The logs were fashioned into rafts and towed to West Coast mills. Access roads were built to open up remote national forest and private timberlands.

FIG. 9. Guayule seedlings were grown by foresters as a source of raw rubber during the Second World War. (*U.S. Forest Service photo.*)

The nation looked mainly to private forest industry, however, for the major production job during the war. The ready market for forest products during the war and prosperous postwar years effected a healthy recovery in private forest industries. It is appropriate now to bring forward the developments in state and private forestry.

State Forestry

The beginnings of state forestry in the United States go back to colonial days. The early ordinances of New Sweden, which became Delaware, of Pennsylvania by William Penn, and of Massachusetts all recognized the importance of the forest and provided regulations for its proper use. With the coming of independence the states continued to take an interest in forestry measures. Agricultural societies in Philadelphia as early as 1791, in New York in 1795, and in Massachusetts in 1804 took notice of the importance of tree planting and urged that steps be taken to encourage it. In 1819 the Massachusetts legislature asked its state department of agriculture to promote the growth of oak suitable for ship timbers.

In 1837 Massachusetts authorized a survey of its forest conditions with a view toward determining measures that might encourage landowners to practice forestry. In 1788 New York passed laws to regulate lumber grading to protect the good name of New York lumber in national and international trade.[20]

The surge of settlers to the prairie states after the passage of the Homestead Act in 1862 made the middle western folk acutely aware of the need for trees. Sod huts for living quarters and buffalo chips for fuel were a real hardship to people used to an abundant supply of wood. Tree planting caught the public interest.

Michigan in 1867, and Wisconsin shortly thereafter, created forestry commissions to investigate timber operations, land clearing, and the importance of proper care of timberlands within the state. Consideration was given to shelter belts and to proper management of forest areas. Similar forestry commissions were set up in New York, Connecticut, New Hampshire, Vermont, Ohio, Pennsylvania, and North Carolina.[8] These states together with Kansas and Maine had state forestry organizations prior to 1885. In that year California, Ohio, Colorado, and New York established state forest administration. With the exception of New York's, these early state forest organizations were temporary, but they laid the groundwork for permanent programs to follow.

In 1872 New York set up a commission to investigate the preservation of the Adirondack forests, which led in 1885 to the establishment of the Adirondack Forest Preserve. This preserve was made up of lands which had reverted to the state for taxes. Later on lands were purchased by the state to add to the preserve. After 1894 citizens became seriously alarmed by the way in which the state preserves were managed. A constitutional amendment was passed prohibiting the cutting of trees in the preserve and has been reenacted at subsequent constitutional conventions.

New York's example of acquiring state land for permanent forest use was followed by Pennsylvania in 1895, Minnesota in 1899, and by many other states thereafter.

State acquisition of land for forestry came about, however, more from default of owners on taxes and land abandonment than from deliberate state policy. It was inevitable that American pioneers clear forests for crops. Hill lands usually had less dense forests, were freer from malaria, and hence were often chosen for settlers' homes and fields; but soils were thin and steep slopes easily eroded. Mechanization of tillage placed the hill farmer at still greater disadvantage in competing with those on fertile level lands. The agricultural depression that followed the First World War brought the problem before the country.

Land abandonment—it began in New England as early as 1850—is a painful process. As family income shrinks, savings are exhausted. Repairs to the dwelling, other farm buildings, and fences are postponed. Equipment wears out. Livestock declines in vigor, quality, and numbers. Crop yields decline from inadequate tillage and fertilizer. Severe family privation may occur. Eventually the farm as a whole is abandoned, or reduced to part-time operation, while the breadwinner engages in some other occupation.

In several counties in the Lake States more than 50 per cent of the land became tax delinquent. Land abandonment spread at an alarming rate, and clearing forests for farm lands virtually ceased. In New York, farms once occupied 80 per cent of the land area. By 1950 this was reduced to 57 per cent and some 4 million acres still were deemed best suited to forest. Legislatures authorized land economic surveys to determine the causes and to suggest remedial measures. Vast increases in state forest holdings were most commonly proposed.

Farmers and rural bankers themselves recognized the economic waste in farming poor lands. In New York they obtained passage of the Hewitt Law to purchase submarginal farm lands for forests. By 1950, 500,000 acres had been acquired and 350,000 acres of open land planted to trees. Other states have followed similar programs. In 1934 the Federal Resettlement Administration entered the field. Many of the lands acquired by this organization are now administered as state forests under long-term lease.

As of 1949 state forest holdings totaled 14.8 million acres. States having the largest areas were: Michigan, 3,860,000 acres; New York, 3,026,000 acres; Minnesota, 2,094,000 acres; Washington, 1,700,000 acres; Pennsylvania, 1,767,000; Idaho, 950,000; Oregon, 570,000; and Montana, 520,000.[5, 17] Most state forest lands, including New York's 500,000 acres of land acquired from submarginal farms, are administered for timber production, recreation, watershed protection, public hunting lands and other multiple-use purposes.

The Weeks and Clarke-McNary laws proved a great boon to state forestry departments. In most states fire-control expenditures have remained the largest item in the budget of state forestry organizations. Federal aid has been a mainstay in providing financial support and in insuring freedom from undue political pressure on forest officers employed in state organizations.

The program in Maryland might be taken as an example of state forestry. Maryland forestry first dealt primarily with protecting the lands from fire. Soon thereafter the state agreed to perform forestry services for landowners on a fee basis. Later a state extension forester was employed to stimulate forest care by farmers. Gradually the state acquired a number of forest properties which, though small in size, were managed for both timber production and for recreation. In 1943 the state passed a law regulating timber-cutting practices on private lands. It cooperates with the Federal government in farm and other private forestry programs.

Similar service programs have developed in New Jersey, Connecticut, Vermont, New Hampshire, and many other states. State efforts to regulate forest practice on private lands seem destined to increase greatly in the years ahead. By 1951, laws providing compulsory regulation had been adopted in six states, and voluntary regulation had been adopted by nine additional states.

Farm Forestry

Farm forestry has been largely a state function. The Smith-Lever Act, passed in 1914, allotted funds through the state agricultural colleges for extension work in forestry. Initial programs emphasized tree planting by farmers. A common method of working was to hold demonstrations and present talks to farm groups brought together by the county agricultural agents. Extension foresters have wide latitude, but their programs are reviewed by their state agricultural extension director and by the Agricultural Extension Service in Washington, D.C.

It became evident by 1924 that considerable additional effort was needed. The Clarke-McNary Law provided for distributing forest planting stock to farmers and for strengthening extension forestry.

Largely as a result of the experience of the extension foresters, the Norris-Doxey Farm Forestry Act was passed in 1937. It authorized appropriations of 2.5 million dollars to provide advice to farmers on timber growing and timber marketing. Farm foresters, employed by the states under this Act but paid in part from Federal funds, visited the farmer, estimated his timber stands, marked trees that should be cut, and referred the farmer to timber buyers who might be interested in the product. This Act, which proved especially popular with farmers and with the states, brought to over 100,000 farmers a clear understanding of what is meant by good timber management practices. By 1950, 220 farm foresters were operating in 39 states. In addition, 65 forest-extension specialists were operating under an enlarged program in 46 states and in Puerto Rico.

The cooperative Forest Management Act of 1950 extended the benefits provided farmers under the Norris-Doxey Act to all woodland owners, and the original Act was repealed.

Private Forestry

The beginnings of private forestry in America are lost in the early history of our country. Certain it is that many private landowners took an interest in their forest and attempted to protect young growth from unnecessary cutting. Farmers, plantation owners, and even loggers and lumbermen have set aside specific forest areas where logging was kept at a minimum or entirely excluded. Some have practiced good forest management over a long period of time with little or no aid from a technical forester.

One of the first men in the United States to undertake the growing of timber as a financially profitable investment was Zachariah Allen.[6] In 1820 he planted some abandoned farm land to oaks, chestnut, and locust and kept an accurate record of his costs and of the income he received from his 40-acre plantation. His total costs for the venture including land, expense of planting, interest at 6 per cent, taxes, and interest on taxes was $3,804. In various sales he removed during a 57-year period timber valued at $4,948. Using the same value for land as his original investment and allowing $800 for timber standing at the end of the period, the enterprise showed a net profit of $2,543 or 6.93 per cent per year for the entire 57-year period. This profit is over and above 6 per cent interest on land, taxes, and planting costs. Zachariah Allen's private forestry venture is of especial interest inasmuch as most of the material he sold was fuel wood. He was selling at a period when fuel wood generally was abundant in the United States, but by taking advantage of local shortages, he was able to make his forest property pay him a very good return.

Among the early large-scale forestry ventures in the United States were those on the Vanderbilt, Webb, and Whitney estates that were begun near the turn of the century. Gifford Pinchot was employed by George Vanderbilt as consulting forester for his 80,000-acre estate in North Carolina, later named the Biltmore Forest. Pinchot brought in a sawmill and started timber operations. The venture was successful financially and attracted the attention of other forest owners. Later Mr. Vanderbilt employed Dr. Schenck to manage the forest property.

Ne-Ha-Sa-Ne Park, owned by Dr. Seward Webb in Hamilton County, N.Y., and the Whitney estate adjoined one another. Gifford Pinchot and Henry Graves were employed to work up plans for management of these two estates. The first silvicultural rules governing cutting practices that were ever written into a logging contract in the United States probably are those prepared for the Ne-Ha-Sa-Ne forest. Ne-Ha-Sa-Ne Park was logged first in 1898 and 1899 and again in 1915 to 1930. A third cutting began in 1947. The Whitney preserve has been under more or less continuous forest management from 1898 up to the present time. Both properties are outstanding among privately managed Adirondack properties in the quality of timber now standing on them and in the timber they have yielded the owners over the long-term period.

The accumulation of forest lands in large family estates began in Maine as early as 1850.[14] Some of these became very large properties, of which the Coe and Pingree estate is an example. The founder, David Pingree, restricted cutting of spruce to trees 14 inches in diameter and larger. After pulpwood logging came to the fore, the estate passed to David Pingree's heirs and cutting was carried to much lower limits. Although these earlier efforts at private forestry had little influence on the attitude of most large lumber and other companies, they did give some impetus to private forestry. The business panic of 1907, however, caused many private owners who had initiated forestry programs with the help of the Division of Forestry to abandon them; hence, forestry efforts became largely a public function from 1907 until almost 1933.

About 1912 Finch, Pruyn and Company started a forestry program on its Adirondack holdings. Trees to be cut were marked by foresters and the cutting budget projected on a sustained-yield basis. Largely because of hardwood competition growth has not come up to expectations. Heavy taxes, windstorm, and other vicissitudes have vexed their program, but the company continues to stand by it.

More recently, interest in private forestry on the part of lumber and other companies has grown at a rapid pace. The Goodman Lumber Company in Wisconsin, which began its first cycle of selective cutting in 1927, has been particularly successful in carrying out good forest practice in the woods and has introduced diversified utilization in its plants. It operates or has operated a sawmill, veneer mill, a pulp mill, a shingle mill, and a wood-chemical plant. It has a market within its own industrial plants for all the timber that grows on company lands and also provides a market for farmers and others who own timberland tributary to the company's mill at Goodman, Wis.[14]

The Urania Lumber Company in Louisiana began operations on a sustained-yield basis about 1917. It has been so successful in the management of its lands that today it has more timber growing on its lands than the company mills can cut, a condition rarely to be found in lumber-company operations.

The Crossett Lumber Company at Crossett, Ark., owns about 500,000 acres of land which it manages on an intensive forestry basis. The land has been divided into 50,000-acre blocks, each in charge of a forester who directs all woods operations. This company also operates a sawmill, pulp mill, and wood-chemical plant.

Other companies that began forestry programs in the 1930's or earlier include the Nekoosa-Edwards Company,[19] the Consolidated Water Power and Paper Company, the International Paper Company, the Patten Timber Company, Von Platen-Fox Company, the Dead River Company, the Minnesota and Ontario Paper Company, the Weyerhaeuser Timber Company, the Long-Bell Company of Washington, St. Regis Company, the Crown-Zellerbach Company, Gaylord Container Corporation, Collins Almanor Forest, Camp Manufacturing Company, St. Paul and Tacoma Lumber Company, and Champion Paper and Fibre Company.

The interest of these and many others in practicing sustained-yield forestry has led to widespread purchase of forest land in the South, the West, and the Northeast by large lumber, pulp and paper, and other companies.

Associations of private forest landowners date back to 1911, when the Clearwater Forest Protective Association was formed. Others followed shortly thereafter. At first, these were concerned with protection of forests against fire, but later on they became interested in good forest management also.

Recent developments in forestry will be outlined in Chaps. 15 and 16.

Conclusion

The following major periods of American forestry might be recognized:

1770 to 1890. This was a period of forest exploitation, of acquisition and disposal of the public domain, and of warnings of threatened timber exhaustion but one of little permanent progress toward establishing forest reserves and basic forest policy.

1891 to 1910. The policy of Federal land ownership and management was established and implemented by creation and management of national forests. Permanent state forest programs began.

1911 to 1932. National forests were consolidated and new lands acquired in the East; states extended protection to most forest lands against fire, insects, and tree diseases. State forestry was promoted and widely developed.

1933 to 1942. Forestry was recognized as vital to the national economy. Interest in private forestry revived, stimulated first by the Lumber Code and later by the American Forest Products Industries, Inc., and other private associations.[19] Forestry work was recognized as a useful measure to absorb unemployment. The value of forest was recognized in control of floods, dust storms, protection of watersheds, public recreation, and stabilization of communities. Working conditions for woods labor improved.

1943 to 1951. Rapid expansion occurred in forestry programs of large timberland-owning companies. Farm forestry expanded. Growing concern was felt by the United States for forestry programs in other lands. Outstanding developments occurred in research, especially in forest-products research. State regulation of private forest practice extended to 15 states.

Foresters and forestry sympathizers have always been in a minority. Forestry has reached its important place in America today because men with physical and moral courage battled against strenuous and often violent opposition. No advances in forestry met with universal acclaim. All were won by struggle. Fire control on grazing and blueberry lands, establishment of national forests from public lands, control of grazing on national forests, purchase of lands for national forests in the eastern states, planting of trees, salvage of hurricane-damaged timber, forest research, shelter-belt planting, public aids to private owners through farm foresters and others, aids to lumbermen in wartime timber production, and finally public regulation of timber cutting on private lands—all have been opposed. Many are still vigorously opposed. Vast inertia stems from public indifference. Forestry measures that require public expenditures are always opposed by economy blocks. Bitter opposition may occur from those who fear their private or corporate welfare will be impaired by public forestry measures.

Few foresters can escape these conflicts.

LITERATURE CITED

1. Allen, Shirley Walter. 1938. "An Introduction to American Forestry," 2d ed., McGraw-Hill Book Company, Inc., New York. 402 pp.
2. Butler, Ovid. 1946. 70 Years of Campaigning for American Forestry, *American Forests,* 52(10):456–459, 512.
3. Chinard, Gilbert. 1945. The American Philosophical Society and the Early History of Forestry in America, "Proceedings, American Philosophical Society," 89(2):444–488.
4. Crafts, Edward C. 1949. Coordination of Forest Survey Data for Large Areas, "Proceedings, Third World Forestry Congress," Helsinki, Finland.
5. Fontanna, Stanley G. 1949. State Forests, "Trees, The Yearbook of Agriculture," pp. 390–394, U.S. Department of Agriculture, Washington, D.C.
6. Holst, Monterey Leman. 1946. Zachariah Allen, Pioneer in Applied Silviculture, *Journal of Forestry,* 44:507–508.
7. Hosmer, Ralph S. 1947. The National Forestry Program Committee, 1919–1928, *Journal of Forestry,* 45:627–645.
8. Illick, Joseph S. 1939. "An Outline of General Forestry," 3d ed., Barnes & Noble, Inc., New York. 297 pp.
9. Kling, John B. 1951. "Cooperative Forest Fire Control-Policy Determination and Administration in the Clarke-McNary Grant-in-aid Program," New York State College of Forestry, Syracuse, N.Y. 106 pp.
10. Lillard, Richard G. 1947. "The Great Forest," Alfred A. Knopf, Inc., New York. 399 pp.
11. Paxson, Frederic L. 1924. "History of the American Frontier, 1763–1893," Houghton Mifflin Company, Boston. 598 pp.
12. Pinchot, Gifford. 1947. "Breaking New Ground," Harcourt, Brace and Company, Inc., New York. 522 pp.
13. Schoepf, Johann David. Trans. by Alfred J. Morrison. 1911. "Travels in the Confederation [1783–1784]," William J. Campbell, Philadelphia. Part I, 426 pp.; Part II, 344 pp.
14. Shirley, Hardy L. 1949. Large Private Holdings in the North, "Trees, The Yearbook of Agriculture," pp. 255–274, U.S. Department of Agriculture, Washington, D.C.

15. Sparhawk, William N. 1949. The History of Forestry in America, "Trees, The Yearbook of Agriculture," pp. 702–714, U.S. Department of Agriculture, Washington, D.C.
16. U.S. Department of Agriculture. 1949. "Trees, The Yearbook of Agriculture," Washington, D.C. 944 pp.
17. U.S. Department of Agriculture. 1950. "Report of the Chief of the Forest Service," Washington, D.C. 69 pp.
18. U.S. Department of the Interior, Bureau of Land Management. 1950. "Brief Notes on the Public Domain," 21 pp. mim.
19. Winters, Robert K., ed. 1950. "Fifty Years of Forestry in the U.S.A.," Society of American Foresters, Washington, D.C. 385 pp.
20. Wolf, Robert E. 1948. "A Partial Survey of Forest Legislation in New York State," unpublished thesis, New York State College of Forestry, Syracuse, N.Y.

CHAPTER 4. *Forest Regions*

THE FORESTS of the United States stand intermediate in floristic complexity between those to be found in northern Europe, which are relatively simple, and tropical forests, which present unending complexity. To make the study of American forestry comprehensible, foresters have recognized a number of forest types. Forest type is a term used to designate stands of similar composition and development. The Society of American Foresters lists 147 forest types, of which 97 are found in the eastern United States and 50 in the western United States. For further simplification, type groups or formations may be recognized.

Major forest formations are shown in Fig. 10. Even formations are too numerous to permit detailed discussion in a book on general forestry, especially since boundaries between formations are irregular and must be arbitrarily drawn. However desirable it might appear to follow the forest types of nature, man must work with human institutions as well. For this reason he divides the various forest regions along political lines that may be used in compiling statistics (Fig. 11) and in relating thereto other economic information. The following discussion deals with both natural forest regions and regions based on state and county lines. A characterization of five natural forest regions will be given followed by a discussion of the forest industries based on economic regions.

The Northern Forest Formations

The northern forest in the United States is characterized by sugar maple, beech, basswood, and yellow birch among the hardwoods; and by spruce, balsam fir, and hemlock among the softwoods. It includes the northern coniferous forest, the northeastern pine forest, and the northern hardwoods.

Five tree species have ranges that correspond more or less closely with the distribution of the northern forest in the United States. These are northern white cedar, eastern hemlock, yellow birch, northern white pine, and balsam fir. Perhaps the latter is the best indicator of the northern forest. It is interesting that sugar maple, which more than other

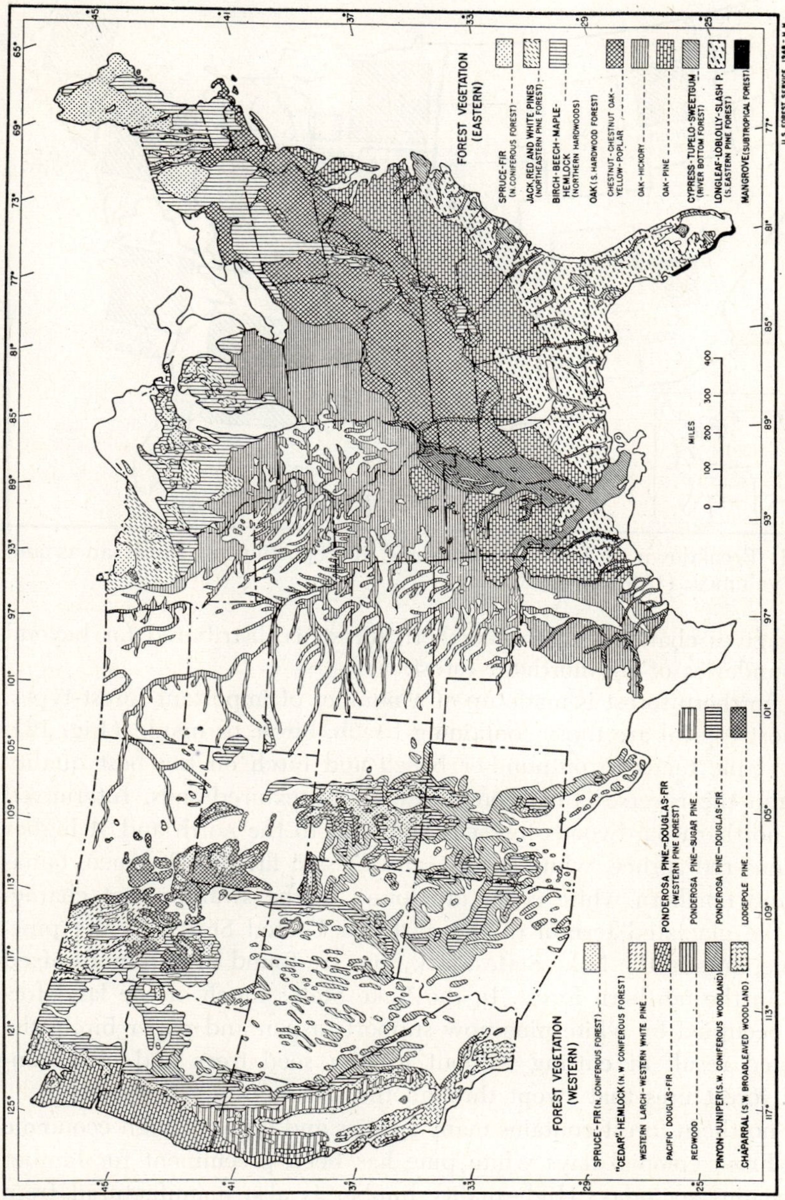

FOREST VEGETATION OF THE UNITED STATES

Adapted from Shantz and Zon's "Natural Vegetation" map of the United States in the "Atlas of American Agriculture"

FIG. 10. Natural forest regions of the United States. (*U.S. Forest Service.*)

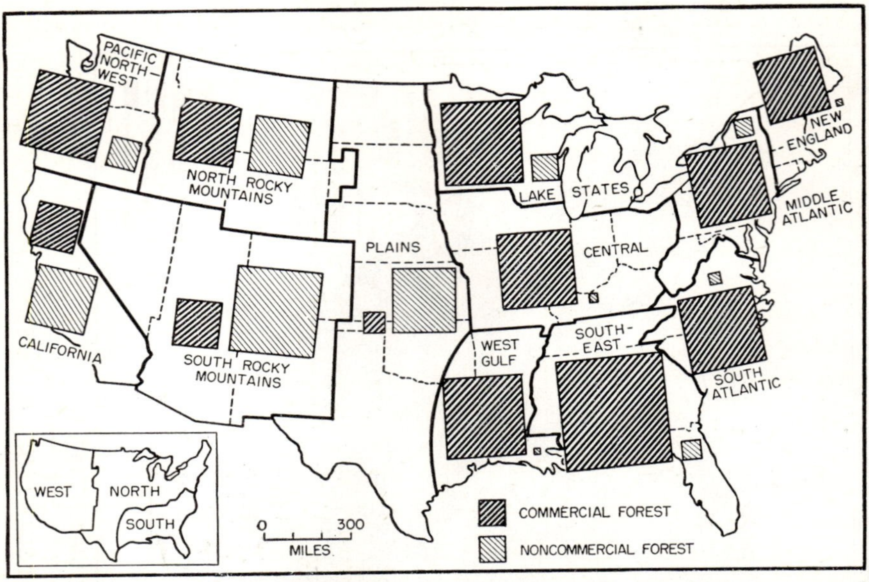

FIG. 11. Breakdown of forest regions used in reporting on forest areas and timber volumes. (*U.S. Forest Service.*)

species gives character to the northern forest, is distributed far beyond the boundaries of the northern forest.

The northern forest is made up of a number of important forest types. The most typical are those containing beech, birch, or maple (Fig. 12). Sugar maple tends to outnumber beech and birch on the best quality soils, and the reverse is true on thin or light-textured soils. Intermixed in the northern hardwoods forest is hemlock; to the north and at higher elevations red spruce, white spruce, and balsam fir. Black spruce, tamarack, and northern white cedar are found in the swamps that characterize the glaciated terrain of the northern United States. White pine, red pine, and, in the Lake States, jack pine are found in pure and mixed stands in the northern forest. In the Lake States much of the land formerly occupied by white pine now supports aspen and paper birch, the combined result of cutting without leaving seed trees and of uncontrolled forest fires that swept through the slashings.

The northern forest contains many species and types of high economic value. From colonial days white pine has been preeminent for lumber and remains so today. High-quality lumber is also manufactured from red pine and from red and white spruce. Hemlock, tamarack, jack pine, and balsam fir produce intermediate-quality lumber. Northern white cedar is usually small in size but produces a durable wood suitable for shingles, boatbuilding, poles, and posts. Among the hardwoods sugar maple and yellow birch are splendid woods for furniture manufacture, flooring, and interior trim. Beech likewise is suitable for flooring, handles, and other uses requiring a hard, dense wood. White ash, black cherry, and basswood produce specialty woods suited for a variety of purposes. Ash is strong, hence useful for handles, athletic equipment, ladder rungs, and other purposes requiring high strength. Cherry is made into furniture and is extensively used for type backing because of its high dimensional stability. Basswood produces a soft, fine-grain lumber suitable for cabinetwork and is also used for veneer, cooperage, and woodenware. Gray birch, paper birch, and aspen are considered temporary types that have come in following fire. Paper birch is especially useful for dowels and stamped veneer products such as tongue depressors and

FIG. 12. A second-growth northern hardwood forest with sugar maple and yellow birch.

novelties. Various turned products such as spools are also manufactured from paper birch. Aspen is widely used for excelsior and meat boxes.

The most highly prized pulp species are red, white, and black spruce and second to these is balsam fir. All may be used for both groundwood and sulfite pulp. Hemlock, jack pine, and also red and white pine may be pulped by the sulfate process successfully. Aspen is used extensively for groundwood pulp.

The northern forest, particularly the coniferous types, grow densely and may shade out the undervegetation. The number of stems per acre for trees averaging 10 inches in diameter is 300 to 500. Volumes in excess of 50 cords per acre may be found on lands where the largest trees are less than 18 inches in diameter. Growth rates of the individual trees are not as high as those found in other sections of the country primarily because of the long winters and short growing season. The volume growth per acre, however, often equals that of the southern and Rocky Mountain forests where wide spacing of trees results from insufficient rainfall.

White pine will grow at the rate of 500 to 800 board feet per acre per year under ideal conditions, though a growth rate of 300 feet per acre is more to be expected. Mixed northern hardwoods made up of maple, birch, white ash, black cherry, and beech will grow at the rate of 300 to 500 board feet per acre. Jack pine, aspen, spruces, and balsam generally grow at ½ to 1 cord per acre per year in well-stocked stands. The growth rates given above apply, however, only to well-stocked stands on good land and are never realized over large areas. These are goals toward which a forester might strive. The growth figures given in Table 1 are estimates of actual growth as it exists over large areas. Because of poor soil, understocking, and large areas supporting timber below merchantable size, these are only one-twentieth to one-fifth that obtainable under ideal conditions.

The northern forest region occupies the major area of forest land in New England, the Middle Atlantic states, and the Lake States. These are the most densely populated regions of the United States and are highly industrialized. Forest and forest products, nevertheless, make up a substantial part of the regional economy in each of these.

New England. The six New England states have a total of 31 million acres of forest land, almost all of which is commercial (Table 1). Nearly half of the forest area of New England lies in Maine. Since much of this is still remote from highway and rail transportation, stream driving remains a major method for transporting both logs and pulpwood. One area in Maine is reported to contain between 150,000 and 200,000 cords of virgin spruce and fir situated on high ground remote from major watercourses. This has been uneconomic to log because the brooks are too small to carry the pulpwood.[10] Forestry in the unorganized towns of northern Maine, therefore, remains on an extensive rather than an intensive basis and is likely to remain so until the country is opened up by highways. In land not opened to roads, the hardwood timber, made up chiefly of paper birch, yellow birch, beech, and sugar maple goes unutilized.

TABLE 1. COMMERCIAL FOREST AREAS AND GROWTH RATES BY REGIONS*

Region	Current growth, billion bd. ft.	Commercial area, million acres	Growth per acre, bd. ft.
New England	1.8	30.9	58
Middle Atlantic	2.7	41.6	65
Lake†	1.2	50.0	24
Central	2.3	44.2	50
Plains	0.2	3.3	61
Total North	8.2	170.0	49
South Atlantic	6.1	42.9	142
Southeast	8.2	89.4	92
West Gulf	5.6	51.0	110
Total South	19.9	183.3	109
Pacific Northwest:			
Douglas-fir	3.7	26.0	142
Pine	0.5	20.2	25
Total Pacific Northwest	4.2	46.2	91
California	1.2	16.4	73
North Rocky Mountain	1.3	29.0	45
South Rocky Mountain	0.3	15.8	19
Total West	7.0	107.5	65
Total United States	35.1	461.0	76

* From U.S. Department of Agriculture, Forest Service. 1946. Gaging the Timber Resource of the United States, Report 1, from "A Reappraisal of the Forest Situation." 62 pp.

† From Cunningham, R. N., and Forest Survey Staff. 1950. "Forest Resources of the Lake States Region," U.S. Department of Agriculture, Forest Resource Report 1. 57 pp.

The unorganized towns of Maine are owned by large timber companies and by wealthy families. The holdings are large, practice extensive rather than intensive, and development has proceeded at a modest pace.

The major forest industry of New England is pulp and paper manufacture. These companies have the largest forest holdings, use more timber than other companies, and exercise a dominant influence on forest-land management.[6] The industry is primarily devoted to specialty products, though they do manufacture some kraft and newsprint. Although the mills were built to operate on spruce and fir, many companies have already begun to use hardwoods in substantial volume and may be expected to increase such use in the future. The industries' requirements for softwood cannot be met without substantial imports from Canada. The last few decades have seen the closing of a number of individual mills in New England and the transfer of operations to the South. If hardwood pulping develops, as seems possible, little further decline in the pulp and paper industry in New England need occur.

New England is also noted for its wood-turning establishments and novelty mills. These manufacture such products as ice cream spoons, shoe shanks, various veneer novelties, spools, bobbins, wooden heels, brush and other handles, chair stock, small cooperage, and related products. Furniture manufacturing also is important.

Southern New England is highly industrialized. Here timberland holdings are mostly small in size and developed more for recreation than for commercial forestry. The large number of small wood-using industries permits rather intensive utilization, diversification of product, and consequent stability of employment and income. Some small wood-using industries have operated in individual communities for 50 years and longer.

The Middle Atlantic States. The five Middle Atlantic states have 42 million acres of forest land (Table 1), much of it occupied by northern hardwoods or oak. This region, like New England, is highly industrialized, but it also has a well-developed agriculture. Population density is high; consequently, markets for wood products are good. Perhaps nowhere in our country is utilization of forests more intensive than in the Middle Atlantic states. New York has many pulp mills. Pennsylvania and West Virginia have a large market for mine timbers as well as for pulpwood. Delaware and New Jersey produce relatively minor quantities of forest products, most of which find a ready market within their own states. The roofing-felt industry and floor-covering industry of Pennsylvania and New Jersey use substantial quantities of small-sized timber for pulpwood. The wood-chemical industry was at one time highly developed in western Pennsylvania and New York. It declined rapidly after the First World War and since the Second World War has become a minor industry. A market still remains for limited quantities of charcoal for curing tobacco and for other purposes.

Only a few large timberland holdings exist in the Middle Atlantic states. These are owned by coal companies, private estates, railways, and other companies. Holdings of 5,000 to 10,000 acres make manageable properties within this region because of the diversity of product and intensity of utilization. The furniture industry, handle industry, toy manufacturing, musical-instrument manufacturing, and a host of others are found in this great industrial section of the nation.

Industrial growth of this already highly developed region is likely to be less rapid than that of other regions. Nevertheless, the peculiar geographic position and abundance of native resources, particularly mineral wealth, seem destined to maintain the Middle Atlantic states in a key industrial position in our economy for decades to come.

The Lake States. The three Lake States have 55 million acres of forest land of which 50 million acres are considered commercial [2] (Table 1). The forests of the Lake States are made up primarily of pine, northern hardwood, and oak, with limited quantities of spruce and fir. The Lake States have been cut and burned more severely than any other one region, and as a result, most of the land is seriously depleted and must be cared for over a period of 20 to 40 years before high production is restored. Current growth is estimated at 24 board feet, or 14.4 cubic feet per acre, annually. The depletion of the standing timber naturally caused a decline in the lumber and related wood-using industries. The pine lumber industry has practically disappeared, but the hardwood lumber business is still active and probably will continue to operate on a moderate scale until young growth becomes sufficiently large to support it on a sustained-yield basis.

The pulp and paper industry is well developed in Wisconsin. This industry originally depended upon spruce but now relies mainly on jack pine, aspen, and the spruce and pine imported from Canada and from the Rocky Mountain region. The freight rate alone on pulpwood being shipped into Wisconsin was as high as $12 to $15 a cord in 1950. The Wisconsin paper industry in common with that in the Northeast and in the Middle Atlantic states manufactures chiefly specialty products; very little newsprint or kraft paper are produced. The kraft pulp made from jack pine goes into a variety of specialty products.

Michigan is one of the foremost furniture-manufacturing states in our country. The industry centered at Grand Rapids is known throughout the land. The furniture mart in Chicago is the major furniture display and marketing center in the nation. Manufacture of novelties such as clothespins, wooden bowls, veneer products, and turned products is also important in Michigan.

The Lake States region is well developed industrially and also has considerable agriculture. The forest-products industries represent a secondary item in the general economy. Large private holdings of land are mainly in the hands of pulp and paper, lumber, and mining companies. The pulp and paper companies have introduced forestry programs and employ many foresters and so have a few lumber companies. Mining companies own considerable land and a few of these also engage in forestry. Public ownership of depleted cutover lands is well advanced in all three states. National forests total 5,921,000 acres in area. State, county, and community forests total by states: Michigan, 3,868,000 acres; Wisconsin, 2,288,000 acres; and Minnesota, 2,668,000 acres.[3] Further expansion of public ownership appears likely because of the cost and time involved in restoring depleted forests. On the whole, the Lake States are seriously depleted of growing stock.[2]

The Central Forest Formations

The central forest is bounded roughly by the northern forest on the north, by the coastal plain on the east and south, and by the prairie on the west. The central forest is characterized by the various species of oak, of which red oak, white oak, black oak, and chestnut oak are most common. It includes the several eastern forest types in which oak occurs as a dominant species. Associated with the oaks are various hickories, shortleaf and Virginia pine, yellow-poplar, white ash, black walnut, and, formerly, chestnut. On the wetter soils silver maple, cottonwood, elm, and sycamore are found. White pine, sugar maple, and beech occur as secondary species in the central hardwood forest and in certain parts of it form important constituents of the total stand. Perhaps no other forest region of the United States contains a greater number of tree species having high technical uses than does the central forest. Mast-producing species such as oaks, hickories, walnut, and beech afford the central hardwood forest great potentialities for wildlife. Squirrels, wild turkey, and deer as well as lesser game animals fatten on the nuts from these trees. The mast may also form a substantial item in the diet of domestic animals, particularly swine.

The characteristic forest types of the central region are those in which one or more species of oak are common, those in which shortleaf pine occurs, and those in which yellow-poplar, beech, silver maple, and river birch occur.

The forests of the central states furnish a wide variety of products. Black walnut, shagbark and shellbark hickories, together with beech produce edible nuts prized by many people (Fig. 13). The chestnut oak produces bark of high tannin content. Hickory, ash, and oaks are used for tool and implement handles. The various oaks, of which white oak is the most valuable, are used for railway ties, tight cooperage, furniture, veneers, and other products. White oak is particularly sought for ship timbers, because it is relatively impervious to moisture, is durable, holds nails and screws well, and is very strong. It is particularly valuable for keels, framing members, and skegs.

The two most valuable woods for cabinetwork, furniture, and a variety of special uses are black walnut and yellow-poplar. Both are straight-grained, uniform in texture, easily worked, strong, and durable. Because of its high resistance to shock, decay, and warping, black walnut is the preferred wood for gunstocks. Highly figured veneer is cut from black-walnut stumps. Yellow-poplar produces a veneer suitable for core stock and for highly technical uses, including aircraft veneer.

Useful lumber is also cut from gum, silver maple, beech, sugar maple, and basswood. Black locust, cottonwood, American elm, sycamore, river birch, hackberry, buckeye, and a host of other minor species have a variety of uses. Locust is valuable for posts. Elm makes a tough veneer suitable for baskets and crating.

The conifers of the central forest region include Virginia pine, eastern redcedar, and shortleaf pine. Virginia pine seldom reaches sufficient size to be valuable for lumber, but it is used extensively for sulfate pulp. Eastern redcedar produces a fine-grained, aromatic wood valuable for

FIG. 13. The central hardwood forests are rich in an abundance of high-quality forest trees. The shellbark hickory or king nut. Turkey Run State Park, Ind.

manufacturing pencils and for cedar-lined chests and closets. Shortleaf pine produces lumber comparable in quality with that of the other southern pines. In general, the conifers occupy the poorer soils on land that had been abandoned for agriculture, and none of the three can maintain itself in competition with hardwoods.

The growth rate in the central forest regions is somewhat more rapid than that in the northern forest region and the stand density about the same or somewhat less. Growth rates of 300 to 500 board feet per acre annually can be attained on well-stocked stands. Even higher growth rates could be expected were it not that agriculture has crowded the forest off the most productive soils. Heavy use is made of the forest for lumber, ties, posts, and for other purposes. The region supports a great number of wood-using industries and consequently provides a good market for the better quality trees that are produced. Next to the more highly industrialized sections of the North, the central forest region probably provides some of the best opportunities for intensive forest management.

The six Central states have a total of 44 million acres of commercial forest land (Table 1). Small ownerships scattered through the more rugged sections of southern Ohio, Indiana, Illinois, Missouri, and Kentucky are the rule. Foresters are used as wood-procurement officers for veneer mills and sawmills looking for high-quality oak, walnut, ash, hickory, and other products. Cooperative marketing is only moderately developed. Publicly owned land occupies a very minor percentage of the total forest area. Forestry opportunities are good in this region, but probably not as good as in the Northeast or in the Middle Atlantic states, primarily because of the small size of holdings and the small size of many of the industrial plants that use forest products.

The Southern Forest Formations

The southern forest region begins with southern Delaware and extends along the coastal plain, including all but the tip of southern Florida, and west to eastern Texas and Oklahoma. The region sends a narrow tongue up the Mississippi Valley to southern Missouri and Illinois. Primarily, the southern forest region is a pine region with swamps occupied by bald cypress, tupelo, red and black gums, swamp oaks, and various other species, including evergreen magnolia, cottonwood, willows, and bays. From the standpoint of timber production the southern region is potentially the most important in the United States. It is a large region, much of which is destined to remain in forest, it is near the great wood-consuming centers, and climate favors rapid tree growth.

The southern region is made up, for the most part, of relatively level, more or less sandy soils, though clays and heavy loams occur in the piedmont region, up which the southern forest advances in several areas. It is characterized by the longleaf-loblolly-slash pine type groups and the cypress-tupelo-sweetgum groups. Upland hardwoods are highly valuable including oaks, hickories, cucumber tree, yellow-poplar, and many others useful for furniture and construction.

The Southern Pine Forest. The southern pine forest is made up predominantly of four species: loblolly, shortleaf, longleaf, and slash pines; but pitch pine and pond pine occur to a limited extent. Loblolly pine, which is rapid growing and the most aggressive in invading old fields, grows in moderately dense stands and attains a height of approximately 100 feet and diameters up to 30 inches. The timber is of medium quality and, when properly manufactured and seasoned, is widely used for general construction (Fig. 14).

FIG. 14. Trucking loblolly pine in Georgia. (*A.F.P.I. photo.*)

Shortleaf pine produces lumber similar in quality to that of loblolly pine, but the tree grows somewhat more slowly and attains slightly smaller size.

Longleaf pine, which grows all along the coastal region, extending into western North Carolina and down along the coast as far west as Texas, is one of the most interesting pines. The needles attain lengths up to 12 inches. For the first four or five years of the tree's life it may remain in the grass stage, but during this time the tree develops a deep taproot and may attain a diameter of almost 1 inch at the ground surface. The huge terminal bud remains close to the ground and well protected by thick layers of bud scales. While in this stage, the plant resembles grass and may go entirely unnoticed by someone unfamiliar with its habits. After it becomes well established, the tree begins to grow in height, shooting up at the rate of a foot or more a year, and it continues to grow rapidly until it reaches maturity. Longleaf pine may attain 24 inches or more in diameter and a height of 90 feet. The trees tend to grow rather widely separated, as considerable soil space is needed to obtain ample moisture.

Longleaf pine is one of the most fire-resistant of all conifers. While it is in the grass stage, fires may repeatedly burn over the area, consuming the needles yet failing to kill the terminal bud. From the time the tree starts to grow from the grass stage until it has attained a height of 10 or more feet, it may readily be destroyed by fire (Fig. 15). After the tree has attained a height of 20 or more feet, however, it develops a very resistant bark. Consequently, fires may burn through at frequent intervals with only limited damage to this tree.

Longleaf pine produces a strong, heavy, dense lumber with a high resin content. It is highly prized for general construction and is the principal producer of gum for naval stores.

Slash pine, the most rapid-growing of all southern pines, resembles longleaf pine in many respects. Slash pine has a restricted natural distribution, but it has been successfully planted outside its natural range. It is almost equal to longleaf pine as a producer of naval stores. Both of these pines reproduce poorly after the stands reach maturity. Undergrowth of palmetto, various grasses, and resinous shrubs may completely occupy the soil and tend to exclude pine seedlings unless removed by controlled burning.

The growth rate of southern forest trees is high. At 20 years of age southern pines may be harvested for pulpwood, at 40 years for piling and lumber. Plantations have given especially high yields in pulpwood. During the past two decades pulp production in the southern states has expanded at a rapid rate. High-grade kraft paper may be made from southern pines by the sulfate process. The fibers are long and the paper is very strong.

The growth rate per acre in southern forests is not greatly in excess of that found in the central and northern forests. Five hundred board feet per acre annually is not unusual, though it is far above the average. Much of the southern forest land supports stands of low density with correspondingly small yields. The growth rate varies widely with the quality of the soil, but a growth rate of 400 board feet per acre per year would be a good goal for the southern forest owner to work toward.

Open southern forests support grass useful for grazing livestock. Cattle can be grown along with pine trees in the South, though tree growth is generally better where grazing is excluded. Hogs are extensively pas-

FIG. 15. Fire is both a destructive agent and a silvicultural tool in southern longleaf pine forests. (*A.F.P.I. photo.*)

tured in southern forests. Where pasturing is moderate, the damage done is less than the value of the meat produced.

The South Atlantic States. Virginia and North and South Carolina have a total of 43 million acres of commercial forest land (Table 1). The forests occur in three topographic subregions: the coastal plain, the piedmont, and the Appalachian Mountains. These subregions support respectively the southern, central, and northern forest associations. The coastal plain region is low-lying and supports primarily pine, of which loblolly and shortleaf predominate. The piedmont region has a mixture of pine and hardwoods, with hardwoods tending to take over the land after the pine crop is harvested. The Appalachian Mountain area is high and rugged and is made up primarily of hardwoods with scattered spruce and fir at high elevations. Ownership is primarily in small holdings in the mountains and piedmont sections and in large holdings in the coastal plain.

Both lumber and pulp manufacture are major industries in the South Atlantic states. A well-developed furniture industry is centered at High Point, N.C. Novelty manufacture is common among the mountain folk. This section of the South is developing rapidly industrially and in forest management and forest industry. The kraft pulp mill at Georgetown, S.C., the world's largest, uses 800,000 cords of wood a year.[9]

The Southeastern States. Tennessee, Mississippi, Alabama, Georgia, and Florida have 89 million acres of commercial forest land (Table 1). This is the real southern pine region. It is characterized by loblolly and slash pine in the south, by hardwoods in Tennessee and the northern sections of adjoining states. This region embraces a large portion of the cotton and tobacco belt of the South. Both of these crops tend to deplete the soil and have been planted on greater acreages than is necessary to maintain an adequate supply for our country's needs. More and more these states are looking to timber as a major item in their economy and as one that can take the place of soil-exhausting crops.

Both large and small ownerships are found in the Southeast. The small ownerships are characteristic of the more rugged sections of Tennessee, Georgia, and Alabama. The large holdings are found farther south on the coastal plain region. The Southeast has many problems in common with the South Atlantic states, that is, a tendency for the pine to be replaced by hardwoods and consequently the need for silvicultural measures to maintain the most valuable timber crop. It also has a considerable volume of hardwoods, particularly the Mississippi-bottomland hardwoods, for which only moderate use has been found. This region is growing industrially and is growing in the importance of its lumber and pulp and paper industries. The naval-stores industry is primarily a small business, which probably has reached its peak of development. Synthetic chemicals are being substituted for the natural turpentine and rosin produced from pine resin.

The land on the whole has been heavily cut and much of it averages only 1,000 board feet per acre in growing stock. Recovery is rapid, however, and yields of 300 to 400 board feet per acre annually are easily possible.

The Gulf States. Arkansas, Louisiana, Oklahoma, and Texas have a total of 51 million acres of commercial forest land (Table 1), made up primarily of southern pine and hardwood. Some of the best pine-producing lands are to be found in southern Arkansas and Louisiana. Sustained-yield forestry has been practiced here by several lumber companies for a number of years. Much of the soil is better than that to be found in the coastal plain of the Southeast; hence forest growth is more rapid. This section has been only moderately industrialized except for the narrow belt along the Gulf Coast, and agriculture is a major activity, particularly the growing of cotton. Both longleaf and loblolly pine do well in this section, but loblolly perhaps occupies a more important place than any other one species. Though some large paper mills are found in the west Gulf states, lumber manufacture rather than pulp and paper manufacture is the major industry. Interest in forestry has been developing rapidly in this section of the South. The value of forest land is relatively high because of the rapid rate of growth, and since many small investors appreciate the value of a good tract of timberland, speculation in it is widespread. This region borders the plains region on the west. Meager rainfall helps to maintain pine in a region that might otherwise go to hardwoods. The hardwoods seem to be much less aggressive than in the Southeast or in the South Atlantic states. The northern section has hardwoods and may be expected to produce much good-quality hardwood timber in the future. Development of wood-using industries in this section is still only moderate, and good opportunities exist for future expansion.

The plains region is not a natural forest area. Trees are confined largely to stream courses and shelter groves. These will be discussed elsewhere.

The Rocky Mountain Forest Formations

The Rocky Mountain forest is bounded by the plains region on the east, the Sierras and Cascades on the west, Canada on the north, and Mexico on the south. The forests in the Rockies are located almost exclusively in the mountains at elevations having suitable precipitation and temperature to support the growth of trees. Throughout the Rocky Mountain region the forests change rapidly in character with changes in elevation. At the lowest levels, where forests are located principally along the stream courses, aspen, cottonwood, willows, and alder prevail. In the foothills may be found the Rocky Mountain juniper, the bur oak, and western live oak and in the South mesquite and the piñon pine. These trees, growing in open groves or as scattered individuals, carry out the struggle between the true forest vegetation found at higher elevations and the desert scrub, sagebrush, and short-grass plains at the lower elevation. Moisture and protection from too severe exposure to sun and wind favor the trees. Droughts, fire, and open exposures favor desert scrub and grasses.

Proceeding up the mountain slopes, one encounters ponderosa pine, blue spruce, more aspen, and oftentimes scrub oaks and maples. At middle elevations lodgepole pine, western white pine, grand fir, western larch, Douglas-fir, and white fir occur. Here are found the most productive forests. In Idaho and western Montana are some of the best forests to be found anywhere in the Rockies. Western white pine, western hemlock, and ponderosa pine form splendid stands with volumes per acre in excess of 20,000 board feet. The trees attain large size, 3 feet or more in diameter, and heights in excess of 140 feet. They produce excellent lumber. From Colorado southward the trees are generally much smaller in size and the lumber value of the timber is modest. These forests do contain many trees of pulpwood size, but water supply is generally inadequate to support pulp mills. Some lodgepole pine and Engelmann spruce from Colorado have been shipped to the Lake States for pulp manufacture.

At higher elevations lodgepole pine becomes more common than ponderosa pine, the Douglas-fir and Engelmann spruce are more common than blue spruce, and limber pine appears. At high elevations mountain hemlock, alpine fir, Engelmann spruce, red fir, whitebark pine, and bristlecone pine are the common species. Just as the forest thins out at low elevations, so also does it thin out at high elevations. Here the trees must find shelter against the high winds that sweep across the mountain crests and against the driving snow which tends to flatten them down to the ground. They must be able to perform their life functions in the short summer period when the mountain peaks are free from snow. Ravines and protected slopes where the snow does not drift too deeply are forested at higher elevations than exposed ridges. Timber line occurs at about 11,000 to 12,000 feet in the central Rockies, at 8,000 to 10,000 feet near the Canadian border.

In the intermountain or great-basin region the forests are found only on the isolated mountain ranges that occur in this area. Their distribution is limited to a relatively narrow belt varying from about 7,000 to 10,000 feet elevation.

The forests of the northern Rocky Mountain region produce trees of large size growing in stands of high density (Fig. 16). They furnish important quantities of high-grade lumber and pulpwood. In Wyoming, Colorado, Utah, Nevada, much of Arizona and New Mexico the trees attain small sawtimber size and low density. Locally the forests furnish much-needed fuel, posts, poles, railway ties, and lumber for general construction. Water being scarce, pulp mills are absent. Pulpwood has been shipped to Wisconsin and other states from these forests.

The chief products of the central and southern Rocky Mountain forests in 1951 were water, forage for domestic livestock, and recreation. Timber operations for increasing water yield were being given serious consideration. Use for recreation interferes little with the forest. The value of these mountain forests as range for domestic livestock is often

FIG. 16. Partial cutting on the Kaniksu National Forest, Idaho. (*U.S. Forest Service photo.*)

of considerable importance to local ranchers. Forage yields are low because of short growing seasons, but the grass becomes available when plains and foothill ranges have passed their productive peak. Where wisely administered, grazing in mountain forests harvests a useful resource without serious damage to water yield, timber, or recreational use.

The southern Rocky Mountain region has a total of 73 million acres of forest land, of which only 16 million acres are commercial. Most of the forest land is in public ownership, and forest industries consist primarily of sawmills and industries preparing forest products for local use. A contract based on establishing a pulp mill in Colorado has been signed by the Forest Service. The sale is for 4,565,000 cords, enough to keep a pulp mill of 200 to 250 tons' daily capacity busy for 30 years.[12] About two-thirds of the timber involved has been killed by bark beetles. Increased use of other timber resources is planned, but water and recreational use remain major factors in forest management of this region.

The northern Rocky Mountain states, Wyoming, Idaho, and Montana and the Black Hills region of South Dakota, have a total of 53 million acres of forest land, of which 29 million acres are commercial. Much of this forest land produces valuable timber with rapid growth rate, and potentialities for sustained-yield forestry are good. The commercial forest land lies mostly in national forests, 18 million acres of it. An additional 4 million acres are found in other forms of public ownership. Private ownership totals 7 million acres, of which half is in large lumber company holdings. This region is being heavily cut over so that the forest industries are declining rather than increasing in activity, and it is unlikely to develop very rapidly industrially or agriculturally. Manufacture of forest products is primarily for lumber, poles, and ties. The use of high mountain country by recreationists is almost certain to grow. Range management is a major activity of the region and will undoubtedly be intensified in the years to come.

The Pacific Coast Forest Formations

The forests of the Pacific Coast are the grandest and most awe-inspiring to be found anywhere in the world. These are exceptionally heavy timber producers. Here are found the nation's most valuable forest trees (Fig. 1): the Douglas-fir, the sugar pine, the western white pine, redwood, Sitka spruce, western redcedar, Port-Orford-cedar, and, in the Sierras, ponderosa pine, white fir, red fir, and the bigtree.

The most valuable commercial stands occur in the North Pacific region, where Douglas-fir and western hemlock are in mixture with Sitka spruce, silver fir, and western redcedar. Here stands of great density occur, up to 100,000 board feet an acre or even higher. Douglas-fir trees

6 to 8 feet in diameter and containing seven to twelve 16-foot logs are not uncommon. The forests of this region attain their density and impressive development because of the high rainfall and favorable temperature. Winter snows are rare along the coast and dry periods infrequent and of relatively short duration. Beyond the coast range are dry interior valleys that tend to be devoid of trees. As the Sierras and Cascades are approached, the forests again resume sway and hold it up to timber line, which occurs at about 12,000 feet in the South and 8,000 feet in the North.

The three Pacific Coast states together contain more than half the reserve sawtimber volume of the United States. From these uncut forests of fir, pine, and redwood comes the major volume of high-grade softwood lumber.

The forests of the North Pacific region require very heavy logging equipment for their utilization. High-lead logging equipment has been destructive, as trees not large enough to log are broken over as the huge logs are skidded out. Such logging, however, removed only about half the woody volume of the logged area. In recent years the large lumber companies have found it profitable to send in a crew with moderately heavy equipment to take out the small-sized trees suitable for pulpwood and small saw logs before the large trees are cut. The heavy equipment is next brought in to harvest the main timber crop. Since a considerable volume of useful material may still be left in tops and broken logs, a third operation may then be carried out. This normally requires use of a portable mill equipped to saw small-dimension material from broken logs and tops and from small-sized trees that have been injured. Prelogging, logging, and postlogging are terms used to designate these three operations.

Logging and lumber manufacture represent a major item in the economy and receive important attention from legislatures and citizens. All three of the Pacific Coast states have adopted forest-practice laws which, by removing the competition of the irresponsible operator, are making the practice of forestry on private lands more economical for all owners. The pulp and paper industry is likewise developed on the Pacific Coast, but not as extensively as in the South and the Northeast. Douglas-fir plywood manufacture is a unique feature of West Coast industry. Veneer in huge ribbons is unwound from giant peeler bolts. Dried, coated with adhesive, crossbonded, and pressed, it forms the plywood of countless uses in American markets (Figs. 46, 47, 48). The West Coast is a rapidly developing region. It is perhaps closer to reaching its industrial maturity than the South. The development of the wood-using industry in the West aside from plywood, lumber, and pulp and paper manufacture has proceeded at a moderate pace. Opportunity exists for additional industries engaged in intensive wood processing in the West.

Many communities in the West that depend upon sawmills for their main source of income are destined to undergo economic decline unless some substitute industry can be found. The virgin timber resource is rapidly being depleted, and second growth and new industries are not in all cases available to fill the gap. The next 20 to 30 years are likely to see major economic adjustments in individual companies and communities. The exploitative phase of forestry has still to run its full course.

The Pacific Northwest. Oregon and Washington have a total of 54 million acres of forest land, of which 46 million are commercial. About 27 million acres are in various forms of public ownership, primarily national forests, and the remainder is held by lumber companies, pulp companies, farmers, and nonfarmers. The total area in private ownership is 20 million acres. Special skill is required in the North Pacific forest to protect the land from fire. Nature is here more friendly to the forester than in the Sierras, but when the Pacific Coast forest does become dry, it forms a tinderbox far more difficult to control than anywhere else in the United States. The huge Tillamook fire of 1933 swept over more than 245,000 acres and destroyed 10 billion board feet of mature timber before it finally burned itself out. This fire paved the way for other serious fires in 1939 and 1945. Over 150,000 acres burned in 1945.

Some of the country's largest private timber companies are located in Washington and Oregon.[7] Favored by the rapid growth of young timber, these are progressively placing operations on a sustained-yield basis. It was here that the tree-farm movement originated, and many companies maintain large forestry departments integrated with all woods-management activities. Second-growth timberland is in high demand by lumber and pulp companies. Integrated use of wood in grouped mills manufacturing lumber, veneer, and pulp is developing rapidly. Banks and other financial institutions take an active interest in the forest-products industries and encourage commitments for long-term operation. The general outlook for forestry is very bright in the Pacific Coast states.

California. California has 45 million acres of forest land of which 16 million are commercial (Table 1). About half of the commercial forest land is in national forests, about 3 million acres are owned by lumber companies, and the other by farmers and nonfarm holders. Exploitation of these forests is primarily by large companies which either own timberland or purchase stumpage from the national forests. This state in common with the Pacific Northwest states and the northern Rocky Mountain region is still seeing a decline in its total volume of standing timber.

Along the Pacific Coast from the Oregon border down almost to San Francisco lies the redwood region, the world's most impressive forest, the sight of which incites a spirit of awe and reverence. Here, magnificent red-barked trees from 6 to 12 or more feet in diameter lift their crowns 200, 300, and even 350 feet above the ground. Shafts of sunlight filter through the crowns like light from the lofty windows of a huge cathedral. The forest floor is covered with a deep bed of needles which cushion the step. Individual trees contain enough wood to build a modern house, and stands on a single acre would build a small village. As one travels through the forest, his view is cut off in almost every direction by the immensity of these huge tree stems. The bases of the trees show a moderate degree of buttressing. In old-growth stands, little vegetation occurs beneath the mature trees, but where the canopy has been broken huge ferns and redwood tree seedlings can be found. This is truly a forest of the giants.

Redwood logging is a difficult and slow operation. The huge trees must be very carefully felled if their stems are not to be shattered when they strike the ground. Oftentimes it is necessary to prepare a cradle made up of branches and small trees as a bed to break the fall of the giants. The bark on these massive trees is 5 to 6 inches in thickness. The logs must, therefore, be handled with sling equipment rather than grabhooks. After logging, a huge amount of debris is left on the ground to be disposed of, ordinarily by burning, before a second crop can be started.

Redwood trees attain ages in excess of 1,000 years. Hence, a forest of these giants once felled will not be regrown short of 500 or more years. Fortunately, redwood can be used in smaller sizes and grows very rapidly. The growth of redwood and Douglas-fir forests may exceed 1,000 board feet per acre annually under ideal growing conditions.

The Sierra forests contain giant Sequoia, ponderosa pine, sugar pine, western white pine, whitebark pine, Jeffrey pine, lodgepole pine, Douglas-fir, white fir, red fir, and incense-cedar. These also are impressive forests. The sugar pine may attain diameters of 6 or more feet and contain five to seven logs 16 feet long. The Douglas-fir and ponderosa pine attain large size and have high value for timber use (Fig. 17). The stands are considerably less dense than those along the coast. Since the Sierra slopes are characteristically steep, logging presents mechanical difficulties. Railway logging with high-lead skidding was characteristic in the past. Today much of the Sierras is logged by truck and tractors and by the use of arch skidding equipment. The Sierra forests are drier than those of the northern Pacific Coast. Hence, protection against fire is more difficult and requires the greatest vigilance. These high slopes once burned over are oftentimes difficult to reproduce to trees, especially if they become overgrown with mountain mahogany, manzanita, *Ceanothus,* and other shrubby species that choke out the young tree seedlings. A high degree of skill is, therefore, required of the forester to maintain

FIG. 17. Felling a large ponderosa pine with an electric chain saw. (*A.F.P.I. photo by K. S. Brown.*)

the Sierra forests in high productivity and to protect the watersheds from erosion and flash floods.

The Forests of Alaska

Extending along the coast from British Columbia as far north as Kodiak Island and the Alaska Peninsula are the coast forests made up of Sitka spruce, Port-Orford-cedar, western hemlock, mountain hemlock, Alaska cedar, western redcedar, and red alder. The forest in commercial sizes is located in an exceedingly narrow belt along the coast and extends up to about 2,000 feet in elevation and for a distance of 5 to 6 miles back from the seacoast or the inland passage. These forests, however, contain valuable logs from which aircraft lumber could be sawn. The timber on the Tongass National Forest is estimated to be 78,500 million board feet.

It is also estimated that 75 per cent of the commercial timber lies within 2½ miles of tidewater. The average stand is 20,000 board feet per acre, of which 74 per cent is western hemlock and 20 per cent Sitka spruce. The large percentage of hemlock makes Alaskan coastal forests primarily valuable for pulpwood. The U.S. Forest Service estimates that the forest will supply raw material to support a permanent pulp industry employ-

FIG. 18. Spruce and aspen forest 40 miles west of Fairbanks, Alaska. (*U.S. Department of the Interior, Bureau of Land Management photo.*)

ing 30,000. Interest in establishing a mill in Alaska arises periodically.[4] American Viscose Corp. started building a mill at Ketchikan in June, 1952.

Behind the coast range lie the interior forests of Alaska made up of white spruce, black spruce, alpine fir, and aspen. These forests, although extensive in area, have trees of only moderate size (Fig. 18). These trees have high value to the local population for construction timber, fuel wood and other purposes, but the forests are unlikely to produce very much timber for commerce because of their remoteness, small size, and slow growth rate. Since these lands have had very little protection against fire, fires have burned over immense areas before dying out, but since 1940 fire control has been attempted. Fortunately, the attitude of the local population has changed. Control of fires is the very essence of land development. Certainly the forests are essential if human beings are to live in the area to trap fur, mine minerals, or engage in agriculture.[1]

Tropical Forests

Tropical forests in the United States are confined entirely to the southern tip of Florida below Lake Okeechobee and are subtropical rather than tropical. The land is mostly swampy and covered with thickets of mangrove and other salt-water shrubs that grow along the seacoast in tropical regions. Palms, including the coconut palm, are found along the coast. The tropical forest in Florida has little commercial value for timber production, though it does serve to protect the soil from erosion when heavy seas and hurricanes occur. The elevation of much of this land is only a few feet above sea level. Were it not for the mangrove and other shrubby species capable of growing in salt water, the southern tip of Florida might be little more than a huge sand bar. Geologists tell us that the coast of Florida is rising. Consequently, the southern tip is extending southward, adding new land over the course of centuries. Within the swampy lands of southern Florida live many large birds and animals, including alligators, crocodiles, flamingos, and other creatures that require the solitude of remote areas for their dwellings.

The United States does have under the American flag some real tropical forests in Puerto Rico, the Virgin Islands, and Hawaii. Puerto Rico, an island about the size of Connecticut, has a great diversity of vegetation. At its eastern end rainfall is in excess of 200 inches annually and dense tropical rain forests can be found. Here are giant tabanucos, guaraguao, ausubo, and the valuable *Magnolia splendens.* The forest is highly diversified, supports large trees, and has a rapid growth rate (Fig. 19). Many of the woods are hard, heavy, durable, have beautiful figured grains, and, once properly seasoned, are suitable for cabinetwork.

In a range of mountains extending along the island from east to west are found most of Puerto Rico's forests. On the rest of the island the demands of agriculture and grazing are so heavy that space does not exist for growing timber. Coffee is cultivated at moderately high elevations under the shade of trees. The wood of the *Erythrinas* and *Ingas,* which are used for this purpose, has only limited value, but the trees, being legumes, improve the soil.

The moderately dry sections of Puerto Rico will grow mahogany, silk cottonwood, Spanish cedar, teak, and other valuable trees. The tabanuco will also grow on moderately dry lands. On the driest sections of the island only cactus and lignum vitae will survive. The latter is one of the heaviest and densest woods that occurs anywhere. It will sink in water, and it is thoroughly impregnated with waxes which make this wood especially valuable as stuffing boxes for the propeller shafts on ships.

The forests of Puerto Rico are intensively utilized for fuel and to a limited extent for construction timbers. The population is so dense and the demands for fuel so heavy that trees, even those as small as 2 inches, are cut down and burned for charcoal to provide cooking fuel. It will be a long and difficult process to build up the forests of Puerto Rico to such an extent that they can supply the needs of the native population for construction timber and export wood. Fortunately, both the U.S. Forest Service and the Insular Forest Service are taking steps to this end.

FIG. 19. A 48-inch tabanuco tree, Luquillo National Forest, Puerto Rico. (*U.S. Forest Service photo.*)

A tropic experiment station maintained at Rio Piedras is accumulating much valuable information on how to protect and grow tropical forests.

The three Virgin Islands, Saint Croix, Saint Thomas, and Saint John total only 133 square miles in area. Saint Croix has considerable agriculture and grazing but little forest. Saint Thomas is rugged and the present forest growth is mostly tropical scrub. The best forests are found on the 19-square-mile island of Saint John. The tree species include those common to the dry sections of Puerto Rico. The wood is used chiefly for charcoal, although limited amounts are converted into tourist novelties by local handicraft industries.

The Forests of Hawaii

The total area of the Hawaiian Islands is approximately 4 million acres. More than half of the area is forested, but only 1.1 million acres is considered accessible. A large part of the forest area is covered with desert scrub, of which mesquite is a typical species at the lower elevations. Most of the forest is found between 2,000- and 8,000-feet elevation. The original Hawaiian forests contained many woods valuable for cabinetmaking, among them sandalwoods and koa, known commercially as Hawaiian mahogany. The forests were badly abused through overcutting and grazing by wild goats, pigs, and cattle; and eventually forest degradation caused failure of mountain streams used to irrigate the valuable sugar and pineapple plantations. At the present time Hawaiian forests are managed mainly for protection and for recreational use. A small amount of timber is harvested locally and sawed in the four sawmills found on the islands. The rainfall on many of the Hawaiian Islands is excessive. One mountain on the island of Kauai has more than 800 inches of rainfall a year. The top of this mountain is perpetually wet and the clay so sticky that it can be climbed only with the greatest difficulty. Vegetation covers most of the mountain, but rainfall is too heavy to permit tree growth. The forest officers in Hawaii are attempting to plant burned-over areas and denuded forest lands with tree species that will be particularly valuable in maintaining the soil in condition to absorb moisture rapidly.[5, 8, 11]

The other United States tropical possessions in the Pacific area are widely scattered, small in size, and inconsequential as timber producers.

LITERATURE CITED

1. Chapman, Oscar L. 1950. "Annual Report of the Secretary of the Interior" (fiscal year ended June 30, 1949). U.S. Government Printing Office, Washington, D.C. 407 pp.
2. Cunningham, R. N., and Forest Survey Staff. 1950. "Forest Resources of the Lake States Region," U.S. Department of Agriculture, Forest Resource Report 1, Washington, D.C. 57 pp.
3. Gill, Tom, and Ellen C. Dowling, compilers. 1949. "The Forestry Directory," The American Tree Association, Washington, D.C. 420 pp.
4. Heintzleman, B. Frank. 1949. Forests of Alaska, "Trees, The Yearbook of Agriculture," pp. 361–372, U.S. Department of Agriculture, Washington, D.C.
5. Judd, C. S. 1935. Forestry in Hawaii, *Journal of Forestry,* 33:1005–1006.
6. Lewison, Wayne C. 1949. What Is Private Forestry Doing in the Northeast from the Viewpoint of an Industrial Forester? "Proceedings, Society of American Foresters' Meeting, Boston, 1948," pp. 189–190, Washington, D.C.
7. Moir, Stuart, DeWitt Nelson, Michael Bigley, William H. Price, and W. D. Hagenstein. 1949. What Is Private Forestry Doing in the West? "Proceedings, Society of American Foresters' Meeting, Boston, 1948," pp. 169–181, Washington, D.C.

8. Rhett, Beverly. 1947. Sugarwater Forests, *American Forests,* 53(1):8–10, 47–48.
9. Seigworth, Kenneth J., Ray F. Bower, Harley Langdale, Jr., and Paul M. Garrison. 1949. What Is Private Forestry Doing in the South? "Proceedings, Society of American Foresters' Meeting, Boston, 1948," pp. 159–169, Washington, D.C.
10. Sewall, Joseph. 1949. What Is Private Forestry Doing in the Northeast? "Proceedings, Society of American Foresters' Meeting, Boston, 1948," pp. 184–186, Washington, D.C.
11. United Nations, Food and Agriculture Organization. 1948. National Situations: Information on Forestry and Forest Products in Asia and the Pacific Area, *Unasylva,* 2(6):330–331.
12. U.S. Department of Agriculture. 1950. "Report of the Chief of the Forest Service," Washington, D.C. 69 pp.
13. U.S. Department of Agriculture, Forest Service. 1948. "Forests and National Prosperity—A Reappraisal of the Forest Situation in the United States," Miscellaneous Publication 668, Washington, D.C. 99 pp.

CHAPTER 5. *Protecting Forests*

FOREST TREES, in common with other long-lived plants and animals, have no fixed life span. Some, such as aspen, rarely exceed an age of 70 to 90 years, but eastern hemlock may live for 700 years and the giant Sequoia to 4,000 years. Few trees die because of age alone. Instead they become broken by storm, are struck by lightning, are weakened by drought, are attacked by insects, or break off from the onslaughts by heart-rotting fungi.

Groups of trees growing together in forest stands also have no fixed age. But stands, like individual trees, reach a stage of maturity in which they contain the maximum of sound-wood volume. Thereafter, the stand thins out. Individual trees become defective. Fallen stems and broken tops indicate decadence. In such stands many trees weaken at about the same time, thus presenting insects and disease organisms an opportunity to build up to outbreak proportions. Open conditions favor wind throws and allow the sun to dry the surface litter so that fires, if started by lightning or other causes, spread rapidly.

The purpose of forest protection is to forestall these and other losses so that man will have plenty of trees to serve his own needs. This is accomplished by direct attack on fires, insects, and other destructive agents and by managing the forest so that conditions favoring damaging outbreaks are minimized.

Controlling Fire

Man has always feared forest fires, and with good reason. During the last hundred years fires have swept over vast areas and taken a heavy toll of property and human life. The Peshtigo fire in Wisconsin in 1871 burned over a million and a quarter acres and took 1,500 lives.[9] The same year a 2-million-acre fire in Michigan took 138 lives. In 1894 the Hinckley fire in Minnesota burned over 160,000 acres, taking 418 lives. The great Idaho fire in 1910 burned 2 million acres and cost 85 lives. The 1918 Cloquet fire in Minnesota took 432 lives. The Tillamook fire of 1933 in Oregon burned 300,000 acres. This same area burned again in 1939 and in 1945. In 1947 a huge fire in Maine burned over 220,000 acres, destroying 1,000 homes and taking 16 lives.[4, 23] Total damage was estimated at 30 million dollars. These are but examples of large forest fires that have taken human lives. Even small fires have been known to take life. On August 5, 1949, Montana's 5,000-acre Mann Gulch fire killed 13 people, 12 of them experienced smoke jumpers. The crew was caught despite exercising every reasonable precaution against a disaster. A sudden shift in wind by almost 90 degrees changed a long flank of the fire into the head, and strong spot fires, started across a creek, drew the fire from the ridge down the gulch, trapping the fire fighters.[6] The same year a district ranger was killed fighting a small fire on the Saint Jo National Forest.

Man's Attitude toward Forest Fires. In early days fires were generally looked upon as a benefit in that they burned off woody growth obstructing agriculture. Even today there are those who feel that fire in the woods is desirable to kill snakes and chiggers, to improve grazing, and to clean out the undergrowth. As the country became settled, however, man's attitude toward forest fires changed from tolerance to respect and fear.

Federal and state forest officers were the first to assume responsibility for protecting wild lands against fires. Expenditures by state and Federal agencies for fire prevention and control exceed by far those for all other forestry purposes. For the fiscal year 1951 the Federal government budgeted 10 million dollars for cooperative fire control. The total expenditures for fire control on Federal, state, and private lands in the United States during the year 1949 were probably in the neighborhood of 45 million dollars.

Despite these large expenditures forest fires still cause heavy damage. In the year 1947, 200,799 forest fires burned over a total of 23,226,000 acres. This is an area the size of the state of Indiana. The damage amounted to more than 55 million dollars to timber, towns, schoolhouses, and scientific laboratories.[2] In 1949, 184,000 fires burned some 15 million acres.[22] Fires on unprotected lands accounted for over half the number and 80 per cent of the area burned.

Forest fires destroy wildlife, mature timber, and young growth. They strip the soil of its protective cover, thereby augmenting floods. A very destructive flood occurred in Salt Lake City in 1945 on a watershed of about 600 acres that had been burning over 11 months before the flood came. The fire itself did little direct damage, but the flood caused a damage of $347,000. A 225-acre fire near San Antonio was responsible for a flood causing damage of $47,000.[10]

Notwithstanding the disastrous effects of uncontrolled fires, there are times and places where controlled fires are useful tools. In areas of the

South grass and pine needles accumulate on the forest floor forming a heavy "rough." Burning this rough under careful control during the winter eliminates fuel that might give rise to disastrous fires in spring and fall. Such controlled burning tends to discourage hardwoods and favors the reproduction of longleaf pine, a very fire-resistant species. It also lessens damage by brown-spot needle disease of longleaf pine.

With forest management come roads, trails, working crews scattered over the forest, and detailed control plans. Fire control becomes integrated with forest management.

Causes and Prevention of Forest Fires. The objective of fire-control planning is to prevent as many fires as possible and to suppress those that do start before they get larger. Only 10 per cent of all forest fires are due to lightning, 90 per cent to man's carelessness. Smokers, incendiaries, and debris burners cause two-thirds of all fires. Railroads, campers, lumbering operations, and miscellaneous causes account for the remainder. Man-caused fires have been reduced by shutting down logging operations during hazardous periods, installing spark arresters on locomotives, requiring permits for debris burners, enforcing state laws against incendiaries, closing forests to sportsmen in severe seasons, and by constantly calling the attention of the public to the necessity for caution with fire in the woods. In 1949, fire-prevention publicity paid for by newspapers, radio stations, and advertisers was valued at 4 million dollars.[22] Improvements in camp sites, roads, and trails prevent many carelessly dropped cigarette butts or matches from falling on inflammable material. Warnings are printed on match folders and cigarette packages. Signs are posted along the highways, and newspapers and radio announcements warn people during hazardous seasons. Motorists in forested areas may be stopped and warned of the danger. Prompt apprehension and strict enforcement of laws against debris burning and incendiary fires are strong deterrents to the careless individual. In 1949, prosecutions numbered 6,880, and 95 per cent resulted in convicting offenders.[22]

PRESUPPRESSION ACTIVITIES. Presuppression activities are closely related to fire prevention. The causes of fire are analyzed, and a major fire plan prepared to provide for prompt detection and suppression. Fire lines and roads are constructed to furnish access and lines from which fires may be effectively fought. Equipment is obtained and located at strategic points. Forest-fuel types are mapped and efforts made to reduce fuel on areas where fires are most likely to start. Fire-danger rating stations are established, at which weather readings are systematically taken and measurements made of the moisture content of forest fuel. Fire-danger meters are devised to integrate the hazards due to temperature, humidity, fuel moisture content, days since last rain, wind velocity, season, and

FIG. 20. Fighting fire in a southern forest. (*A.F.P.I. photo.*)

time of day. A reading of fire danger serves as an index to the fire-control officer of the steps he should take.

The key to quick action is prompt detection. Trained watchers in lookout towers are on constant alert during hazardous periods. They spot the fire, determine its location and probable size, and telephone the information to the district ranger. Foot and aircraft patrols may be used when visibility from lookouts is poor.

FIRE SUPPRESSION. A small fire can oftentimes be controlled by one man or by a very small crew. The smoke chaser, the first man sent to a fire, determines its size, cause, if evident, and rate of spread. By the time the secondary suppression crew arrives the smoke chaser has made his reconnaissance and reports to the fire boss.

Small fires are fought directly with shovels, rakes, fire swatters, water from back-pack pumps, and other simple tools. The crew first knocks down the hot spots, cuts off the head of the fire, and then proceeds to build a line around its flanks and back (Fig. 20).

The fire dispatcher's job is to send enough men and material to the fire to get it put out as quickly as possible. On large fires camps must be established with a supply system, workshop, communication center, and fire-camp headquarters. A relatively large organization may be involved with a thousand or more men working on the fire line. Coordination becomes a major item, and communication must be established between the fire boss and the foremen who are responsible for the several sectors. Heavy equipment will oftentimes be used on such fires. These include tractors with bulldozers, heavy fire plows, and tank trucks with hose for spraying water on hot spots.

Kinds of Forest Fires. GROUND FIRE. The ground fire occurs where the fire burns in a heavy layer of duff or peat such as accumulates in a swamp.[11] Such fires may burn to a depth of 6 or more feet and are extremely difficult to put out without the actual digging out of the burning material. They burn slowly and are not a serious menace in themselves, but they may smolder for weeks and then suddenly set surface material afire.

SURFACE FIRES. These fires, which are the commonest of all, burn the dry vegetation and forest litter. Where tree growth is sparse, such fires may travel rapidly. Wind accelerates spread.

Surface fires can be fought with all types of equipment and by many stratagems. Small fires can be fought on the advancing front, quickly knocked down, and the flanks then extinguished. Fire fighters usually attack large fires on the flanks, working toward the head. On very large fires lines are prepared in advance of the fire. As the surface fire approaches, backfires are set and are drawn by the draft of the head fire down toward it. The two fires then meet some distance back of the fire line and burn themselves out, but constant vigilance must be exercised to extinguish any spot fires that occur from sparks thrown over the fire line.

CROWN FIRES. These burn in the crowns of coniferous trees, are exceedingly hot, travel rapidly, and can run many rods ahead of the accompanying ground fire. Usually the heat from the surface fire dries out the needles and heats them to the ignition point. Crown fires are very spectacular. They throw sparks for long distances, and great care must be exercised in fighting them lest the crew be entrapped in the fire. Crown fires can be best fought by putting out the surface fire during the night and early morning, when the crown fire generally dies out. The only other available method may be to fight such fires by means of backfires that burn off the surface material that would provide the heat necessary to keep the crown fire going.

After any fire has been brought under control, mop-up operations begin and continue until the last spark is out. Mop-up involves digging out burning stumps, burying logs, felling burning snags, and digging out fire in peat or heavy litter.

Use of Aircraft. Aircraft are used for patrol, observation, and transport of men, equipment, and supplies. Air photos can be developed in flight and dropped to the fire boss within 15 minutes from the time the exposure is made. Experiments have been conducted in the dropping of bombs containing water and chemicals on small fires before the fire fighters actually arrive. Attempts to extinguish fires by bombing have been successful, and they have slowed down the rate of spread, but the method is very costly and of only experimental value as of 1950. Attempts have also been made to dissipate electric charges causing lightning and to induce rain by seeding clouds with dry ice.[4] Helicopters have been used for scouting, liaison, and landing men near the fire line.

The most spectacular use of aircraft is transport of smoke jumpers. These skilled fire fighters are flown over the fire, then parachute down, arriving fresh within a short time after the fire is detected. Specially designed parachutes that can be steered by pulling on the shrouds are used, and experienced smoke jumpers can drop within a short distance of the target they set. The initial supply of food and equipment is dropped to them by parachute. Smoke jumpers signal the pilot that they have landed safely, take off their padded jumping suits, collect tools, and attack the fire. They have been particularly successful in extinguishing fires in remote sections that could not readily be reached by motor travel. Thousands of jumps had been made by 1950, and no one had been permanently injured or killed by smoke jumping.

Fire-control Organizations. A fire-control organization is of necessity large and far flung, and a maximum of responsibility must rest on the men on the ground. On national forests the district forest ranger is usually the fire boss. In state forest organization the district foresters and rangers have this responsibility. The state forester backs them up with extra equipment and manpower if required but does not attempt to take over the job unless the district organization breaks down.

In the organized towns of Maine, responsibility for forest-fire control rested with the town fire wardens. This system proved entirely ineffective in the huge fire that occurred in the fall of 1947, for fires soon got beyond wardens' control. During this same dry period the unorganized towns of Maine, which include by far the greater forested area of the state, underwent only minor fire damage. The state-control system in authority here functioned with full effectiveness. Maine's forest-fire laws were changed after 1947 to give the state forest commissioner supervision over the town wardens.

Taking a lesson from the Maine disaster, the northeastern states have enacted a fire-control pact with central coordination which permits states to furnish men and material to a neighbor state if a fire threatens both states or is too large for one state organization to handle.

Despite the great progress that has been made in nationwide fire control, there still remains in the South some 70 million acres without organized control.[22] On this area 105,500 fires occurred in 1949, burning

15.6 per cent of the land. This rate is 24 times as high per unit area as that on protected lands.

Review. After a large forest fire has been extinguished, there remains the final job of reviewing the fire. This includes determining where it started and the cause of its starting and a review of the initial action taken, of all subsequent actions, and of their efficacy in hastening the time when final control was achieved. These reviews have been particularly valuable to the men who fought the fire and to the review board in determining what additional training of personnel may be needed, what new equipment could have been used effectively had it been available, and what precautions must be taken to avoid, or at least minimize, the chances of future fires getting beyond control.

Protecting the Forest from Insect Attack

It is estimated that fire destroys 460,000,000 cubic feet annually and that annual losses from insects, disease, windstorm, and other causes amount to 1,019,000,000 cubic feet. These two sources of loss together account for 1,479,000,000 cubic feet of timber annually. Cutting accounts for 12,182,000,000 cubic feet.

The amount of money spent annually for controlling insects, diseases, and other destructive agencies is only a small fraction of the amount spent to control fires. Not until 1947 did Congress authorize the Bureau of Entomology and Plant Quarantine to make adequate surveys to detect incipient insect outbreaks and to undertake prompt control work. Before then specific appropriations from Congress were required before a major outbreak could be attacked.

Insects causing damage to forest trees may be grouped under the following types: sucking insects such as aphids and scales, defoliators, bark beetles, and insects that destroy wood. A few insects such as the pine shoot moth and the white pine weevil destroy the terminal buds or terminal shoots and hence deform the trees.

Aphids and Scale Insects. These do not directly kill forest trees but they may so weaken the trees that they succumb to attacks by other insects or to disease. An example is the beech scale which in conjunction with the Nectria fungus disease has caused widespread dying of beech in Maine and New Hampshire. The aphids and scale insects suck plant juices from the stems, leaves, and other succulent tissue of the tree, and a few attack roots. Often the aphids may be transported from tree to tree by ants that are fond of the honeydew that is excreted by the aphids. Aphids and scale insects are subject to predation by the ladybird beetle and other insects and to a limited extent by birds. They are not considered a major pest to forest trees under ordinary circumstances, though locally they may cause considerable damage from time to time. Sprays are effective on small areas, but no practical method for controlling these insects in the forest has been devised.

Defoliators. The serious defoliators are the larvae of certain moths and sawflys and, to some extent, the adults and larvae of some beetles.

Examples are the tussock moth, forest tent caterpillar, the larch sawfly, Leconte sawfly, gypsy moth, brown-tailed moth, and spruce budworm. Most are native insects, though the gypsy and brown-tailed moths were introduced from Europe. All have caused serious damage when outbreaks occurred. In an outbreak that began in 1909 and continued for almost a decade, the spruce budworm destroyed 250 million cords of spruce and fir pulpwood. It was active in Quebec, Maine, New Brunswick, and Minnesota. A second outbreak, which began in Ontario in 1935, had spread over 125 million acres by 1944. By 1947 most of the mature fir and a considerable part of the white spruce stands had been killed over an area of over 20,000 square miles. The amount of timber killed was enough to supply the pulp mills of Canada for more than three years. This outbreak reached New York in 1945 and Maine in 1946. Considerable defoliation occurred in New York in 1946, but the budworm population dropped off during 1947 and 1948 so that serious damage did not occur. During the winter of 1946–1947, approximately 75 per cent of the insects died. Parasites together with the winter kill reduced the population by almost 83 to 98 per cent. Insect-killing birds destroyed additional larvae and pupae. As a result, the population in New York State was restored almost to normal by 1949. The spruce budworm larvae show a preference for feeding on balsam fir buds that develop the pollen cones.

The seven-year locust, or cicada, may sweep over considerable areas of forest, feeding heavily on leaves and damaging branches in which they deposit eggs. Brood years are usually well separated so that damage is in the form of reduced growth rather than death of trees. May beetle adults cause some defoliation, though this is rarely serious. The larvae, white grubs, feed on roots and cause heavy losses in nurseries and young plantations.

Bark Beetles. These beetles are the most destructive of all the insects in terms of the quantity of timber destroyed. They burrow beneath the bark, feeding on the inner bark and cambium, where they find considerable quantities of sugars and starches. Although the healthy trees oftentimes produce a vigorous flow of sap that kills the insects, trees infested in late summer, and those which have a limited moisture supply or are unable to resist the attack for other reasons, are girdled and killed. These beetles provide avenues for the entrance of fungi and bacteria and may carry spores on their bodies from infected trees to healthy trees.

Defoliators and bark beetles have a tendency to fluctuate widely in population. They may be present in the forest for a long period of time and cause no appreciable damage and then within a period of two or three years develop a huge population that causes a devastating outbreak.

THE PINE BARK BEETLES. The pine bark beetles are made up of several species of small engraver beetles. Similar beetles attack spruce and other conifers. The western pine beetle working on ponderosa pine from 1921 to 1946 is estimated to have killed 25 billion board feet of timber in the Pacific states; the Black Hills beetle working in the Rocky Mountain section from 1895 to 1946, 2.5 billion feet; and the mountain pine beetle, 5.5 billion feet. The latter insect working in California has caused damage to sugar pine, western white pine, and lodgepole pine totaling 14.5 billion board feet.[3, 19] The Engelmann spruce beetle killed 4 billion board feet of spruce timber during the period 1942 to 1948.

Wood-destroying Insects. Wood in service is attacked by various wood-destroying insects, the most serious of which are termites, carpenter ants, and powder post beetles. Termites and carpenter ants can work only in wood that has a fairly high moisture content. These and other wood borers work beneath the surface and offer little outside evidence of their activity. The powder post beetles will work in tool handles, furniture, flooring, and barn timbers where the moisture content is considerably below that required by termites. No insect can work in wood dried to 6 per cent moisture content.

A recent survey of insect work in lumberyards in New York State showed an annual damage estimated at $44,400. Powder post beetles damaged 235 thousand board feet of ash, oak, and hickory valued at $41,000; solitary wasps caused a loss of $2,200, and carpenter ants $1,200. These estimates, made during a period of rapid lumber movement, are believed to be a conservative statement of annual loss.

Degrade caused by insects attacking the green log results in annual losses in lumber of $63,200, for which round-head borers and ambrosia beetles are mostly responsible. Insects damaging buildings in New York include termites, anobiids, carpenter ants, and wood-boring weevils.[20]

Control of Insect Damage. Insect damage is controlled by preventive measures brought about by good forest management, reduction of favorite foods or habitat conditions, introduction of parasites and predators, and direct control. Pure plantations of trees over large areas are an open invitation for insect or disease outbreaks. Where mixtures of several kinds of trees are used or the plantations are small in size and interspersed with natural stands, the damage is less likely to be severe. Severity of white pine weevil attack can be reduced by planting pine in mixture with hardwoods, but, unfortunately, the hardwoods oftentimes prove to be a more serious menace to the pine than the weevil. A more successful practice is to plant the pines close together. Since weevil tend to attack the tallest trees, those just below the main canopy may escape attack and develop into good crop trees. In any case, close spacing prevents the formation of heavy branches that render severely damaged trees unfit for saw timber. Spruce budworm damage can be minimized by removing the mature balsam fir before the insect reaches outbreak proportions. This is possible only on forests that have well-developed road systems or other means of transportation so that partial cutting is possible.

Good forest management has proved to be the most effective over-all method that we have of combating both insects and diseases. Control of western bark beetle damage has been possible by learning to recognize the types of trees that the beetles will attack and removing these in early logging operations.[17] A great deal of effort has been spent in attempting to build up parasites, predators, and diseases of destructive insects. These have oftentimes given good results and are particularly effective in controlling a normal population, but they can rarely be relied upon to halt a serious outbreak.

Damage to wood in service can be prevented largely by keeping the wood dry and preventing direct contact between the wood and the soil. Certain timbers which must be located closer to the soil than 18 inches can be protected by impregnating them with various preservatives.

DIRECT CONTROL OF INSECT OUTBREAKS. Attempts to control insects by direct methods have proved costly in the past. Huge sums have been spent to control bark beetles. Infested trees were cut, stripped of bark, and the bark burned to kill the insects before they reached adult size. Where the outbreak was small and effective work was started promptly, control was oftentimes achieved. But, once an outbreak reaches large size, this method is extremely expensive.

Efforts were put forth to control the gypsy moth by direct attack. The females of this species do not fly. Research discovered that the males which do fly could be trapped readily by attracting them through the odor given off by the tips of the female abdomen. If the males could be trapped, there was indication that the insects were present within a relatively short distance of the trap. A diligent search of the surrounding area would reveal the egg masses. Eggs could be killed by painting the egg mass with a poisonous chemical. Despite serious efforts at control by this means, the insect continued to spread. When DDT and other powerful insecticides became available, however, airplane spraying of this chemical proved exceedingly effective in eliminating the insects. Entire populations were destroyed.

One of the most spectacular jobs was the control of the tussock moth in Idaho. An outbreak began in 1944 or 1945 and in 1946 whole mountainsides appeared brown because of defoliation by this insect. Through joint efforts of the Federal and state governments, the wood-using industries, and other agencies, a huge spraying job with aircraft was undertaken in the spring of 1947. The job began on May 22 and was completed on July 16. A total of 413,000 acres were covered at a cost of $1.57 per acre. After the spraying was completed, no live tussock moths could be found anywhere on the treated area.

What the future holds for direct control cannot be foretold. Incipient insect outbreaks, if detected early, can readily be controlled by airplane spraying or dusting or even by ground control methods. Meanwhile, as our forests are opened up to intensive management, it becomes possible to salvage more of the timber that is killed by insects or diseases and to keep the stands in healthy condition.

Tree Diseases

Various diseases of trees kill or damage seedlings, young timber, and mature forest trees. Diseases are of two main types, nonparasitic and parasitic. The nonparasitic tree diseases include those caused by drought, sunscald, winter injury of various types, improper nutrition, injuries from smoke or gases, and injuries resulting from flooding. Some nonparasitic diseases such as sunscald and smoke injury are of local occurrence. Others such as drought damage and wind burn may be widespread.

Few soils of the United States are so deficient in mineral nutrients that trees cannot survive. Many are infertile, however, and support only small-sized trees, as do the sand plains of New Jersey. Glacial outwashed sands of sections of the Lake States and New York, following cultivation, have been seriously reduced in their capacity to support vigorous tree growth. The poorest soils are often found on rocky ridge tops and along lands recently elevated from the sea.

Locally, trees may be damaged by chemical wastes, such as the acid water from mines, smokes, and smelter fumes, and by other toxic chemicals in heavy concentration. The outstanding example is the copper basin of Tennessee. Here smelter fumes, mainly sulfur dioxide, have resulted in the death of all trees for a distance of about 8 miles from the smelter.[14, 15] Trees are also readily killed by flooding caused by dams. Unfavorable weather will be discussed later.

Parasitic Diseases. Parasitic diseases of trees are caused by bacteria, fungi, viruses, nematodes, and mistletoes.[7, 12, 16, 18, 21] Fungi cause more losses to trees than any other types of disease, particularly in the larger trees. Nematodes are particularly destructive of seedlings and of newly germinated seed. The phloem-necrosis disease of elm and brooming of black locust are caused by virus infections. The parasitic organism causing a disease is called a pathogen.

FUNGAL DISEASES. Many tree diseases, of which the chestnut blight is a good example, are spread by spores carried by the wind. The chestnut blight, which was first reported in New York City in 1904, spread rapidly, killing all trees as it went. The blight is the death of the bark on small stems. Many trees have sent up sprouts following the killing of the main stem, but these soon become infected and die also. No effective control for the disease has been found. Consequently, the chestnut has been wiped out as a commercial tree throughout its range. The elimination of the chestnut from the eastern forests of the United States has meant the loss of the country's chief local source of tannin and of a useful wood for many purposes.

The rusts are highly specialized fungi. A classical example is the white pine blister rust fungus, which was first found in North America at Geneva, N.Y., in 1906. It occurred on cultivated gooseberries and currants, which were promptly destroyed. By 1909 a new infestation was found on eastern white pine seedlings which were imported from Europe and shipped widely over the Northeast and the Lake States. Attempts to destroy these were unsuccessful. By 1915 all hope of eradicating the fungus from the United States was abandoned, since it was learned that new infections had been introduced on the West Coast. These have spread on the western white pine and sugar pine.

The fungus attacks all five needle pines. It enters the trees through the needles, spreads down the twigs to the main branches, ultimately to the main stem,[13] and kills the tree by killing all the bark and thus girdling the stem. The rust cannot spread directly from pine to pine but must spread from pine to currants or gooseberries. On these it causes only moderate damage and may spread from one currant bush to another. In the fall of the year spores are produced which can infect only pine. The white pine blister rust may, therefore, be controlled by eliminating currants and gooseberries which enable the infection to spread to pine stands. Approximately 28 million acres in the eastern and western United States are within the white pine blister rust control areas. Here periodically the *Ribes* (currants and gooseberries) are eliminated by pulling them up or by killing them with herbicides, but *Ribes* produce abundant seed which live for a considerable length of time in the soil. Hence, when white pine is harvested the currants and gooseberry seeds germinate and produce dense stands which may become infected with blister rust fungus. It is believed that *Ribes* can be eliminated on areas that have been logged of white pine by two or more broadcast burns or by other appropriate methods. This rust is one of the most seriously destructive diseases of forest trees in our country.

The Dutch elm disease, an excellent example of a disease in which the causal agent is carried by an insect, broke out in Europe some 30 years ago and was found in the United States in 1930. The fungus is parasitic on living trees and saprophytic on dead trees. Its yeastlike spores produce toxins which are carried by the sap stream throughout the tree and so cause death. The disease organism is carried from diseased to healthy trees by the elm bark beetle. There are two elm bark beetles, one native to America and the other to Europe. The European beetle is the more important in the spread of the disease. Beetles were introduced at the same time as the disease in various importations of elm logs used for producing high-grade furniture veneer. Since the adult beetles breed only in recently dead or dying trees, the insect is benefited by the association of the fungus. Efforts have been made to control the disease by destroying infected trees and by protecting healthy trees through spraying with DDT to control the beetles. Whether the disease will ultimately wipe out the elm in the eastern United States is still open to question.

Year in and year out heart rots are responsible for more damage to timber than other diseases. The fungi are spread by wind-borne spores that infect the tree through injuries caused by logging, ice breakage, fire scars, insect attack, or through other injuries that expose the heartwood of the tree. The fungi causing heart rot are of several types, but the group known as the polypores are the most common and widespread. They produce bracket-shaped, spore-bearing conks on the tree trunk. Heart rot has caused heavy damage in western stands and is particularly prevalent in eastern stands where the best trees have been removed (Fig. 21).

No feasible method has been found for preventing heart rot in forest trees. The only effective control is the practice of intensive forestry. Care should be taken to eliminate fire, since much heart rot gets its start through fire scars. Where selective logging is carried out, care should also be taken to avoid injuring the base of trees in tractor operations or other skidding methods. The annual loss for heart rot is estimated at 1.5 billion board feet.[12]

The fungi that cause heart rot in trees also cause decay of slash and fallen trees. This is a valuable function in the forest. If the wood did not decay promptly, we would soon have the forest cluttered up with highly inflammable debris making the growth of future trees almost impossible.

Wood-rotting fungi cause considerable loss to wood in service as well. It has been estimated that approximately 10 per cent of the annual production of lumber in the country is used each year merely to replace wood which has decayed in service.[1] Control of decay in wood is obtained by keeping it dry, for wood which does not contain free water in cell cavities is safe from fungus attack. Examples of commonly accepted building practices for limiting decay by keeping wood dry are keeping timbers at least 18 inches off the soil, providing adequate ventilation, avoiding joints or pockets where moisture can collect, and applying vapor barrier materials on inner (warm) walls of homes to prevent water

FIG. 21. In the virgin forest wood-destroying fungi and other destructive agents harvest mature trees.

condensation. Many uses of wood, such as for railroad ties, telephone poles, and piling, necessitate its locations in places where it becomes wet. In these cases, wood decay is arrested by treatment of wood with poisonous chemicals such as creosote, zinc chloride, or pentachlorophenol.

MISTLETOES. The eastern mistletoe and, to some extent, the western mistletoe are carried by birds. These are parasitic seed plants that grow on the branches of pines and hardwoods, sending their rootlike haustoria down into the conductive tissue of branches. The eastern mistletoe attains fairly large size. Since it produces seed covered with a sticky, gelatinous material that adheres to birds' feet, the birds may spread the infection in carrying the seed from one tree to another. The dwarf mistletoe causes witches'-brooms and other damage to conifers, particularly to ponderosa pine. The diseases are spread by birds, but seeds of this plant are small and forcibly ejected from the sack in which they mature. They may be shot out for a distance of 20 to 40 feet horizontally. Once a tree becomes infected with the dwarf mistletoes, the infection can be spread by the ripened seed.

No effective control has been found for the dwarf mistletoe, even though it is believed to cause the heaviest losses of all diseases of the ponderosa pine in the Southwest. The mistletoe can be removed by pruning, and where logging is feasible, infected trees can be harvested, leaving the healthy ones.

Avoiding Introduced Pests

Many serious tree pathogens and insect pests have spread from one country to another. The maintenance of a plant quarantine and inspection service at all ports of entry has been immeasurably helpful in keeping to a minimum the number of serious diseases that have been introduced, but with rapid air travel the hazard has been immeasurably increased. Spraying the interior of commercial aircraft with insecticides prior to landing can be of considerable help in keeping out unwanted insects. Plants to be imported should be grown under quarantine until such time as it is clear that they are free from disease. Quarantine can also be more vigilant when it is known what diseases will attack the native trees and shrubs. This can be determined by establishing plantations in various sections of the world where plants similar to those that grow in the United States are found. Special precautions can then be taken against importing material from areas where such diseases occur.[8]

Protecting the Forest against Animals

Domestic Animals. No measure is available of the total damage caused to American forests by domestic animals. It seems safe to state, however, that next to man himself, his domestic animals cause more damage to the forest than any other destructive agent. Domestic animals are grazed in forests throughout most of the United States. Grazing is practiced the least in the Northeast and in the heavy forests of the Pacific Coast since these regions produce little forage in the forest to support domestic animals. Wood lots are commonly grazed throughout the Middle Atlantic and central states, the South, and all the West except in the heavy forests along the coast.

Grazing is defensible in the West where timber stands are of low density and a considerable amount of grass can be found between the trees. This is true to a certain extent of the pine forests of the South also. Throughout the rest of the United States little grazing value would exist in the woods, providing the trees were growing as densely as they are capable of growing. Where forage is available in eastern forests, the stands are thin as a result of past grazing, fires, improper logging practices, or deliberate attempts on the part of the owner to keep the forest open so that it will support forage for domestic animals.

The fact that cattle, sheep, horses, hogs, and goats do not destroy trees beyond the seedling stage leads many people to assume that they cause little damage to the forest. Actually, they cause a great deal of damage. All compact the soil, reducing infiltration of water, inhibiting root growth, and reducing aeration. Erosion is a particularly serious menace in western forests subject to heavy grazing by sheep and cattle.

Hogs consume seed when it first falls, and they continue to destroy young seedlings until these have passed the cotyledon stage. Hogs may root up seedlings of longleaf pine and eat the roots.

A white oak forest grazed over a period of 40 years has been observed to deteriorate progressively. One by one the old trees have dropped out through lightning, disease, and timber cutting. Only a few scattered trees now remain. The soil became very compacted so that new seedlings did not become established. Erosion was common, especially along animal paths that led to water and to the barn. Bared roots were injured by trampling. Eventually, what was once a fine white oak forest became a woodland pasture. Grazing improved somewhat as the trees disappeared, but the soil being only moderately fertile tended to support worthless shrubs, blackberry briars, honeylocust, and other thorny thickets that prevented the animals from grazing. Within these thickets seedlings of hickory grew up and also a few black walnuts, which were not relished by the animals. Grazing had in this case completely changed the composition of the forest and resulted in a relatively poor forest of wide-spreading, scattered trees where a splendid forest of well-spaced trees originally existed. A study in Ohio revealed that grazing caused a substantial reduction in the flow of maple sap which, when translated in terms of maple syrup production, amounted to $10.67 per acre annually.[5]

BENEFICIAL EFFECTS OF GRAZING. If properly regulated, grazing has some beneficial effects. Grazing animals can consume much of the grass and other herbaceous vegetation that would otherwise become a serious fire hazard, and grazing can control the composition of a forest by killing out unwanted hardwoods. It is seldom used, however, as a conscious silvicultural measure.

Protection against Wildlife. It is a maxim that any animal that lives in the forest must of necessity affect the forest. It must get its food from the forest, and this often comes in part from the woody vegetation, either directly in the form of leaves, twigs, buds, and bark or indirectly in the form of nuts, fruits, and fungi that grow on the live or dead trees.

No satisfactory estimate is available of the total damage caused by wildlife throughout the United States. Compared with that caused by domestic animals, it is believed to be small. It is certainly comparatively light except where overdense populations exist, but these become a menace to both vegetation and the wildlife itself.

Squirrels are fond of pine seed, acorns, hickory nuts, walnuts, butternuts, hazelnuts, and other nut-type foods, which they gather in autumn for winter use. Where squirrel populations are dense, they may destroy an entire seed crop, but if the squirrel population is kept within reasonable bounds, plenty of seed will be left to perpetuate the forest. Moreover, squirrel caches are good sources of cones for the seed collector. Mice, chipmunks, and birds eat tree seed in large quantities. Grosbeaks have destroyed the entire seed crop of certain southern pines. Birds also eat the young seedlings when they first emerge. In coniferous nurseries these must be screened lest the birds destroy entire seed beds. Direct seeding of forest trees is almost precluded in many regions, because the seed that is planted is promptly destroyed by rodents and birds.

Rabbits, mice, deer, and other browsing animals will eat young tree seedlings. Plantations and even natural reproduction of pine in the Lake States have been completely destroyed by snowshoe hare during peak periods of their population cycle. Where deer are overabundant, they can be as destructive as domestic animals. Deer, rabbits, and other animals also tend to influence the composition of a forest by selective feeding. Deer and rabbits both tend to browse more heavily on hardwoods than on softwoods, thereby exerting generally a favorable influence on the reproduction of conifers. Where they are overly dense, however, they effectively keep down seedlings of both hardwoods and softwoods.

The bark of trees is eaten by rabbits, mice, squirrels, porcupines, deer, and elk. Mice are likely to be particularly destructive of sugar maple and other hardwood seedlings and saplings. They work beneath the snow during winter. Rabbits gnaw the bark off seedlings and saplings that are too large for them to cut.

Porcupines, where abundant, may cause extensive damage to both hardwood and softwood trees by gnawing bark. They may girdle trees at almost any height, and they also cut off twigs. Ordinarily, however, these animals are not dense in the forest, and the space occupied by the trees they kill is quickly taken up by surrounding trees.

Elk gnaw the bark from young aspens and alders when other feed is not available during the winter, but this damage is highly localized and is seldom severe to trees of merchantable size.

Ruffed grouse, squirrels, porcupines, deer, rabbits, and many other animals feed on the buds of trees. The actual damage caused by this feeding is difficult to appraise. The ruffed grouse is a solitary bird, normally commanding an area of from 1 to 5 acres, which it holds against other birds of the same species. The amount of damage it causes by the buds consumed, therefore, is negligible. Squirrels cut off terminal shoots and may later return to the ground and eat the buds. Such cutting during the winter and late spring months is beneficial to animals that cannot reach the high branches in the trees. It is believed that none of the animals does serious damage through destruction of the buds once the trees have passed the sapling stage.

Beaver are the only animals that fell large-sized trees. They may also girdle trees that they do not completely cut down. Beaver cutting is normally limited to trees of 2 to 4 inches in diameter, but they have been known to cut down trees as large as 18 inches and even larger (Fig. 22). Beaver eat the bark from the felled trees and store the branches under water for use during winter. They also kill a considerable number of trees by flooding low areas. Their dams cause only limited damage along watercourses, and the benefits from the dams to fish, to water storage, and particularly in checking erosion are generally much greater than the damage they cause.

CONTROL OF WILDLIFE DAMAGE. Wildlife-damage control can best be effected through good wildlife management combined with good forest management. The wildlife population is likely to be most destructive during those periods of the year when food is least plentiful. If forest-

FIG. 22. An 8-inch red maple tree partially severed by beaver. Note recent bark removal at base.

cutting operations are carried out at this time, deer, rabbits, and other animals that inhabit the forest can browse on the branches of felled trees, which provide them with much-needed winter food. It is obvious that the forest manager should make concessions to favor wildlife. Not only is wildlife of value for recreation and sport, but it also performs useful service to the forest.

From time to time wildlife populations in local areas may become excessive and result in severe damage to the forest. This will require direct action. Hunting and trapping are usually the most effective for game and fur animals. Live trapping and transport to other areas has rarely succeeded in reducing population. Fencing is costly and ineffective for large areas. Nature eventually will bring about an adjustment through predators, disease, and starvation, but such adjustments are often delayed until severe damage to the wildlife range occurs. Furthermore, natural adjustments may severely deplete the seed stock.

Mice can be controlled on small areas by the distribution of poisoned bait. This is an extreme measure and may not be undertaken unless the forester has the permission of the game authorities, but areas that are to be reforested by direct seeding must almost always be protected from the rodents. When bait is properly distributed, nothing other than rodents will be affected.

Unfavorable Weather

Native forests, though not immune to injury, withstand normal vicissitudes of weather. They may be seriously injured by abnormal weather. Plantations are likely to be particularly susceptible. Drought injury can be minimized by planting drought-resistant species, by eliminating vegetation that competes for soil moisture, and by preserving moderate overhead shade. Tree seedlings hardened in the nursery by exposure to soil dryness and not overfertilized are likely to survive better on dry sites than unhardened stock.

High temperature sometimes causes loss to tree seedlings, particularly in nurseries. If the soil surface is dark, the damage is likely to be more severe than on light-colored soils. The sun will cause heat lesions on the south and west sides of the seedling which may cause its death. This can be prevented by watering the seedlings during the hottest part of the day. Heat loss is rare in forest plantations but has been known to occur on occasions. Usually it is spotty and confined chiefly to places where air drainage is poor. Air temperatures that kill seedlings must be as high as 120 to 130°F., and such temperatures rarely occur in regions where trees normally grow.

Frost action on heavy soil may heave fall-planted trees from the soil or lift them so far out that they die. Frosts may also stunt growth and actually exclude trees from "frost pockets," which have poor air drainage. These may occur near the head of shallow valleys and at other places where free movement of air is impeded, and it is also possible to create frost pockets by cutting mature timber in a low-lying area such as a swamp. Once air drainage has been blocked, it is almost impossible to restore a good tree cover. Such areas tend to be restored naturally, but the process may take many decades. Frost that extends over wide areas may also cause severe damage to trees, especially late spring frosts after the buds have opened and vegetative growth has started. Ordinarily, however, these do not result in the death of large trees.

Sunscald and winter burning of forest trees occur on unusually warm days with bright sunshine when the ground is still frozen. Winter burning is due to severe loss of water from needles when the roots are un-

FIG. 23. Remains of a virgin spruce-fir forest in the Adirondacks after the November 25, 1950, windstorm.

able to replace it. Sunscald appears to result from freezing of tissue newly activated by solar heating.

Wind. The most destructive of all atmospheric influences is wind. High winds will uproot trees or break them off. Moderate winds will break the tops. Cyclonic storms of tornado type will sweep through a forest, leveling trees in their paths. Hurricanes cause severe damage to forests. Certain trees of which the black and red spruce are good examples have relatively shallow root systems and are easily overthrown by wind when left in exposed situations. This is particularly true in wet sites where the soil is thin and saturated with moisture (Figs. 23 and 24).

Although forests cannot be completely protected against damage by windstorm, damage can be minimized by proper density and by main-

FIG. 24. A windstorm in July, 1940, left this badly bent survivor of a beautiful virgin red pine stand, Chippewa National Forest, Minn. (*U.S. Forest Service photo.*)

FIG. 25. Severe ice damage to northern hardwoods of central New York. (*U.S. Forest Service photo.*)

taining good borders at the edge of the forest to prevent the wind from sweeping under the crowns. In thinning operations it is best to take out those trees that are likely to be thrown down by the wind, provided they have commercial value. Since trees with weak crotches will always be subject to wind damage, they should be removed.

Rain. Heavy rainfall sometimes causes land slips, gullying, flooding, and heavy washes that may wash trees out by the roots. Most forests, however, keep the soil sufficiently open so that the water soaks in and causes little erosion or damage. Snowslides or avalanches may harm forests in high mountains. Only limited steps can ordinarily be taken to prevent these, such as keeping vegetation on the upper slopes and promoting the growth of woody vegetation wherever possible. Hail may strip the leaves from trees, but such storms are generally of local occurrence and recovery from such damage usually is rapid.

Snow and Ice Damage. Wet snow, and particularly rains in freezing weather, may overload tree crowns and cause extensive breakage. A single ice storm may damage trees over thousands of acres (Fig. 25). Trees planted out of their natural range are especially susceptible. Hardwoods are usually damaged most by ice, conifers most by wet snow. Trees with narrow crowns usually resist damage better than broad-crowned trees. Ice and snow breakage can be minimized by proper care in selection of the trees that are planted in regions subject to snow breakage and also by maintaining proper densities. Trees in stands that are neither too dense nor too thin usually have sturdy stems which enable them to resist snow breakage. If they are in the open, they get heavily laden with snow and ice, and wind may then cause breakage.

In protecting the forest against unfavorable weather, the forester has as his main tool intensive forest management, which is also the best measure available to him for safeguarding the forest against fire, insects, and diseases. Combined with game management, forest management will keep the forest free from excessive wildlife damage. Actions by man taken with full knowledge of natural tendencies and natural events in the forest can be so planned as to keep the forest in healthy condition and to improve the yield of both forest crops and wildlife crops with increasing human use. Poorly planned operations, on the other hand, cause a progressive decline in forest yield, wildlife yield, water yield, and the services of the forest. They tend to make fires more frequent and devastating and open the forests to attacks by destructive insects and diseases. Unwise forest management, even though aimed at creating good conditions, can oftentimes prove disastrous, as the extensive planting of conifers in sections of Europe has well demonstrated. Fortunately, we have rich and varied forests in the United States, and given reasonable care, they can be kept productive and free from excessive losses due to the various destructive agencies that are always present. Intensive management with a good road system is essential to good protection and salvage of damaged trees.[11]

LITERATURE CITED

1. Boyce, John Shaw. 1948. "Forest Pathology," 2d ed., McGraw-Hill Book Company, Inc., New York. 550 pp.
2. Brown, A. A. 1949. Progress, but Still a Problem, "Trees, The Yearbook of Agriculture," pp. 477–479, U.S. Department of Agriculture, Washington, D.C.
3. Craighead, F. C., and John M. Miller. 1949. Insects in the Forest: A Survey, "Trees, The Yearbook of Agriculture," pp. 407–413, U.S. Department of Agriculture, Washington, D.C.
4. Crocker, Clayton S. 1949. Fighting Fires from the Air, "Trees, The Yearbook of Agriculture," pp. 508–516, U.S. Department of Agriculture, Washington, D.C.
5. Dambach, Charles A. 1944. Comparative Productiveness of Adjacent Grazed and Ungrazed Sugar-maple Woods, *Journal of Forestry,* 42:164–168.

6. Forbes, Robert H. 1949. The Smoke Jumpers Carry On, *American Forests,* 55(10):18–19.
7. Gill, Lake S., and Jess L. Bedwell. 1949. Dwarf Mistletoes, "Trees, The Yearbook of Agriculture," pp. 458–461, U.S. Department of Agriculture, Washington, D.C.
8. Gravatt, G. F., and D. E. Parker. 1949. Introduced Tree Diseases and Insects, "Trees, The Yearbook of Agriculture," pp. 446–451, U.S. Department of Agriculture, Washington, D.C.
9. Guthrie, John D. 1936. "Great Forest Fires of America," U.S. Forest Service, Washington, D.C. 10 pp.
10. Hammatt, R. F. 1949. Bad Business; Your Business, "Trees, The Yearbook of Agriculture," pp. 479–485, U.S. Department of Agriculture, Washington, D.C.
11. Hawley, Ralph C., and Paul W. Stickel. 1948. "Forest Protection," 2d ed., John Wiley & Sons, Inc., New York. 355 pp.
12. Hepting, George H., and James W. Kimmey. 1949. Heart Rot, "Trees, The Yearbook of Agriculture," pp. 462–465, U.S. Department of Agriculture, Washington, D.C.
13. Hirt, Ray R. 1933. "Blister Rust, a Serious Disease of White Pine," New York State College of Forestry, Syracuse, N.Y. 5 pp.
14. Hursh, C. R. 1932. The Forest Legion Carries On, *American Forests,* 38:16–19.
15. Hursh, C. R., and C. A. Connaughton. 1938. Effects of Forests upon Local Climate, *Journal of Forestry,* 36:864–866.
16. Hutchins, L. M. 1949. Diseases and the Forest, "Trees, The Yearbook of Agriculture," pp. 443–445, U.S. Department of Agriculture, Washington, D.C.
17. Keen, F. P. 1949. Pine Bark Beetles, "Trees, The Yearbook of Agriculture," pp. 427–432, U.S. Department of Agriculture, Washington, D.C.
18. Martin, J. F., and Perley Spaulding. 1949. Blister Rust on White Pine, "Trees, The Yearbook of Agriculture," pp. 453–458, U.S. Department of Agriculture, Washington, D.C.
19. Rohwer, S. A. 1949. The Key to Protection, "Trees, The Yearbook of Agriculture," pp. 413–417, U.S. Department of Agriculture, Washington, D.C.
20. Simeone, J. B. 1949. "Damage by Wood Borers in Lumber and in Buildings," unpublished manuscript, New York State College of Forestry, Syracuse, N.Y. 3 pp.
21. Swingle, R. U., R. R. Whitten, and E. G. Brewer. 1949. Dutch Elm Disease, "Trees, The Yearbook of Agriculture," pp. 451–452, U.S. Department of Agriculture, Washington, D.C.
22. U.S. Department of Agriculture. 1950. "Report of the Chief of the Forest Service," Washington, D.C. 69 pp.
23. Wilkins, Austin H. 1948. The Story of the Maine Forest Fire Disaster, *Journal of Forestry,* 46:568–573.

CHAPTER 6. *Forest Care and Use*

MAN IS WILLING to care for what he expects to use. He will care for a forest provided such care is necessary for his use of it. The type of use he expects to make of a forest determines the type of forest care needed. Forests to protect municipal watersheds should be such as will maintain the soil porous so that it readily absorbs rainfall. Forests for game production must have an abundance of animal food, water, and coverts. Forests for recreation need impressive old trees, trails with interesting vistas, screening from vehicular traffic, parking lots, and areas for concentrated use. Forest for timber production should contain species of high timber value and individual trees with tall, straight, clear stems.

To use a forest, man must protect it, care for it, harvest its products, and process them for use. This is true whether its products be water, wildlife, timber, or human recreation.

Needs of the Growing Tree

Most forest trees start out as seedlings. The seed of a forest tree may be a very light, tiny seed such as that of the trembling aspen. This can be carried miles by the wind and must germinate within four or five days or the seed dies. Other seeds are large such as those of the oak, chestnut, walnut, and largest of all, the coconut. These fall directly to the ground. They may roll a short distance, but unless transported by man or animals, they germinate near the mother tree. These large seeds may also not be long-lived. They can be killed by excessive drying and, consequently, if they are to germinate, they must find a moist place on the forest floor and be covered with leaves or litter. Rarely do these seeds live through the first summer unless they germinate. Seed of hawthorn, basswood, and junipers are long-lived, rarely germinating before the second summer following ripening.[8]

Each seed has stored within it reserve food to support the life of the seedling—for three or four days in the case of aspen, as long as one year in the case of oak. As soon as the reserve food in the seed is exhausted, the seedling is strictly on its own. It must manufacture its own food in its leaves and absorb through its roots water and the mineral elements necessary for its livelihood. The organic food which is converted into stem, root, branch, and leaf tissue must come from the photosynthetic activity of the leaves. Photosynthesis is the process by which the leaves use the energy of sunlight to combine carbon dioxide from the air with water from the soil to form carbohydratelike substance. Simple sugars, starches, oils, lignin, and cellulose are formed from the primary and secondary products of photosynthesis. Simple sugars are further combined with nitrogen to form proteins. All are used by the tree for growth. The seedling must, therefore, have access to water, carbon dioxide, and light to provide its food substance, and its roots must obtain from the soil nitrogen, magnesium, phosphorus, calcium, potassium, sulfur, and iron to support the life of the tree. Roots also absorb copper, boron, zinc, silicon, magnesium, sodium, chlorine, and other elements some of which may be necessary for the life of the tree. Each of these has been found useful in very small amounts for the growth of at least some plants.[5] Favorable temperature is also essential for photosynthesis and growth.

The rate at which a tree grows depends largely on the rate at which it can be supplied with its basic growth needs—mineral nutrients, water, carbon dioxide, and light. If any one is absent, growth slows down or ceases entirely, and if it remains absent for long, death will ensue. Fortunately, throughout the tropics and temperate zones most elements are available in adequate amounts for satisfactory tree growth. The one that is most likely to be limiting is moisture supply. The mineral elements needed by the trees are abundant in the earth's crust and are usually present in ample concentration to support vigorous growth in most forest soils. Occasionally, a soil is found that is lacking in calcium, potassium, sulfur, or some other one element. Tree growth can be improved on such soils by applying a small amount of the deficient element. An interesting case of deficient supply of potassium occurred on formerly heavily farmed sand soils on the Pack Demonstration Forest at Warrensburg, N.Y. A red pine plantation badly stunted quickly resumed normal growth where pine or hardwood slash from healthy trees was distributed and allowed to decay. This led to a series of experiments applying various individual and combinations of mineral salts until the deficient element, potassium, was revealed.

Nitrogen in the form of nitrates, nitrite, or ammonium compounds, which trees can use, may also be below the optimum for maximum tree growth. Experiments have shown that trees may make substantial response where nitrogen is applied to poor soils. Carbon dioxide is always present in the air in sufficiently large amounts to support tree growth. Sunlight is always ample in the temperate and tropic zones except beneath the shade of a dense forest overstory. If the sunlight that penetrates to seedlings is much below 15 per cent of full sunlight, they will not grow satisfactorily.

When any one element necessary for the growth of trees is present in minimal supply, the rate of growth is determined largely by the supply of this one element. If two or more elements are low in supply, an increase in growth rate can usually be obtained by increasing the supply of either one of the deficient elements. Optimum growth, of course, requires a complete supply of all the several elements.

Many tree seedlings survive best if protected from the hot rays of the sun that heat and dry the soil. Usually, seeds that germinate best in the shade produce seedlings that are capable of surviving for a considerable period of time in low light intensities. These are said to be shade-tolerant. Those that germinate best in the open may require abundant sunlight if they are not to weaken and die. These are said to be shade-intolerant seedlings.[1]

Hazards to Survival. From the very beginning the young tree seedling must face not only conditions that favor its growth but conditions that tend to destroy it. A dry hardwood leaf that is blown accidentally over a tender pine seedling may, if packed down by rainfall, be sufficient to cause its death. A small insect may clip the stem or destroy the growing point before it can form leaf tissue to support life. An animal may tread on the seedling or a bird eat it. A hailstone may break it off, and even rain may beat it into mineral soil. Literally hundreds of seedlings must start for each tree that grows to maturity.

When the seedling gets larger, a rabbit may eat it, a disease may strike, or insects consume it.

When the seedling reaches sapling size, it must compete with its fellow trees and sometimes with older trees that shade it and send their roots into the same soil areas. Growth may be slow for an extended period. Eventually weaker trees drop out. Wind, lightning, fungi, or insects cause death of the older trees, and so the sapling gets its chance to go ahead. When it attains pole size or larger, wind, sleet, snow, and ice all present hazards to survival.[7]

Climates for Trees. In the temperate zones trees must have about 15 to 20 inches of rainfall each year to survive; in tropical regions, 25 to 30 inches. Trees require generally a minimum growing season of approximately 100 days, that is, 100 days free from frost. Those adjusted to northern climates, however, can complete their growth in a shorter period, and many northern trees are capable of surviving satisfactorily in climates where frost may occur any month in the year. Active growth may take place at temperatures above 40°F., though the most rapid growth generally occurs at 60 to 80°F. Evergreen trees may carry on photosynthesis at temperatures as low as 32°F., and a few have been reported to perform photosynthesis at even lower temperatures, though at a very slow rate.[5]

The length of daylight affects tree growth in that it tends to limit the range over which an individual species will extend. It may determine the times at which buds open in the spring and leaves are dropped in the fall. By this means tree functions are regulated so that buds do not open during the first warm days of early spring when they might be killed by late frosts. Leaf casting in the autumn is also controlled in part by the length of day which stimulates the tree to prepare for winter by hardening the tissue in buds and stem.

The growth of leaves, stems, and roots is regulated by plant-growth substances called plant hormones, or auxins, which are complicated chemical substances that tend to inhibit growth in one region and stimulate it in another. Growth substances are largely responsible for the expression of dominance in terminal buds and in the main root system. Growth substances are influenced in part by the light intensity, temperature, and other factors which are favorable to growth.

Maturity and Death. As the tree grows to maturity, it reaches the period when it begins to produce flowers and fruit containing seeds that serve to perpetuate the species. Seed production is generally favored by an abundance of sunlight and moisture. Consequently, after partial cuttings remaining trees tend to develop a larger crown and to produce abundant seed. Seed is not produced regularly every year by most trees. Good seed years may occur annually in some species, biannually in others, and at irregular intervals of three to eight or nine years in many species. The causes of good seed production are not accurately known, but it is known that certain factors tend to stimulate production. Among these are major injuries that tend to weaken the tree, an abundant carbohydrate supply, and a proper ratio between the supply of carbohydrates and the supply of nitrogen in the form of proteins or amino acids.[8]

Seed production becomes more and more bountiful as the tree approaches maturity. When abundant seed is formed, less food reserves remain to support the growth of the tree. When growth rate declines, the tree is unable to extend its root or crown system fast enough to meet its needs. The new stem and bark tissue formed may be inadequate to transport the large amounts of water and nutrient elements up the stem and the food down the stem. So, gradually, the overmature tree weakens. Its branches may be broken by windstorms. Heart-rotting fungi grow in wounds caused by broken limbs, fire scars at the base, or injuries by man or animals. Eventually, the weakened tree is either thrown by wind or dies where it stands (Fig. 21).

Establishing Forests on Open Land

Lands unwisely cleared, badly eroded fields, and extensive fire-swept areas in the Lake States, South, and West must be planted to tree seedlings if they are to serve a useful purpose to man. So must shelter groves be planted to protect fields and farmsteads in the prairie-plains region. Altogether, the Forest Service estimates that 75 million acres of land are in need of forest planting.

Forest Nurseries. To establish a forest by planting, seed of forest trees must be collected, seedlings grown in a nursery, and the seedlings set out on properly prepared planting sites and protected until they are able to take care of themselves.

Tree seed may differ radically from agricultural seed. It is borne high in the trees where it is difficult to observe and collect. Since much of it is dispersed by the wind, it must be collected as soon as ripe but before wind dispersion occurs. Careful judgment and test are necessary to determine the exact time at which the seed is mature enough for use so that seed-collecting activities can begin promptly. Tree seed is adjusted to germinating in the natural forest habitat and also is adjusted to many different means of dispersal. Some of it is well protected against destruction. Basswood for example must lie for one to two years on the forest floor while the seed coat is softened up by action of microorganisms before sufficient water will penetrate to enable the embryo to swell and emerge. Cottonwood, willow, elm, maple, and most of the seed that ripens in early spring is short-lived. The forester must know how to collect cones and fruit, how to extract the seed, how to hold it in storage until ready for use, and how to pretreat it so as to get satisfactory germination in his nursery beds. Through study a forester can evade many of the limitations that nature places upon seed (Fig. 26). Under proper conditions many kinds of tree seed can be stored for several years without serious decline in viability.[8]

Seed may be sown directly in the forest where the seedlings are to grow. This is desirable practice for heavy-seeded species such as oaks, walnuts, and hickory. Care must be taken to protect such seed against rodents, lest they dig up and devour every seed. If a proper seedbed can be prepared, the seedling will spring up in the place where it is to grow permanently.

The majority of tree seeds are so delicate that if the forester attempts to sow them directly on the planting site, however, survival is exceedingly low. To plant successfully such species as northern white cedar, white spruce, hemlock, many of the pines, larch and fir, it is desirable to grow the seedlings for one to three or more years in a nursery.[6]

Fig. 26. The modern tree seed extraction plant is fitted with automatic temperature and humidity controls. Cass Lake Extractory. (*U.S. Forest Service photo.*)

Seedbeds are carefully prepared, and the seed, after appropriate pretreatment, is sown in rows or broadcast in the beds, covered to protect it against soil drying, and watered daily until germination occurs. The nursery beds must be weeded, watered, and properly fertilized to produce a satisfactory crop of tree seedlings suitable for forest planting. The nursery must also be protected against diseases and insects, birds, mice, and other destructive agents. The job of a nurseryman is an exacting one. He must be on duty daily to discover early and correct promptly any unfavorable factor that may be operative in his nursery. The task of the nurseryman is not to grow fine-looking seedlings, but to grow seedlings that are vigorous and rugged enough to stand up under field conditions when they are set out on the land (Fig. 27).

Field Planting. Field planting of tree seedlings requires proper preparation of the planting site to receive the seedlings. For many areas very little preparation is needed. For others, existing vegetation must be destroyed, at least in the neighborhood of the individual planting spots. This can be done by furrowing, scalping, burning, or complete cultivation of the soil. The latter has proved the best way of converting lands invaded by brush to forest. It requires the use of heavy equipment and is expensive, yet cheaper than what could be done by hand methods.

FIG. 27. Forest nurseries are highly mechanized. A tree lifter in operation, Superior National Forest. (*U.S. Forest Service photo.*)

Each seedling must be carefully set in the ground at the proper time of the year so that it has good opportunity for survival. Spring is normally the best time to plant, but trees can be successfully planted in the fall in many parts of the country. This is a desirable practice in that it extends the period over which forests can be established. The modern tree-planting machines have greatly facilitated field-planting operations on sites on which such machines can be used efficiently. Two men with a machine can plant 8 acres a day and on favorable sites even more. This is about four times as fast as hand planting. It has also been found that good machine planting is better than average hand planting (Fig. 28).

Plantations must be well protected against fire and must also be protected against destruction by browsing animals, such as cattle, rabbits, deer, and other animals that may destroy the seedlings during the first two or three years of their life.

Competition from herbs, shrubs, and weed trees must not be excessive lest the seedlings be choked out completely or so stunted that they never develop into satisfactory trees. Protection of newly planted trees by an overhead canopy tends to enhance first-year survival, but if older trees are left standing after the seedlings once become established, they will

FIG. 28. A mechanical tree planter in operation. With such machines two operators plant 6,000 to 8,000 trees per day. (*A.F.P.I. photo.*)

stunt the growth. The plantation, therefore, should be released by removing these older trees as soon as the seedlings are well enough established to take care of themselves.

Plantations may require thinning at least once before the trees are of sufficient size to bring in enough income to cover the cost of the thinning operation. Where a market for Christmas trees and other small-size products is available, such thinning operations may be made at a substantial profit.

The Tree and the Forest

The foregoing sections have dealt with the life needs of an individual tree and with the establishment of plantations of trees. In supplying the needs of individual trees and establishing plantations, the forester does work that parallels that of the orchardist and arborist. But a forest, as stated earlier, is much more than a group of trees occupying an area of land. It is a community made up of trees, shrubs, herbs, birds, mammals, and myriads of lower forms of plants and animals. All play an interdependent part in the life of the forest as a whole.

Shrubs use space not occupied by trees. Many produce valuable food for wildlife. They protect the openings and forest borders against sweeping winds. When dense, they may impede the growth of seedlings, but shrubs, like trees, tend to go through a life cycle. Groups get overdense and overlarge in size as they reach maturity. They may then die out over wide areas and make the land available for tree seedlings and other vegetation.

Herbs also serve as food for wildlife. Many of them such as the wild vetches and peas serve to nitrify the soil. They provide foods for insects many of which burrow in the soil opening up channels for roots and water. These, with burrowing mammals, earthworms, and other forms of life, cultivate the forest soil.[9]

Animals browse on shrubby thickets, preventing them from getting too dense; they also help to keep insects in check. The insects, in turn, break down wood and kill weak trees, thereby preparing the way for drastic changes which will occur in time. Bacteria help to store up nitrogen in the soil. Bacteria and fungi also break down wood, leaves, and other plant material to the basic inorganic compounds usable for new plant growth.

Climax and Virgin Forests

Change occurs in the forest from year to year, decade to decade, century to century.

The forest is a splendid builder of soil. As soil accumulates, conditions favor more and more complex forms of forest. Pioneer plants such as lichens, mosses, algae, and ferns prepare the way for grasses, other herbs, and small shrubs. These prepare the way for trees. The first trees may be light-demanding trees such as redcedar, juniper, gray birch, or piñon pine. Gradually, the forest passes through various stages in a succession toward a climax forest vegetation that will occupy the soil for an indefinite period.

The climax forest needs to be distinguished from the virgin forest. We may have a virgin forest of paper birch, white pine, or Douglas-fir, but these are not considered climax forests. They owe their origin to some fire or other catastrophe in the past. A virgin forest is one that has been undisturbed by man's activities, has not been cut for timber, intentionally burned, or heavily grazed by domestic livestock. It may represent almost any stage in the succession from the pioneer trees to the climax forest.

The climax forest, on the other hand, is composed only of those species that have the capacity to maintain themselves indefinitely in possession of the land without the aid of man, fire, windstorm or other major destructive agents. Neither the climax forest nor the virgin forest is necessarily ideal for man's use. Examples will make this clear. On Star Island in Cass Lake, Minn., no commercial timber cutting has been permitted. This island was used by Indians as a stopping point on their canoe trips along the upper Mississippi, but the forest was little modified by them. The soil is essentially uniform throughout. On the island can be found every stage in natural forest succession for the region. The pioneer jack pine forest occupies the lands most recently burned, but even though this forest is near maturity, few jack pine seedlings and even fewer poles can be found. The longer-lived red pine, where mixed with the jack pine, is taking over instead. The red pine also reproduces only moderately well; so it is replaced by the still longer-lived white pine. The white pine forest is grandest of all. The trees reach heights of 120 or more feet, ages of 400 or more years, and produce a forest of high aesthetic value and utility. But white pine also is giving way to a maple-basswood climax. These trees, being ill adapted to the sandy soil, attain a height of only 40 to 50 feet and ages of barely 150 years. This hardwood forest is neither impressive to look upon nor does it offer any temptation to the logger. It can, however, hold the soil against the pines and other species and therefore represents the climax.

From the standpoint of the timber user the virgin forests which our ancestors discovered in America were also far from ideal. The average timber volume standing in the Adirondacks was 10 to 12 cords per acre. For the country as a whole, the average timber volume was estimated to be 6,300 board feet per acre. Stands of this density represent good merchantable timber but they are far from the ideal. Good pulpwood stands oftentimes run 30 to 50 cords per acre, and sawtimber stands even in the eastern United States may run as high as 20,000 to 40,000 board feet per acre.

In the virgin forests heavy stands were interspersed with open areas, burned areas, windfall areas, and areas where the timber died from attack by insects, tree diseases, or other causes. Droughts undoubtedly had a great deal to do with the dying of timber over substantial areas. The popular conception of the virgin forest is one in which trees of all ages are found. As soon as a veteran dies, a young tree is presumed to be ready to take its place. This may be true of some types of virgin forest, but it was not true of large areas of timberland in the United States. Instead of having all ages represented, the virgin timber stands tend to be either even-aged or made up of a series of even-aged groups. Even in the northern hardwoods of the Allegheny Plateau which are often cited as examples of the all-aged virgin forest, careful study has revealed that a high percentage of the trees owed their origin to some particular catastrophe in the past. It might have been a fire, drought, heavy windfall, perhaps an ice storm, or some other catastrophe that killed off many of the older trees and made room for the young trees to become established and grow. Practically all of our pine forests in the South, the Douglas-fir forests of the West Coast, and particularly the jack pine and the red pine forests of the Lake States, owe their origin to fires or to other destructive agents.

Making the Forest Useful

The forester's job, therefore, involves more strictly technical work than planting trees and protecting forests from fire and other destructive agents. He must mold the forest stands as they grow into forms that best serve his purposes. To produce valuable timber, he should make certain that the forest growing space is used by straight, clean-boled trees of useful kinds. He will take advantage of markets for small-sized material to dispose of injured, defective, crooked, diseased, and poor-quality species. Thereby he diverts the growing potential of the soil from poor to valuable trees.[4] This is the exact reverse of common past practice—harvesting the useful trees and leaving the unwanted ones to occupy the soil.

Caring for Seedling and Sapling Stands. Stands of natural seedlings and saplings like those of plantations improve with care. They may require weeding, that is, the cutting away of undesirable tree and shrubby growth that is tending to choke out good trees; release cutting from older trees, after these have seeded in the area and protected the seedlings during initial establishment; and thinning so as to favor those seedlings which will make the best crop trees. In most hardwood stands thinning can be avoided in the seedling and sapling stage, but this may not always be desirable. Young conifer stands, particularly pines that become established on open areas following fires or logging, may be so dense that thinning is necessary if the trees are to develop sufficient sturdiness to withstand snows, wind, and other natural hazards to which they are subject. If they are not artificially thinned, they will be thinned by nature through extensive blowdown, snow and ice breakage, or other natural means. Rabbits, deer, porcupines and other animals sometimes participate in the thinning operation, but normally the type of job they do is far from perfect from the forester's viewpoint, for they are likely to concentrate their feeding in small areas.

Pole-size Timber. When the forest reaches pole size, the individual tree stems are large enough to yield merchantable products if a fuelwood, chemical-wood, or pulpwood market exists. The forester will tend to select those trees for early cutting that are least desirable in the final harvest. He may either pick out the small-sized trees—this method is called a thinning from below—or he may select the large-sized trees, a thinning from above. The latter method has the advantage of removing material of sufficient size to be merchantable early in the life of the stand. It therefore reduces the investment in the stand and brings in substantial income to help meet carrying charges. A further advantage is its tendency to remove those trees that have the heaviest limbs and therefore will produce ultimately the lowest percentage of high-grade lumber. Usually it is best to combine the two methods, taking some of the large trees and some of the small ones that would die naturally if not removed. The growing space can then be made available for full use of the trees that will produce the highest values in subsequent cutting operations. Usually two or more thinnings can be made in the stand while it is in the pole-timber stage. Best results are obtained if thinnings can be made lightly and frequently. The Danish stands that yield as high as 1,000 board feet per acre annually are thinned at about three-year intervals.

Management of Sawtimber Stands. Stands of sawtimber size may still be thinned and improved by cuttings. Improvement cuttings remove the poor-quality trees and those injured by wind, falling trees, fungi, or other causes. If damage has been severe, such operations are called salvage cuttings. Salvage cuttings should be made, if possible, before the tree dies, as the lumber may otherwise be low in value.

Harvest cutting may be either on a clear-cutting or partial-cutting basis. Clear-cutting is done by strips, by patches or blocks, which may be regular or irregular in size, or it may extend over very large areas. Clear-cutting large areas of timber is usually bad practice in that it exposes the soil to the drying action of sun and wind, creates a huge fire hazard unless the slash has been disposed of, and removes all source of seed for natural restocking. Much of the 73 million acres of understocked timberlands in our country today has resulted from clear-cutting over large areas followed by burning. Strip cutting and cutting in blocks or patches is especially to be recommended in that it leaves ample trees surrounding the clear-cut area to protect the area and to provide seed for a second crop. Such cutting practice has been widely adopted by many of the large timber-operating companies in the Douglas-fir region. Here, even-aged stand management appears to be the best practice, and consequently a clear-cutting system is desired.

PARTIAL CUTTING. The simplest type of partial cutting is the seed-tree cutting. From one to ten or more trees are left per acre to seed up the surrounding land. This method works well with species, such as southern pines, that require plenty of light and soil space for establishment and growth. The trees left standing after logging should be sound, wind-firm, good seed producers, and well distributed. Seed trees should also be of good genetic strain so that they produce desirable offspring. Poor results may follow seed-tree cutting if these principles are neglected. Seed trees should be removed when the seedlings are well established, that is, 2 to 6 feet in height.

Fig. 29. Marking a mature Jeffrey pine for cutting. Plumas National Forest. (*U.S. Forest Service photo.*)

A shelterwood cutting resembles seed-tree cutting, but usually more trees are left standing. These provide seed and light shade to protect the young seedlings from direct sunlight and frost injury. If a stand has been very dense up until the time of harvest cutting, it may be desirable to make a preliminary cutting, known as a seed cutting, to stimulate the production of seed. Following this, the shelterwood cutting is made. Once seedlings are well established, the final cut removes the old trees, and the new forest takes over the area. The shelterwood system is not widely applied in the United States nor has it been used extensively in Europe. Locally it has been used very effectively for the reproduction of northern white pine.[4]

The selection system promotes an all-aged forest in which trees are harvested on an individual basis as they reach financial maturity. Young seedlings take over the space relinquished by the veteran (Fig. 29). A type of selection forestry known as Dauerwald (continuous forest) was advocated in Germany and practiced for many years prior to 1940. The true selection forest, however, is not adapted to trees that grow best in even-aged stands, of which pines are an outstanding example. This is true despite the fact that the Dauerwald system was developed for Scotch pine forests. The true selection system is best adapted for long-lived trees that endure shade, such as the northern hardwoods. Some foresters think it unsuited even to these.

Selective cutting or selective logging are expressions loosely used to describe any system of cutting in which part of the stand is harvested at a time, whether the selection is based on species, age, size, or even area. It is not a clearly defined concept and consequently has led to much debate. When properly employed, it embraces features of the low and high thinning method, features of group cutting, and features of the shelterwood system. Selection cutting can be skillfully applied with even-aged stands until the final harvest is made. This must then take the form of a shelterwood cutting or one or more clear-cutting methods.

REPRODUCING FOREST STANDS. After harvest of the mature timber crop, the forest may be reproduced by artificial planting or seeding or by favoring natural reproduction. Because of the expense involved in planting, American foresters normally try to get natural reproduction. This may require preparation of the seedbed by burning, scarification of the soil, or other measures. Sometimes woody shrubs and undesirable trees must be poisoned or destroyed by still other means.

Improving Value with Use. The more intensively man uses the forest, the more rapidly he may shape it to fit his needs, provided his use is properly guided. At least half the woody volume produced by European forests is removed as intermediate cuts before the time of final harvest. The same practice is used on intensively managed forests in the United States. Instead of the ax being the enemy of the forest, as it was in the past, it can be man's most useful tool for improving the forest. It is by judicious use of the ax that the best individuals and species are favored, that overdense stands are thinned, and that adequate light is admitted to the forest floor. This is necessary to maintain vigorous activity by the soil fauna and flora that break down litter, converting it to useful humus, and release mineral elements needed by trees.

The extent to which the individual tree species will respond to release from the competition of their neighbors and the rapidity with which the thinned forest closes ranks so as to put on high wood volume is a measure of the response to management. This response in turn determines the type of silviculture that may be used, the length of the cutting cycle, and the monetary yield that can be expected from efforts expended on the forest.

Silvics, Silviculture, and Forest Management

The forester uses the term silvics to include those phases of soil science, botany, zoology, and protection that deal with the fundamental nature of forest plants and their relationships to one another, to soil, and to the animals that live in the forest.[7]

Silviculture draws upon all fields of knowledge included in silvics for its basic, scientific information. It is not, however, a science in itself, but rather the art of using these sciences so as to produce high yields of valuable crops from the forest whatever the nature of these crops may be. Silviculture is limited by the physical and biological factors of environment, by the genetic capacity of trees, and by the expenditures man can afford to make to favor tree growth. Obviously, forest trees could be made to grow more rapidly if they were selected individuals of superior genetic strain, if they could be cultivated and fertilized, if they could be watered periodically, and if whatever steps are necessary to eliminate competition of other trees could always be taken at the right time. Such steps are generally uneconomic as well as highly artificial, and it is doubtful if man knows enough at present to create an ideal environment for the growth of a forest crop.

Practice of silviculture is an art in that it requires shrewd compromise between what is lowest in cost and what science shows to be the best practice. It is an art also because it must combine knowledge of many sciences in a way that can be skillfully done only through exercise of well-informed judgment.[4] The art of silviculture embraces all of those activities that have to do with the care of the forest and its culture for human use.

Forest management is the application of silviculture to the business of growing timber. Management, therefore, embraces not only silviculture but the other factors that must be combined to make the growing of timber a successful venture in terms of the owner's desire, whether this be for money, protecting watersheds, recreation, or game. The silviculturist is guided in his plans and activities by the aims of management, which reflect the wishes of the owner. There can never be a perfect compromise between the desires of the landowner and the needs of the forest trees for best growing conditions. It is for this reason that it will always require men of thorough training and keen understanding of potentialities to manage forest properties and to mold forest vegetation so as to achieve the ends desired by the timber owner.

Units for Buying and Selling Timber

Any commodity used by man in commerce must be measurable. In the case of a forest this means measurement of area, timber volume, and growth rate.

Forest land is measured in terms of *acres*, *townships*, *sections*, or *square miles*. In countries using the metric system, *hectares* are used as units of area.

An *acre* of land is 43,560 square feet or 10 square chains. (A chain is 66 feet.) An acre is roughly equivalent to a square 208 feet on a side.

A *section* of land is a subdivision of a township of the national land survey system. It includes the land as originally surveyed no matter how accurate that survey happened to be. An accurately surveyed section, except where correction lines for the earth's curvature are made, is a square 1 mile on a side and contains 640 acres.

One *township* is a square of 36 sections.

One *hectare* is a plot of land containing 10,000 square meters. It is equivalent to 2.47 acres.

Timber volume is measured in terms of units in which the wood is used. As more timber is sold for lumber than any other one use, the board foot is the most commonly used unit.

A *board foot* is a unit of lumber measuring 1 inch in thickness and 12 inches in length and breadth or the equivalent thereof. It may be determined by multiplying the length and breadth of a piece of timber in feet by its thickness in inches. A piece of lumber 12 feet long, 6 inches wide, and 2 inches thick contains $12 \times \frac{1}{2} \times 2$, or 12 board feet.

Pulpwood, fuel wood, chemical wood, and wood used in tanning is usually sold by the *cord*. The *standard cord* is a stack of wood 4 feet high, 8 feet long, and made up of sticks 4 feet in length. It contains a gross volume of 128 cubic feet. The net volume may vary from about 70 to over 100 cubic feet, depending upon the size and straightness of the wood in the cord and the way in which it is piled. Many pulp companies buy wood in 5-foot cords, in 50-inch cords, and even in 100-inch cords. The last is spoken of as a double cord. The measurement in each case refers to the lengths of the billets of wood that are cut. Fuel wood may be sold by the face cord with sticks 12, 16, or 24 inches in length.

Crossties, posts, poles, shoe-last blocks, blanks for baseball bats, and a host of similar products are sold by the *piece*. Piling is sold by the linear foot. The value of piling varies with its length and diameter. Price per linear foot for a 50-foot piling is considerably more than for one 40 feet long. Railway ties, likewise, vary in price with the size. A 7- by 9-inch tie is worth considerably more than a 6- by 8-inch tie. Since switch and bridge ties are longer, they are worth more than the standard ties.

Mine timbers may be sold either by *piece* or by *weight*. In the bituminous region in West Virginia and Pennsylvania, the timbers are commonly sold by the piece; in the anthracite region, by weight. The price varies with the size of timber being sold. Heavy mine timbers bring a higher price per unit weight than small mine timbers and lagging. Products sold by the piece and ton can be accurately counted or weighed.

Veneer bolts may be sold on the log-scale basis. If of small size, like birch bolts, they may be sold by the cord.

Measuring Lumber. As most lumber is cut square-edged with parallel faces, the number of board feet in any board, plank, or timber can be easily determined. Lumber rules are used having a hook at one end for turning boards. The board-foot contents is stamped on the rule for boards of various lengths (Fig. 38). Since unfinished hardwood lumber is sold in random widths and lengths, it must be measured and graded when sold. Softwood lumber is cut to standard widths and lengths so that the volume may be computed when the number of pieces of each dimension is known.

Log Rules and Log Scaling. Logs cannot be measured directly in terms of squared lumber they will produce. No log is perfectly circular in cross section or cylindrical in shape. Because of taper in the log one 16 feet in length may saw out lumber 8, 12, and 16 feet in length. The amount of waste in the form of slabs and sawdust will depend upon the skill of the sawyer, the care with which the sawmill machinery is adjusted, and the type of head saw used. Band head saws cut about ⅛-inch kerf; circular head saws, ¼-inch kerf. Saws badly adjusted or aligned may cut even wider kerf. Kerf is the channel cut by the saw as it removes a board from the log. Logs cut into 2-inch and thicker planks will yield more board feet than logs of the same size cut into 1-inch lumber, because less saw cuts are required.

To measure the contents of logs in terms of squared lumber, log rules have been devised. Many types exist. In common use are the Doyle, the Scribner, and the International. The Doyle rule underscales the lumber volume that can be cut from small logs. The Scribner rule underscales the volume to be cut from large logs. A combination of the two, the Doyle-Scribner, underscales both small and large logs. The International rule gives accurate volumes for any diameter of log. As the log buyer or sawmill man is usually the one who determines the unit on which he will buy, the Doyle rule is in widespread use in the small timber of the eastern United States. Inasmuch as the value of logs increases as their size increases, the Doyle rule, when used to buy ungraded logs, is not as unfair as it might at first seem.

Determining the volume of a log in board feet is called log scaling. This is done by measuring its diameter usually at the small end, and determining log length (Fig. 30). These two figures are looked up in appropriate tables which give volume in board feet for logs of various sizes. Volumes are printed directly on special scale sticks and can be read from them directly. The log scaler must make appropriate deductions for defects such as crook, sweep, decay, checks, or splits. The usual method is to scale the log as if sound and then scale the defects and subtract this from the gross scale.

Neither the board foot nor the cord is an exact unit of measure when applied to round products. For this reason the cubic foot and cubic meter are substituted when precise measurement is desired. It is often useful to be able to convert from one unit of measure to another, but this can never be done with any degree of exactness unless size of product is known. For general purposes it is assumed that 1,000 board feet of lumber can be cut from approximately 204 cubic feet of logs, which is the equivalent of about 2.5 cords. One cord is about the equivalent of 400 board feet.

Estimating Standing Timber

Volume estimates of standing timber are important to those who buy or sell it, operate it, or appraise it for taxes or damage. Such estimates were made long before foresters were active in America. The old-time timber cruisers used various methods of arriving at their estimates, some accurate, others inaccurate. It was not until foresters systematized the

Fig. 30. Scaling birch logs in Maine. (*U.S. Forest Service photo.*)

methods that any good crew after appropriate training could produce a usable estimate.[2, 3]

Timber estimating requires determining the volume to be found in trees of various sizes, the number of such trees per unit area, and the total area of trees of the given forest type and condition class. A first step, therefore, in preparing to estimate timber is to obtain tables that will show for trees of various sizes the number of units of product they contain.

Volume Tables. Volume tables have been prepared that give for trees of all sizes the volume in board feet, in cords, in cubic feet, in ties, and residual board feet, in other piece products, and in tons of such products as mine props. The volume tables are prepared by cutting sample trees, measuring their diameters, heights, and merchantable length, and computing the units of products they contain.[3]

Volume tables formerly were based on the diameter and total height of the trees. For each species a separate volume table was prepared. In recent years James Girard, formerly of the U.S. Forest Service, has determined that more accurate estimates can be made by basing volume on the tree form class and the merchantable length of the tree. The form class, as defined by Girard, is the ratio between the diameter inside bark at the top of the first 16-foot log and the diameter at $4\frac{1}{2}$ feet from the ground, where it can readily be measured by means of calipers. Girard has shown that form class can be readily estimated in the field after a man has a little practice. To aid in this estimation, tables of bark thickness for various species have been worked out, and special devices have been constructed for estimating diameter outside of bark at the top of the first 16-foot log by using the angle subtended by the tree. The form-class system enables the forester to use a single set of tables for all species of trees. For short trees of the eastern United States that have their greatest volume in the first log, form-class volume tables are especially useful. Only form class need be determined in the field to adjust the tables for use.

Estimating Timber on Small Tracts. A forester usually takes with him tables of tree volumes and bark thicknesses, calipers or tapes for measuring tree diameters and distances, a hypsometer to measure tree heights, tally forms and holder, and possibly chalk to mark lines and a device to measure diameters at the top of the first 16-foot log. On small tracts the timber cruise may be 100 per cent; that is, every merchantable tree may be recorded. It is a standard method in use on small-sized properties throughout the central states, especially for such high-value trees as black walnut, white oak suitable for veneer and tight cooperage, ash, yellow-poplar, and cherry (Fig. 31).

After all trees have been measured and tallied in the woods, the volumes are determined for each diameter class, height class, and species. These are summarized by species. Volumes may be shown as board feet of lumber, ties, cordwood, and other products. A stand table may be prepared that shows for each species and diameter the number of trees and total merchantable volume. These are particularly useful to sawtimber buyers who are concerned about the size of timber they are buying.

Estimating Timber Volumes on Intermediate-sized Tracts. For timber tracts over 100 acres in size, it may be too costly to measure and tally each individual tree. This is particularly true if the timber is small in size or low in value. Some method of sampling the timber then becomes necessary. Two new tasks then must be performed by the timber cruiser: he must determine the area of the timber tract and of each stand on it, and he must decide on a method for sampling the trees on each.

If a map exists, the area can readily be determined. If no map exists, it may be necessary to make a survey of the property and from this construct a map.

The next task is to divide the timber tract into timber size or condition classes and into forest types. Condition class indicates the size of timber on an area as seedling and sapling, pole, or sawtimber in size, and if

FIG. 31. Estimating timber in a central New York farm woodland. Tallies are being kept by log grades. Otsego Forest Products Coöperative Association. (*U.S. Forest Service photo.*)

of sawtimber size, as light stands or heavy stands. If aerial photographs are available and a good road system, it may be possible in the field to delineate on the photographs these types and condition classes for later sampling. This is the system now used by most men who are obliged to estimate timber on large areas. When photos are not available, condition classes and type are determined on cruise lines.

For the areas that are to be sampled on the ground, the forester must determine the intensity of cruise he will need; that is, how many plots of definite area (1/5- or 1/4-acre plots are usual) he must measure and where these are to be located. This requires experience with similar timber or preliminary sampling and statistical computations to determine variability. He then goes on the ground, following a compass line, and, stopping at regular intervals, he lays out square or circular plots of definite area. On these he tallies all trees by diameters and some for heights. From the latter he determines an average height to apply to trees of each diameter class. After sampling one plot, he moves on to the next. Usually a cruiser works with a helper who runs the compass line and tallies the trees. Separate tallies are kept for each forest type and condition class.

It is desirable to keep the tally on the individual plots separate or at least in groups of five or ten plots. These then can be worked up as units and used to determine the variability of the timber stands. From this it is possible to calculate the probable accuracy of the timber estimate for any intensity of cruise that is used. A 20 per cent estimate may be made of tracts 100 to 1,000 acres in size, and a 10 per cent estimate of tracts 1,000 to 5,000 acres in size. Field tallies are worked up in the office. The total number of trees of each species, diameter, and height class is determined. The volume per tree is looked up in the appropriate table, and the total volume in board feet, cords, or other desired units computed. The process is repeated for other diameters, heights, and species and the volumes summarized by species. Volumes may also be summarized by diameter classes in special cases. The volumes so summarized are multiplied by the appropriate area factor to convert the sampled volume into total volume in the type and condition class.

Cruising Large Forest Tracts. For estimates of large areas it usually pays to use aerial photographs. A system of timber cruising is adopted that involves dividing the area up into types and condition classes and determining the degree of sampling needed in each to arrive at a satisfactory, over-all estimate. Careful statistical control should be used throughout. A competent survey party will include a man familiar with statistical and sampling procedures. It is his task to design the survey so that the greatest over-all accuracy results for the effort expended. Skilled photograph interpreters are employed, equipped with stereoscopes and special measuring devices that enable them to measure crown width and tree-shadow lengths on photos and from these to compute tree diameters, heights, and volumes. By a proper combination of extensive and intensive use of photos and by field checking on estimates of volumes, species distribution, and cull, estimates of timber volumes on vast areas can be made with surprising accuracy and a minimum of field work. Cost of such surveys is correspondingly reduced. The forest-survey leader, however, must know vastly more than how to run a compass line and estimate volumes of individual trees.

TREE GRADES. A refinement that is being introduced into timber cruising is to estimate timber volumes by tree grades. Tree grades are based on the percentage of high-grade lumber that may be cut from them. These grades are a measure of the value of the tree. To date only limited use has been made of tree-grade estimating.

Estimating Timber Growth. The simplest way to determine the growth rate of a forest property is to compare the growth rate with that existing on a similar property bearing the same types of timber and having similar soil and climatic conditions. This is the method generally used in Europe where records have been kept over a long period of time. These values are corrected for the density of stocking, and yield tables are made up which show the volumes that may be expected at any 10-year period in the life of the stand.

Yield tables of the same pattern have been made in the United States. The so-called normal-yield tables, however, do not represent the yield from any one area over a period of years. Instead, the tables are based on the volumes that are found on current stands of various age classes from the youngest to oldest available. These must have normal stocking, that is, enough trees to occupy the area completely. Obviously, opinions differ as to how dense trees should stand to represent normal stocking.

The tables so built up have proved only moderately useful. The main difficulty is that timber stands are variable in stocking, and few areas within any one stand approach normal stocking as thus defined. Yield tables can be successfully applied to stands having less than normal stocking only after corrections are made. Understocked stands as they increase in age tend to approach normal stocking. The rate at which they approach normality is dependent upon the uniformity of tree distribution.

A new method of determining growth rate has been devised based on stocking as it actually exists in the field. This involves determining the average variation in height with diameter for the several species and determining from increment borings the growth rate during the past 10 years and the past 20 years. It is assumed that trees now 14 inches in diameter will grow during the next 10 to 20 years as much as have the trees that were 14 inches in diameter 10 and 20 years ago. In this way the growth rate for each diameter class is projected forward. The volume 10 and 20 years hence is assumed to be equal to this projected volume less the volume lost through mortality, which is likewise projected from the past. Mortality is estimated from the number of dead trees present and their condition. The method obviously has flaws, since agents causing mortality, such as insect attack, disease outbreak, windstorm, and ice damage, may vary widely decade by decade. Also, since trees, as they approach a certain density, slow down markedly in growth, an overestimate is easy to make. This method, however, is checked against growth of trees on permanent sample plots, of which there are many throughout the United States on the various experimental forests. Appropriate safety factors are thus introduced.

How Forest Management Practices Affect Forest Growth

The growth on a given forest area will vary widely with the intensity of forest practice. An example will make this clear. On the Pack Demonstration Forest at Warrensburg, N.Y., a special truck-mounted sawmill has been built. This can be taken into the woods and set up by two men within 90 minutes and will saw timber at the rate of 1,500 board feet per man-day. The mill is adapted to handling small, short logs that will not pay their way in a large sawmill. The use of this mill enables the owner to harvest trees and to saw small top logs that would otherwise be allowed to rot in the woods on a usual forest operation. The yield of the forest, therefore, is increased over that of other forest properties by the amount of timber that is handled by this small sawmill.

The shorter the cutting cycle that can be adopted, the greater the possibility of harvesting all timber that is grown. A striking example of this was worked out some years ago for spruce in the Northeast. Pulp and paper companies in Maine formerly managed their lands on a 60-year cutting cycle. Balsam fir trees which were 20 to 30 years of age when the land was first cut over became defective and died before reaching 80 to 90 years of age at the time of the second cut. Even a considerable volume in spruce also decayed if the stands were dense. It was found that by shortening the cutting cycle from 60 years to 20 years, it would be possible to increase the yield by 90 per cent. The more frequently timber stands can be cut over, the better. Weakened trees, those that have been damaged by wind, ice, or other causes, can thus be harvested to augment the total yield from the property. If the trees to be removed are carefully selected each time, the stand remaining will keep thrifty and grow rapidly.

Trees will thin out naturally when the volume according to normal-yield tables is at about 115 per cent of normal. Such stands can be thinned down to about 60 per cent of normal volume without seriously decreasing the total growth rate. By judicious and frequent thinning, the forest at Friesenburg in Denmark has stands yielding as high as 500 cubic feet per acre annually. The average for the whole of Denmark is but one-fifth this amount.

LITERATURE CITED

1. Baker, F. S. 1934. "The Theory and Practice of Silviculture," McGraw-Hill Book Company, Inc., New York. 502 pp.
2. Bruce, Donald, and Francis X. Schumacher. 1942. "Forest Mensuration," 2d ed., McGraw-Hill Book Company, Inc., New York. 425 pp.
3. Chapman, Herman H., and Walter H. Meyer. 1949. "Forest Mensuration," McGraw-Hill Book Company, Inc., New York. 522 pp.
4. Hawley, Ralph C. 1946. "The Practice of Silviculture," 5th ed., John Wiley & Sons, Inc., New York. 354 pp.
5. Miller, Edwin C. 1938. "Plant Physiology," 2d ed., McGraw-Hill Book Company, Inc., New York. 1201 pp.
6. Toumey, James W., and Clarence F. Korstian. 1942. "Seeding and Planting

in the Practice of Forestry," 3d ed., John Wiley & Sons, Inc., New York. 520 pp.

7. Toumey, James W., and Clarence F. Korstian. 1947. "Foundations of Silviculture Upon an Ecological Basis," 2d ed., rev. by C. F. Korstian, John Wiley & Sons, Inc., New York. 456 pp.
8. U.S. Department of Agriculture, Forest Service. 1948. "Woody-plant Seed Manual," Miscellaneous Publication 654, Washington, D.C. 416 pp.
9. Wilde, S. A. 1946. "Forest Soils and Forest Growth," Chronica Botanica Company, Waltham, Mass. 241 pp.

CHAPTER 7. *Harvesting Timber Crops*

LOGGING EMBRACES those activities involved in harvesting timber crops and transporting them to the place of manufacture. While strictly speaking it might apply only to the cutting and transporting of products in log form, it is used as a broad term to cover the cutting of pulpwood, veneer bolts, crossties, piling, poles, and many other products.

A major aim of silviculture is to produce crops of timber. The major aim of logging is to harvest such crops. Both silviculturist and logger desire to perform their jobs at minimum cost. If their activities are uncoordinated, the silviculturist may insist on operations that increase logging costs unduly, and the logger may so damage the forest as greatly to increase the costs of silviculture. The only sound approach is to coordinate both activities so that combined costs are minimized.

Logging changes the forest more drastically than any activity other than fire. Under the skillful hand of a competent forester logging is a means of molding the forest from one of mediocre quality to one of high quality. The character of logging therefore furnishes one of the best indices to the level of forestry being practiced. If care is taken to preserve young growth, if seed trees and immature trees are left to restock the area, if heavy accumulations of slash are disposed of, and if the area generally is left in condition for producing a second forest crop, it is evidence that conscious effort to practice forestry is being exercised.

Stumpage, the price of trees standing in the woods, made up in 1948 only 10 to 22 per cent of the lumber price f.o.b. mill,[1] and logging costs made up 30 to 50 per cent of lumber price. Labor represents the major item of expense in logging.

The Evolution of Logging in America

Logging has gone through four stages in the United States. Early logging was primitive. It was done with the simplest of tools, the ax, pry pole, and the swing dog, a forerunner of the peavey. After horses or oxen had skidded logs to water, they were floated to the mill, driven down the river as individual logs or collected together into log rafts.

The cross-cut saw, peavey, and other improved tools and equipment helped usher in the second stage, characterized by winter sledding of logs in the North and wagon hauling where snow for sledding was undependable. Huge lumber camps were operated to house workmen and horses.

The third stage was the introduction of railway logging, the dominant system used in the Lake States. Logs were loaded on huge sleds that were drawn by horses over iced winter-haul roads across swamps and lakes to railway spurs. There the logs were loaded on cars and hauled to the mill. Where large lakes or streams were involved, many of the logs were driven after they reached the main watercourse. Railway logging reached its peak on the West Coast, where it was combined with high-speed cable skidding.

This high-lead system of logging has proved very destructive in that small trees that stand in the path of the onrushing logs are badly scarred or broken off completely. Where the logs are drawn along the ground, huge furrows are made. This method of logging is still used, however, particularly in rugged topography where tractors can be worked only with difficulty.

The fourth stage of logging, which is currently widespread, is truck and tractor logging. The logs are skidded by tractor to a landing along a woods road and there loaded onto trucks that transport them to the mill. The great network of public highways throughout the country has greatly stimulated truck logging. A good truck road is more expensive to build than a logging railway spur, but it is also more permanent and admits of greater flexibility in the logging operation. When timber becomes smaller in size and more scattered in location, as is the case throughout the eastern United States, railway logging becomes far too expensive. Here truck and tractor logging is almost universal.

The fourth stage has also been marked by the introduction of power-driven saws for felling and bucking. Power-driven drills, road-building and maintenance equipment, loading devices, and small power skidders have also been introduced. The bulldozer, power shovel, dump truck, and other road equipment are essential items in a modern logging operation.

Planning and Layout

Logging operations may be grouped under four headings: planning and layout, log making, skidding, and hauling. For logging a woodlot or some other small area, relatively little attention needs to be given to planning, but where large areas are to be logged, planning is very important. Planning a large western logging operation may involve these steps: (1) The head cruiser explores the timber area and decides whether or not it can be logged profitably. (2) Other cruisers or timber estimators determine the volume of timber and where the heaviest stands are located. (3) The logging superintendent plans the location of his main haul road and sends in a survey party to stake it out. (4) Secondary roads or logging spurs are laid out and construction begun. (5) Spar trees and landings are located, sidings and turn-arounds established. (6) The entire road system or railway system is built or at least enough of it for logging operations to start. (7) The forest is divided up into cutting blocks to be assigned to individual cutting crews. (8) Repair shops, logging stores, and a logging camp are established if necessary. (9) Spar trees are rigged, logging equipment moved in, and the actual operations begun. While these steps are listed more or less in sequence, they are actually all being carried on at the same time in most logging operations. While one area is being logged, the cruisers and surveyors are preparing for the next year's operation. While road making is going on at one end and logging operations are in active progress farther back, clean-up crews are taking out equipment and railway steel where logs have been removed. Relogging or other clean-up operations complete the logging job (Fig. 32).

FIG. 32. A portable sawmill cutting lumber from logging waste left by the main operation. Re-logging Forest Products Company, Oregon. (*A.F.P.I. photo.*)

The logging foreman must understand the cost of the several operations, and it is his obligation to plan each phase so that the total cost will be minimized. He must know how far logs can be skidded before it pays to build additional road. He must lay out the operation in such a way that it moves in an orderly fashion without bottlenecks and without one crew endangering or being in the way of another. Coordinating all these activities and supervising them is a difficult task calling for special administrative skill.

Logging Operations

Felling. Timber is felled by using the ax or the ax and saw. The felling crew usually has single- or double-bitted axes, a crosscut saw, wedges, a hammer for driving the wedges, and a bottle of kerosene to cut the pitch from the saw. First, the undercut is made perpendicular to the direction in which the tree is to fall. This should be cut straight. Fallers often use an ax placed in the undercut to sight the tree. Skillful timber fallers can drop a tree almost anywhere they wish unless the tree is decayed, leans badly, or has a one-sided crown. Even then considerable latitude exists as to where the tree can be dropped. The fallers begin to saw on the opposite side of the tree parallel to the undercut until approximately 1 to 3 inches of wood remain, depending upon the size and species of tree. Usually, the tree will then start to fall. If necessary, wedges are used to start it tipping in the direction of fall. As the tree starts to fall, the saw is removed and the fallers step back out of the way to avoid being struck by debris or being injured by the butt of the tree, since it may jump back should the crown lodge part way down.

Limbing. Limbs must be cut flush with the stem from the merchantable portion. This ax work is hazardous. A skilled limber stands in such a position that the ax will not injure him even though it glance off the limb he is cutting. He also takes care to clear away brush that might deflect it.

Bucking. Bucking, that is, cutting the tree into measured log lengths, is done with a crosscut saw. The bucker carries wedges and a hammer to release the saw when pinched.

The power chain saw has been introduced for both felling and bucking to take the place of the one- and two-man crosscut saw. The chain saw is used in essentially the same way as the hand-operated crosscut saw in both felling and bucking. The chain saw may also be used for preparing the undercut. With chain saws the undercut may be cut out of the stump instead of the first log, as is the case when the ax is used.

Power chain saws are operated by compressed air, electric motors, and gasoline motors. The more popular types are the gasoline motors inasmuch as these may be transported around the woods without the necessity of dragging cables or air hoses. Two types of cutting teeth are in use, the gouge-shaped tooth that cuts a ribbon of wood all the way across the log and teeth modeled after the ordinary saw tooth which cuts the wood into coarse sawdust. The chain saw cuts far more rapidly than the two-man crosscut saw, but it is heavy to carry in the woods, saves little time on small-sized timber and is cumbersome to use on huge Pacific Coast timber.

The bow saw is particularly effective in timber 11 inches in diameter and smaller. It has a very thin blade and consequently cuts rapidly and easily. In *southern* forests where land is level and there is a minimum of undergrowth, various types of power-driven circular saws mounted on wheels and pushed by hand through the woods are used (Fig. 100). These are very effective for bucking logs at a landing and can be used for felling timber. A similar type of saw has been mounted on the front of a tractor.

Swamping. In heavy timber and particularly with large operations, the tops of trees may lie so close together that it is difficult to extricate the logs without cutting away tops and oftentimes smaller trees to enable the tractor operator to get in with his logging arch to remove the logs. This operation is called swamping. The swamper also cuts out the skid trails along which the logs are dragged and prepares the skid trails for use. On small logging operations the felling crew may perform all of these operations.

Skidding. Greater diversity is to be found in the methods by which logs are moved from the stump to the deck than in other phases of logging operations. The simplest method is to loop a chain around the log and drag it in by horse or tractor. This is almost a universal method for small operations. Unless the terrain is rough or undergrowth bothersome, the horse or tractor makes its own skid road as it goes. At the landing the chain is unhooked and the logs rolled by hand tools onto a deck. The skidding trail is located so that the pull is mostly downhill or on the level. For very steep country several logs may be chained together end to end, thereby serving to brake one another on the slope. Sleds used for logs and pulpwood may be held back on steep slopes by snubbing with a rope around a tree or by special braking shoes. In Norway the sleds have loosely attached shafts with special pointed iron shoes. When the sled glides forward, the runners ride on the shafts driving the shoes into the ground.

To lighten the load on horses or tractor, various supports are used to hold the front end of the log off the ground. The "go-devil" consists of a bunk attached to two runners. A simple sled may be used. The "dray" is a type of bobsled. The logging arch is a device that straddles the logs holding the front end up. It may consist of high wheels attached to an

FIG. 33. Tractor and arch skidding in a western pine forest. (*Western Pine Association, and A.F.P.I. photo.*)

axle and drawn by horses or of elaborate track-laying sulkies with elevated fair leads through which choker cables attached to the logs can be connected to the tractor winch (Fig. 33).

A modern modification of the logging tractor is a combination tractor and arch all built together. Such a tractor, called the tomcat logging tractor, was built in the machine shops of the U.S. Forest Service a few years ago. The tomcat differs from the ordinary logging tractor in having far wider treads, which make it possible to operate on muddy soil. It is much wider than the ordinary tractor and therefore has greater stability on hillsides. The tractor is equipped with a winch which leads through a sheave fastened on a short mast that extends above the tractor. The cable from the winch may be taken out at right angles to the tractor, fastened to a large log, and the log skidded in along the ground by means of the winch until its end can be lifted off the ground. The tractor then moves with the log to the log deck.

Where topography is too steep or where the ground is swampy so that neither horses nor tractors can work, various types of cable logging may be used. In the high-lead system logs are dragged along the ground by a cable that leads through a block on a tall spar tree. This provides enough lift to clear normal obstructions such as stumps. The sky-line system has a heavy cable suspended between two spar trees with a carriage riding on the cable. The skidding cable runs from the winch to the carriage and down to the ground where logs are attached by choker cables. This system is useful in removing logs from steep canyon walls. A simplified version known as the Wyssen cable, which can be set up for small volumes of logs, has recently been introduced from Switzerland.[1] Details on cable and other logging methods are given in textbooks and pamphlets on logging.[1, 2, 4, 6]

Various modifications of the high-lead and sky-line systems of cable logging are used throughout the United States and Canada. The heavier and more elaborate systems are used in the big timber of the West and the simpler systems in the East. The simplest sort of sky-line logging is used in Norway and has been proposed for use in Canada.[3] It requires extending a cable from the base of the steep slope up the mountainside and making it taut with a winch. The pulpwood is fastened to a forked stick that rides on the wire carrying the load to the base. This method is suitable for handling fuel wood, pulpwood, and other small-sized products.

Hauling. At the log landing the logs may be placed on a rollway whence they are rolled onto trucks or freight cars for hauling to the mill. If the high-lead system is used, they may be piled up in a random manner at the base of the spar tree. From here they must be loaded onto the cars or trucks. This requires a special loading device. Various types of log loaders are in use, including jammers, gin poles, sky-line loaders, swinging booms, self-loading trucks and crosshaul loading by means of a team or tractor (Fig. 34). Mechanical loaders include types that pick up the logs in huge iron fingers and drop them on the truck. Hand loading is done with pole skids and peavey. Oftentimes conveyors are used for loading pulpwood.

The logging trucks may be conventional motor trucks fitted with special bunks on which the logs rest, or they may be the tractor and trailer combination, if used for heavy loads (Fig. 16). All sorts of modifications exist, depending on the heft of the load to be hauled, the type of road over which it is to be drawn, and the skill and ingenuity of the manufacturer or builder who has prepared the truck for logging use. For soft ground special wagons equipped with track-laying tread and drawn by a crawler-type tractor may be used. Pulpwood is usually hauled on conventional trucks or special trucks with long racks on which the wood is piled. In railway logging the logs are loaded on gondola-type cars, or on flatcars, which are the easiest to unload.

FIG. 34. A modern log loader that can be easily mounted on a skidding tractor or bulldozer.

Logging in the Northeast

The northern section of the United States has very little large sawtimber left, and much of the cut comes from farm woodlands and other small holdings of second growth. Consequently logging operations are on a small scale and use relatively simple equipment. The common logging operation involves a two- or three-man crew for felling, using the power chain saw. The felling crew limbs and bucks the logs, and a tractor is used to skid them to the landing. Logs may be rolled by hand from the log deck onto a truck.

Trucks are usually unloaded by removing binding chains and stakes and rolling the logs off onto the ground or onto a rollway by use of a peavey or cant hook. Various types of power devices are sometimes used to push the logs off the truck.

In Maine, the Adirondacks, and other sections of the Northeast, pulpwood logging camps are operated. The French Canadians, who supply a high percentage of the labor, work mostly with hand tools such as a bow saw, single- or double-bitted axes, and bark-peeling spuds. In pulpwood production one man will produce 1 to 2 cords per day. Production of saw logs runs 1,000 to 2,000 board feet per man-day, depending on the species, size, and skill of the men.

A considerable amount of tree-length logging is being done in the Northeast. The trees are skidded in and bucked at a landing by power saw. This has increased the production of pulpwood cutters by ¼ to ½ cords per man-day. Simple types of log loaders and pulpwood loaders are in use.

River driving is common in Canada and Maine. Logs to be river-driven are landed on the river ice or along the banks of the stream, whence the logs can be pushed into the river by means of a bulldozer. Most river driving is of pulpwood, though some long log driving is still carried out. River driving often requires heavy capital outlays for construction of splash dams, special flumes to carry the logs around rapids, booms to prevent the logs from being stranded along swampy sections of the river, and various other improvements to avoid log jams or loss of logs through stranding on bars and along swampy banks. Log floating where it is feasible remains the cheapest form of log transportation.

Water transportation is rapidly going out of use, however, despite its cheapness. One of the major problems in Maine is that hardwood logs cannot be floated and that much of the forested area contains a mixture of hardwood and softwood. Repeated cutting of softwood has gradually converted good softwood timberlands into hardwood lands, thereby decreasing softwood yield. Roads are now being built into some areas to take the timber out on motor trucks.

Both pulpwood and logs are transported by water in barges, rafts, and in booms. Barges are used mostly for coastwise shipping and, to a limited extent, on lakes. During the Second World War rafts were used to move Sitka spruce logs from Alaska to West Coast mills, but they have been in little use in recent times. Booms towed by tugs are extensively used to move pulpwood and logs across inland lakes.

Logging in the South

Many southern operations are relatively small in size. The two-man crosscut or chain saw is widely used in felling and bucking. Logs are usually bunched by horses or mules and are skidded to the landing by animal or tractor power. Logging of pine on the coastal plain is more highly mechanized than hardwood logging in the mountains and small woodlands. Production per man-day may be as high as 2,500 board feet, and the crews using power chain saws may produce as much as 5,000 board feet per man-day. Where the stands of timber are large in size and operations are conducted by large companies, heavy tractors are used. The tractors are connected to logging arches to increase their capacity. Cable logging is used in swamps. At the landing the logs are usually loaded on trucks or tractor and trailer combinations, using A-frame jammers or various other types of power loaders. A number of new devices have been put on the market recently to facilitate the handling of logs at a landing.[4, 6]

Western Logging

Logging techniques are more highly developed in the West than in the South and East. The West has the largest trees, heaviest stands of timber, and roughest topography. Here logging is a major industry, methods are complex, and equipment highly mechanized. It is here that a new profession, that of logging engineering, has been developed to cope with the multitudinous problems of logging occasioned by rugged topography and large timber.

Power chain saws are in common use except in the heavy timber along the coast. In the California pine region fallers work in pairs and may do no limbing or bucking. One pair of fallers may cut from 60,000 to 100,000 board feet per day, and one limber can usually keep up with a pair of fallers. Buckers are able to cut from 30,000 to 50,000 feet per day and usually work alone. With the increased use of chain saws, both felling and bucking have become mechanized. The logs are bunched and hauled in by huge tractors equipped with logging arches. A large tractor may yard as much as 60,000 board feet per day in 32-foot logs. Loading is done with specialized cranes or A-frame jammers. Formerly, railway hauling was usual in the pine region, but it has now been replaced primarily by tractor and trailer hauling.[6] Ponderosa and western white pine logging is similar, though logs and equipment are smaller.

Rugged topography, dense undergrowth, and fallen logs limit the usefulness of chain saws in the Douglas-fir region, but here also a choice may be made between the chain saw and the hand crosscut saw depending upon the timber stand, topography, and density of undergrowth. The felling crew in the Pacific Northwest may cut from 25,000 to 30,000 board feet per day, those using power saws 75,000 to 80,000 board feet. Tractors equipped with logging winches are used in bunching and arches in skidding. High-lead logging and sky-line logging are still in use in certain areas where other types of logging are difficult because of topography. Almost all loading is done by power loaders using converted power shovels or special A-frame jammers. Though railways are still in wide use and some timber is still water-driven, a great deal of the transportation is now by truck.[6]

Waste in Logging

Waste occurs in logging because of careless felling, which may break the log or injure valuable trees that otherwise would be left standing. Trees may also be injured or destroyed in skidding and slash disposal. If large enough to be cut for lumber, these are ordinarily salvaged during the logging operation, but otherwise they remain standing and decay in the woods. Bark constitutes about 10 per cent of the volume of a tree; yet at the present time little use is made of bark except for tanning and for special purposes. The bark from pulpwood is peeled and left in the woods or burned in the mill, and that from saw logs is cut off with the slabs and edgings and ordinarily burned to provide fuel for the sawmill and dry kilns.

The Forest Service estimated that in the year 1944, 26 per cent of all timber cut in logging operations was wasted.[5] Of this waste, 23 per cent consisted of parts of trees cut and left in the woods and 3 per cent of trees that were destroyed in logging operations. Only about 2 per cent was salvaged for fuel. The remainder was not used at all. The logging waste referred to in this case does not take into consideration cull trees that were left standing or the waste in stumps.

Waste occurs following logging when cull trees remain to occupy soil, when injured trees are left to become culls before the time of next logging, and when the forest is in such condition that young trees do not promptly become established to replace those harvested. Waste occurs also if camps and other improvements needed to carry on logging are immediately removed or allowed to deteriorate. This prolongs the time and waste that will occur before loggers return for a second operation, because the entire costs of improvements must be met anew.

Waste also occurs if logging occurs too early or too late in the life of the stand. If timber is harvested before the stand is mature, the total volume cut is small and the value of products low. An example will make this clear.

In western Pennsylvania timber on the stump in the early 1940's sold at about $0.50 per cord for chemical wood, which could be harvested from trees averaging 8 inches in diameter or less, about $1.00 a cord for pulpwood, which required trees 7 to 15 inches in diameter; and about $2.00 per cord for sawtimber, which required trees 14 inches and larger in diameter. Corresponding returns per acre annually would be $0.25, $0.55, and $1.05, respectively. Nevertheless, much of the privately owned timber of the region was sold as chemical wood. Considerable restraint is required on the part of the owner to hold his trees long enough to produce the best quality products, which in turn yield the highest prices. It takes only 40 years to produce chemical wood, 60 years for pulpwood, and 80 years for sawtimber.

At the time of logging there are often many defective or cull trees in the forest. These may contain as much as 50 per cent sound material, but it is so expensive to harvest and process such trees that they are left standing in the woods. The stumps and tops of trees are seldom used in logging operations. Exceptions are black-walnut stumps and those of other highly figured woods that are useful for veneer and cabinet purposes and pine stumps used for recovery of naval stores in the South. Tops of sawtimber trees are sometimes used for fuel wood or for pulpwood, but this is the exception rather than the rule. A few companies use top wood for chemical wood or low-grade pulp.

The Place of the Forester in Logging

The forester who understands logging is likely to hold a far stronger position in his company than one who does not. Logging is the first step in the chain of operations by which the product of the forest finds its way to the ultimate consumer. It cannot be by-passed; therefore logging jobs exist whenever timber is utilized.

When forestry is integrated with the timber-operating work, top management is unlikely to consider it a luxury that can be curtailed or discontinued during periods of adversity. Also, roads built for fire protection can also be built for logging. Timber cruising and management activities are definitely integrated with the logging program of the company. In this way forestry work becomes an indispensable part of the over-all timber operation. It is, therefore, highly important that foresters who expect to work for large corporations learn logging. The woodland-manager positions in large corporations are well paid. The men occupying these exercise a high degree of authority and spend vast sums of money for labor, equipment, and technical services.

Careful logging is indispensable to intensive forest practice. Until the loggers understand and become sympathetic toward the aims of forestry, good forest practice becomes largely a matter of chance rather than intent. In Europe logging is commonly considered a function of the forester. That it has not been so considered in America was largely the result of foresters finding employment in government agencies that did little or no logging. Training in logging consequently received scant attention in most colleges of forestry. Fortunately for the profession, the forest industries, and the forest, this situation is rapidly changing. Industrial foresters are becoming the woodlands managers who direct all forestry and logging operations.

Planning logging activities on a large operation requires a great deal more than simple woods experience. It requires knowledge of timber volume and quality, of engineering to survey and build roads, of the uses and limitations of logging machinery, and of cost control so as to minimize over-all expenditures. It is a job for the fully trained forest officer who has been well seasoned by several years of practical experience.

Logging involves training the various men who take part in the operation. This training must emphasize care in protecting the forest, skill in the various operations, and, above all, safety. Logging is one of the most hazardous occupations in America. On operations where intelligently planned safety training has been given and safety constantly stressed, accidents have been no greater than in industries considered to present only normal industrial hazards. In this activity alone the forester can more than earn his pay with a timber-operating company.

LITERATURE CITED

1. Brown, N. C. 1949. "Logging: The Principles and Methods of Harvesting Timber in the United States and Canada," John Wiley & Sons, Inc., New York. 418 pp.
2. Bryant, Ralph C. 1923. "Logging: The Principles and General Methods of Operation in the United States," John Wiley & Sons, Inc., New York. 556 pp.
3. Koroleff, Alexander M., and C. R. Silversides. 1948. "Transport of Wood by Gravity over Suspended Wire," Woodlands Research Index 43, Pulp and Paper Research Institute of Canada, Montreal. 8 pp.
4. Simmons, Fred C. 1951. "Northeastern Loggers' Handbook," Agriculture Handbook 6, U.S. Department of Agriculture, Washington, D.C. 160 pp.
5. U.S. Department of Agriculture, Forest Service. 1947. Wood Waste in the United States, Report 4, from "A Reappraisal of the Forest Situation," Washington, D.C. 45 pp.
6. Wackerman, A. E. 1949. "Harvesting Timber Crops," McGraw-Hill Book Company, Inc., New York. 437 pp.

CHAPTER 8. *Lumber Manufacture and Sale*

LUMBER, A BASIC raw product, ranks high in our national economy on the basis of men employed, values created, and its importance in our daily lives. Among industry groups lumber and lumber products ranked tenth in national employment in 1947 with an average of 636,000 workers. These workers created values of 2,497 million dollars. The lumber and lumber-products industry outranked in employment the chemical industries, industries manufacturing stone, clay, and glass products, and the paper and paper-products industries. It employed fewer persons than found work in printing and publishing, electrical machinery manufacture, and metal fabrication.[7] In values created the lumber industry outranked petroleum and coal products, rubber products, furniture and fixtures and leather products. Sawmills and planing mills alone employed 383,000 workers and created values of 1,492 million dollars.

Of wood products harvested from the forest for the period from 1934 to 1943, logs for lumber represented more than half of the total volume. Of all the many products and services forests provide, timber for lumber manufacture has by far the greatest direct monetary value. It brings direct tangible returns to the timberland owners that far surpass returns received for watershed protection, wildlife, and recreation. Important as the latter may be, they bring in few dollars with which the landowner can pay taxes, or costs of maintenance and silviculture. Ignoring this basic fact leads to misunderstanding of the entire basis for private forest management in our country.

To speak of lumber manufacture only in terms of workers employed, values created, and timber consumed gives an entirely inadequate picture of what a generous lumber supply has meant to the American people. It has meant first of all homes for families. Who can estimate the value of homes to our nation in many satisfactions, family harmony, thrift, mental composure, and loyalty to country? It has helped to make possible independent farmsteads on which single families could pursue their livelihood unbeholden to a landlord class. It has helped make possible the creation of a host of small business enterprises from the crossroad stores to toy manufacturers. It has made possible the erection of summer homes, tourist cabins, and resort hotels so that city dwellers may enjoy our country's scenic grandeur in comfort and at moderate cost. And it has provided a cheap and easily worked material suitable for the expression of man's creative urge, whether as a bookshelf, a pigeon cote, or a fine wood carving. These are intangible values that cannot be expressed in dollars, but they are high among those things that give meaning to human life and solace to untrammeled spirits.

Lumber also plays an indispensable role in our country's military defense. Poles, logs, and lumber were used in frontier blockhouses in colonial and early post-Revolutionary times (Fig. 5). During the First World War the 10th Engineer Regiment was engaged in cutting logs and sawing lumber to supply our fighting forces in France. Lumber was also in high demand at home.

In the Second World War lumber in huge amounts was used for barracks, military shops, for overseas warehouses, bases, and docks, and to repair bomb damage in cities. Over 16 billion board feet of lumber, half the lumber cut that year, were used in 1944 to crate and box military supplies and equipment for shipment overseas.

Development of Lumber Manufacture

The use of wood for houses, boats, fences, and barricades began in America with the arrival of the first white man. Leif Ericson is said to have built ships from timbers harvested along the New England coast. Lumber manufacture was under way at Saint Augustine before the arrival of the Pilgrim Fathers and began at Jamestown, Va., in 1625 and at Berwick, Me., in 1631. The early use of wood was in the form of twigs, small poles, and bark such as the Indians used for their baskets, tepees, long houses, and canoes. When white men brought in the ax, poles and hewn timbers were used to construct cabins, churches, trading posts, and forts. The first lumber was probably cut with the old-fashioned whipsaw. The first power sawmill was a modification of the whipsaw, substituting water power for human power to move the saw up and down. This gave way to steam power early in the nineteenth century. With the application of power came the development of the circular saw, which increased the speed with which logs could be converted into lumber. Toward the end of the nineteenth century the band saw was introduced in the larger and more efficient sawmills.[1, 3]

The Trend in Lumber Production

Trends in lumber production for the United States as a whole and for four states are given in Fig. 35. It was not until 1905 that peak production in Maine occurred. The peak lumber production in New York State occurred much earlier, and there has been little recovery from the low point in lumber production reached in recent years. The graphs of lumber production for Minnesota and Mississippi show even more interesting trends. Production rose rapidly to a sharp peak in 1895 in Minnesota and in 1925 in Mississippi. From this peak, production in Minnesota dropped to less than one-tenth that at the peak; in Mississippi the drop was only one-half. The contrast indicates that Minnesota forests were more severely depleted by logging and fire than those of Mississippi and that second-growth stands in Mississippi have been more successful in sustaining lumber production than they have been in Minnesota. This is due in large part to the more rapid growth rate of southern trees.

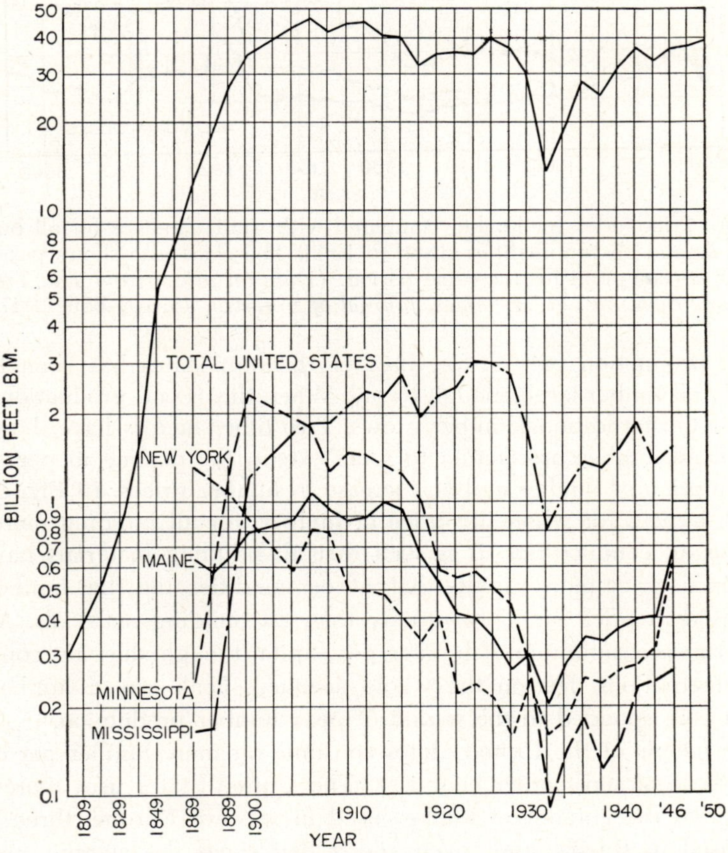

FIG. 35. Lumber production in four states and in the United States from 1799 to 1946. (*Data from "Lumber Production in the United States 1799–1946" by Henry B. Steer. U.S. Department of Agriculture Miscellaneous Publication* 669. *October,* 1948.)

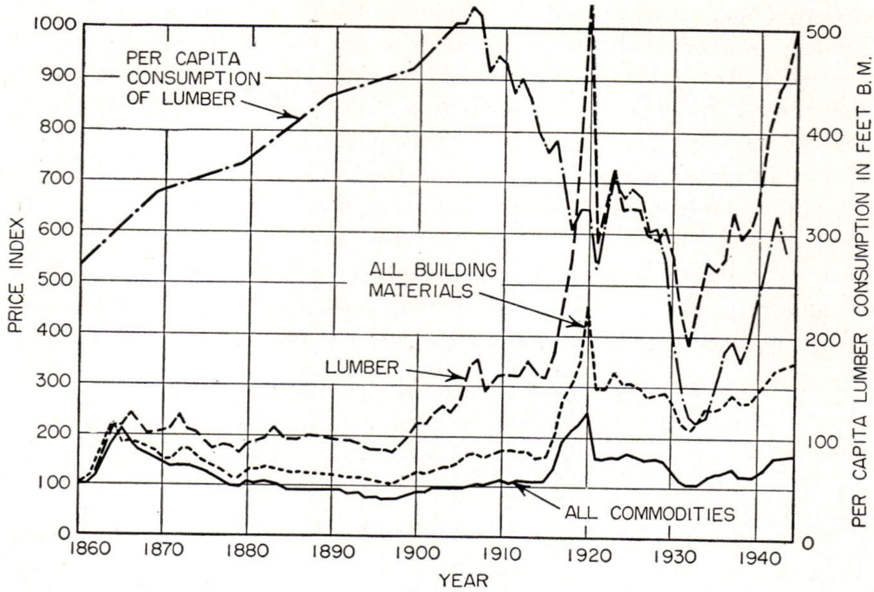

FIG. 36. Price index for lumber compared with similar indices for all building materials and all commodities, 1860 to 1944. The chart also shows per capita lumber consumption for the same period. (*Data on prices from The Trend of Lumber Prices, by Lee M. James, Journal of Forestry,* 45:646–659, 1947.)

The production pattern for three of the four states has been to cut rapidly all timber large enough to sell. When this is cut, production must decline until additional timber grows. Two other factors have, however, influenced lumber production to some extent, the decline in per capita consumption of lumber and an increase in lumber prices. In Fig. 36 are shown indices for wholesale price of lumber, for all building materials, and for all commodities. It is seen that all building materials have increased in price more rapidly than all commodities and that lumber has increased in price far more rapidly than all building materials. At the same time consumption of lumber per capita, though showing considerable fluctuations, has on the whole declined. Peak use of lumber per capita corresponded to the period of peak lumber production.

The people of the United States consume far more lumber per capita than those of most other countries. The United States uses more than four times the amount of lumber used in western Europe, three times that used in Russia, and seven and a half times the amount used in Central and South America and in the world as a whole. Some people argue that because high standards of living are maintained in western Europe with low lumber consumption per capita, we might get along well in the United States with a meager lumber supply. This, perhaps, is true. But whenever one major element in the economy is withdrawn, the economy is reduced not only by the contribution that this particular element made but by its indirect effects on the use of other materials. A high standard of living rests ultimately upon abundant use of many resources.

Whether per capita consumption of lumber will continue to decline is open to speculation. The Forest Service in making its reappraisal of the potential timber requirements for the United States anticipates that it will not. They estimate potential requirement at 39 billion board feet by the year 2000 with population estimated at 185 million. This would provide 210 board feet per capita, which is slightly in excess of the 1939 consumption rate. Such long-term predictions mean little, unless our country is to have the lumber to consume and the need for its consumption.

Lumber Production by Species

The most important species in total lumber production in the United States for the year 1945 are listed below.[5]

	Production, million bd. ft.
Softwood species:	
Yellow pine (southern yellow pines chiefly)	7,210
Douglas-fir	6,237
Ponderosa pine	3,056
Hemlock	1,039
White pine	1,269
White fir	499
Redwood	443
Spruce	358
Total softwood production	21,139
Hardwood species:	
Oak	2,859
Red gum	971
Yellow-poplar	759
Maple	522
Tupelo gum	439
Beech	352
Cottonwood	209
Hickory	159
Elm	159
Total hardwood production	2,982

The Sawmill

Lumber manufacture includes those operations that take place from the time the log is unloaded at the sawmill until the lumber, in either seasoned or green form, finished or rough, is loaded on the car for shipment to the consumer. A typical stationary, medium-sized sawmill has a log pond into which the logs are dumped directly from the truck or railway car, and some may have a temporary log-storage yard. Where it is possible to place the logs directly in the millpond, one operation is saved and damage by insects and fungi that degrade the lumber is reduced. From the pond the logs are carried into the mill by means of a jack ladder. This consists of a power-driven chain with lugs running in a V-shaped ramp that elevates the log to the top floor of the mill. Here the logs are rolled onto a sloping log deck. From the deck they are placed directly on the saw carriage either by hand or by a mechanical loader. The log is pushed on the head blocks of the carriage and held against the knees by heavy pointed dogs while it is passed successively through the head saw.

The head saw may be either a circular or a band saw, but in either case, the saw will have swaged teeth. The circular saws now in use have inserted teeth that may be removed when broken. A circular saw may have an auxiliary saw known as a top saw to cut through the large-size logs. The first pass by the head saw removes a slab. The second pass may remove a 1-inch board. After the second cut the log is normally turned, but it may be turned after the first or after the third or fourth. It is turned on the log carriage so that the flat face is either down or placed up against the knees. Additional slabs and boards are then removed, and the log re-turned for cutting off the third slab and boards. A fourth turning is made to remove the last slab, and the remainder of the cant is then sawed into lumber (Fig. 37). To produce "round-edge" lumber for box boards, logs are sawed without turning.

The term lumber is generally applied to sizes cut 1 or 2 inches in thickness with square edges. It may be used for thicker material, but anything 4 inches in thickness or larger is generally classed as squared timbers rather than lumber. The slabs go to a slasher or swing saw where they are cut up into short lengths for fuel wood, but in large sawmills they may go through additional saws where they are cut into laths and other small-sized material. The round-edge lumber passes through an edger which trims the two edges square. The material removed, called edgings, may be passed through additional saws for lath manufacture or else cut up for fuel or other purposes. After the boards have been edged, they pass to the trim saw, where the ends are cut square, and then onto the green chain, where the lumber is graded and sorted (Fig. 38).

The lumber may go directly from the green chain to the dry kiln or from the green chain to an air-seasoning yard. Usually the better grades

FIG. 37. A modern band saw in operation. The sawyer (left) is signaling the setter (riding the carriage) the depth of the next cut. (*Western Pine Association, and A.F.P.I. photo.*)

FIG. 38. Grading lumber on the green chain before transporting to kilns and yards for seasoning. (*A.F.P.I. photo.*)

FIG. 39. Loader-stacked lumber for air seasoning. (*U.S. Forest Service photo.*)

are kiln-dried and low grades are air-seasoned.[4, 6] In the dry kiln the lumber is dried in a period of two to four weeks. It may then be surfaced, regraded, and shipped, or graded and shipped directly.

Large sawmills commonly saw the lumber into thicknesses of 2 inches or more on the head saw and then send it through a resaw for further cutting. A few sawmills in the United States have adopted the Swedish gang saws for sawing small logs. Either one pass may be made through the gang saw to cut round-edge lumber, or the logs can be passed through one frame of gangs that square the cant on two sides. The cant is then turned down and sent through a second gang saw which removes the other two slabs and saws the cant into 1- and 2-inch lumber.[1] The gang saws are highly efficient for sawing small logs of uniform size. Large mills differ from the one described, mainly in the amount of heavy machinery used. In so far as possible, handling is mechanized. Rubber-tired timber carriers straddle small lumber piles, then lift and carry these bodily to the seasoning or storage yard, where mechanical cranes or other hoisting machines lift them to the yard piles. Seasoned lumber may be handled in similar ways. Methods vary widely from mill to mill (Figs. 39, 40).

The Portable Sawmill. The descriptions given above of large and medium-sized sawmill operations are vastly different from that of typical portable mills. These may consist only of a head-saw rig. Usually an edger and a swing saw that serves to trim and cut up slabs are added. Power is furnished by a tractor or stationary power unit. Logs are delivered to the mill by truck and rolled directly onto deck and carriage. The lumber may be yard-piled or loaded directly onto a truck for sale. Such mills may be moved frequently, and some can be set up to saw as little as 100,000 board feet.

Even more portable rigs have been devised. The Jackson timber harvester is truck-mounted and may be set up for only one day's sawing of 5,000 feet. A two-man truck-mounted mill in use at the Pack Demonstration Forest at Warrensburg, N.Y., may be set up in two hours' time. It cuts 3,000 to 4,000 board feet per day from 8-foot logs less than 13 inches in diameter.

Size of Mills in Relation to Lumber Production. The small portable sawmill common in the eastern United States cuts about 5,000 to 8,000

FIG. 40. Mechanical handling of lumber reduces labor costs. (*Western Pine Association, and A.F.P.I. photo.*)

board feet of lumber per day and employs from four to eight men. Many operate intermittently and may be closed down for months at a time. In contrast, the large mills of the Pacific Coast cut from 100,000 to 1,000,000 board feet per day.

The total number of sawmills operating in the United States exceeds 40,000. It was estimated that there were about 1,700 in New York State in 1948. The small mills, however, account for only a small percentage of the total production. Approximately 3 per cent of American sawmills produce over half the total output of lumber. Gradually, however, the small mills are assuming increased importance as the supplies of heavy stands of virgin timber are depleted.[1]

Technical and Skilled Jobs in Lumber Manufacture. Handling logs and lumber is heavy work and, consequently, expensive work. It requires rugged equipment and workmen who are skilled in sawing, edging, trimming, grading, and seasoning. The most important workman in the mill is the head sawyer, for it is he who sets the pace in the mill and who decides how the log is to be cut up. He stands beside the head saw and works the lever that controls the movement of the log past the saw. On the carriage may be a "dogger" and a "setter." Both men operate the dogs that hold the logs on the carriage, and the setter, on signal from the head sawyer, sets the logs forward for the proper width of cut and, upon further signal, turns the log as the sawyer directs (Fig. 37). The tail sawyer catches the slabs and boards as they come from the head saw and drops them onto rollers whence they are carried to the swing saws and edgers.

Lumber is sold by grade; the higher the grade, the higher the price it brings. The grade of lumber that can be sawed from a log depends on log defects and on the skill of the head sawyer. He will usually place the log on the carriage with the poorest face down. This face is sawed last after he has removed as much high-quality lumber as he can from the other three faces.[8]

The machinery in the sawmill is kept in shape by the millwright, whose task it is to oil and service the various pieces of equipment in the mill.

The machinery inside the sawmill consists of the jack ladder, the head-saw rig—which includes the head saw, the carriage, the driving mechanism for the carriage and log-turning machinery—the edger, the trim saw, and the slab saw. In addition, there are various power-driven rolls, dead rolls, chain conveyors, waste conveyors, sawdust blower, and sometimes resaws. Power may be furnished by steam or a stationary Diesel-power unit, or the machinery may be driven by individual electric motors. A large sawmill must have a burner or other waste-disposal unit for handling sawdust, slabs, edgings, and other waste material.

Seasoned lumber that is to be finished before shipping is sent through a planing mill. The modern planers, which consist of rapidly turning cutterheads on which knives are mounted to cut away the rough surface, will plane four sides of a board in a single pass through the machine. Planing also removes irregularities due to sawing or seasoning. The lumber is ordinarily loaded directly in the car from the planer to avoid additional handling.

Keeping all this equipment in adjustment and good mechanical condition is an exacting job, requiring vigilance and skill. The saws are kept in shape by the saw filer, whose job ranks second only to that of the head sawyer. In large mills saw filing is done with complicated machines. Band head saws must be changed about three times each eight-hour shift; other saws require less frequent changing. Each type requires a special filing machine.

Lumber is graded on the green chain as it comes from the trim saw and after seasoning before it is shipped. Softwood lumber is commonly graded on the basis of its suitability for structural purposes, *i.e.*, its strength. The various classes and numbers of defects present indicate the grade of the board in question. Hardwood lumber is graded on the basis of the yield of clear material above specified minimum sizes that may be obtained by ripping or crosscutting the board one or more times. Although grades are highly standardized, the lumber grader has no time to lay out cuttings or make other detailed inspection. He must quickly determine and mark the grade of each board as it passes by him, turning each one in the process (Fig. 38).

Freshly sawed lumber, called green lumber, has a high moisture content. As this moisture leaves the wood, the lumber shrinks in width and thickness. Shrinkage is a serious fault in lumber to be used for exacting work. Also unseasoned lumber is subject to decay. Seasoning is therefore necessary to remove excess moisture and to bring the moisture content of the wood to equilibrium with that of the air to which it is to be exposed.

Lumber seasoning is a technical operation requiring careful control to avoid degrade due to warping, staining, checking, splitting, or casehardening.[1, 2, 4, 6] Lumber to be air-seasoned should be carefully piled on foundations 18 inches to 2 feet off the ground. Three or more "stickers," 1- by 2-inch strips, are placed between each course of boards. Boards should be spaced wide enough apart laterally to permit free circulation of air, and the lumber pile is sloped from front to back to permit free drainage of rain water. A roof of old lumber elevated 12 or more inches above the top course of green lumber protects the boards from rain and from rapid sun drying with the inevitable cupping (Figs. 39 and 41).

FIG. 41. Hardwood lumber poorly piled for air seasoning. Contrast with Fig. 39. (*U.S. Forest Service photo.*)

Lumber for the dry kiln is carefully "stickered" and piled on kiln trucks. The piles need be neither covered nor sloped. The dry kiln is a chamber in which prescribed rates of lumber drying are maintained by control of temperature and humidity. Kiln operation requires special skill. Prescribed kiln schedules calling for definite periods of time at set temperatures and relative humidities must be maintained, and the operator must also make moisture tests to determine when drying is completed. Special precautions are taken to avoid seasoning defects.

Moderate-sized mills must also have accountants to keep books, analyze costs, and furnish material for price determination, but lumber prices as such are set in the large market places rather than at the sawmills.

Supervision is required for mill operation, handling lumber in the yard, and finishing and loading. Constant care is necessary to keep each operation moving with a minimum outlay for labor. Mill layout must therefore be periodically scrutinized for minor improvement in flow of material and economy of labor.

Concentration Yards

As sawed lumber is expensive, the portable-mill operator rarely can afford to keep a sufficient lumber inventory on hand to supply customers with seasoned lumber of the different grades and sizes they require. Many portable-mill operators sell their lumber mill run to a concentration yard or jobber. The concentration-yard operator performs a valuable service in collecting lumber from widely scattered mills, grading it, seasoning it, holding it until he has carload lots of the types consumers demand, carefully grading it, and shipping it out. Concentration-yard operators oftentimes finance portable-mill operators.

Other operators, who run merely a financing and jobbing operation without moving the lumber to central yards, serve as commission salesmen for a group of mills. Such businesses tend to thrive in the eastern states where small mills operated by owners with limited capital abound.

Wholesale Selling

Much of the lumber may be sold by the large western and southern manufacturers directly to large retail dealers. Even medium-sized retail dealers can use a full carload of lumber at a time. These and small retailers may buy from sawmills or from commission sellers. The Weyerhaeuser interests have their own sales company with wholesale and retail outlets in many cities. Lumber trade associations, of which there are many, establish standards and furnish information to manufacturers and customers that facilitates the lumber trade. All these organizations offer opportunities for useful employment to men who understand lumber and its uses.

Exporting of lumber is carried out by large manufacturers, particularly those having port facilities or located on navigable waters. The British trade sets standards for international lumber dealing. Lumber imported from Canada is handled on about the same basis as that from large United States mills. Tropical timber and special veneer logs are imported by companies operating in the nation's chief ports. These distribute foreign timbers to furniture and other manufacturers throughout the United States.[5]

Retail Lumber Handling

The construction industry uses more lumber by far than all others. Lumber is the chief material used for erecting one- and two-family dwellings, garages, farm buildings, small stores, and warehouses. It is widely used for fences, scaffolding, concrete forms, temporary flooring, and other purposes in heavy construction of steel, concrete, and brick (Fig. 6). The agency furnishing lumber to builders is the retail lumberyard.

The country's average annual lumber production from 1945 to 1950 was about 35 billion board feet. Almost half of this reached the ultimate user through the retail yard. About 25,000 retail yards are found in the

United States. Their principal business is supplying the needs of residential construction. The U.S. Census of 1940 lists more than 37 million dwelling units, mostly one- and two-family size and mostly of wood-frame construction.

New dwelling construction in 1950 approximated 1.3 million units. Home construction for 1951 was estimated at 850,000 units. The median postwar cost per unit was in excess of $8,000, and realtors estimate an annual cost of 5 per cent for maintenance and repairs, probably a generous estimate. Even so, it is easy to appreciate the importance of building materials and the retail lumberyards that supply them.

The retail yard is, generally speaking, small business. Usually the owner serves as general manager. His principal customers are light-construction contractors, and he usually handles builders hardware, cement, roofing, paint, and other builders supplies as well as lumber.

In order to progress very far, the retailer has to be familiar with general construction practices as well as with the technical facts about his own merchandise. He must be prepared to sell "packages" for complete buildings as well as pieces. Traditional methods of construction are constantly being modified to fit changing demands. Different living habits bring about different needs. Although wood as a basic material for residential construction will not soon be replaced, new and more efficient uses for the same material are daily developments. Many new materials reflect sound advances in all the sciences. The task of promoting better means to good homes represents an economic obligation which may not be adequately met by the traditional lumberyard. The field is always open for improvements when the handling costs amount to 22 per cent of the cost to the consumer. By better methods of handling, almost every lumberyard could improve its efficiency. Today there are about four times as many machines and devices available for moving lumber as there were only a few years ago. It is the job of the retailer to determine which is best to meet his particular needs.

The successful retailer must be able to handle all the management details that affect a corporation by himself. A sound understanding of the principles of economics, finance, accounting, marketing, and salesmanship is useful. The typical small businessman does not have a staff of experts at his elbow. Without some skill in using knowledge, he may find even the information which is normally available more confusing than helpful.

Retail lumber dealers are often called on by both builders and the public to work up a bill of materials from simple blueprints. They may be asked to estimate costs of a garage or dwelling. Some engage directly in the contracting business as a side line.

Designing and constructing homes is important work and should be done well. Planning is the key to success in the contracting business. Time spent at the drafting table, shifting partitions, changing dimensions, fitting and balancing building costs against livability, pays off well. Through application of engineering principles to house design, materials may be saved.

Almost anyone who has worked on a house can learn how to put in a foundation, erect framing, and build strong, watertight walls. It is quite another task, however, to combine materials and labor into an attractive home designed for functional living. This problem offers the trained man his real opportunity. Building suppliers are constantly producing new materials that call for modified construction techniques, plumbing and heating are being functionalized and simplified, and kitchens are being built as mechanized workshops. Mobility of population means minimizing heavy furniture. Disappearance of household help places a premium on compactness, adequate storage space, and ease of cleaning.

Perhaps no basic need in America has been less satisfyingly met during the past two decades than housing. With population growing at 1.5 million per year and with a heavy increase in new families destined for the early 1960's, opportunities for imaginative builders appear bright.

Prefabricated Buildings. The principle of prefabrication is used to some extent in almost all construction, as, for example, in window and door framing. The degree of prefabrication employed varies from such uses to complete precutting and section assembly. In 1948 there were 34,000 housing units built through use of ready-made sections of floors, walls, and roofs. These prefabricated houses made up only a small part, but an important part, of the total home construction for the year. In 1946 the Prefabricated Home Manufacturers Institute listed 55 members, 11 of which were national distributors. The next year there were more than 80 producers.

Although the industry seems to be permanently established, there are many problems yet to be solved which cause instability. Large business has the problems of overhead, shipping, and storage. The small operator lacks purchasing power and proper advertising, engineering, and architectural services. The solution of these problems calls for continued changes in the industry as progress is made in lowering the cost of production (Figs. 42 and 43).

There is a trend in the industry toward custom prefabrication, through which the prefabricated units are fitted to the buyer's own floor plans. Custom work calls for an increase in time spent on planning and designing.

FIG. 42. Erecting a roof truss fastened with modern timber connectors. (*A.F.P.I. photo.*)

FIG. 43. Plywood saves time and cost for mass housing projects. (*A.F.P.I. photo.*)

Waste in Lumber Manufacture and Use

Waste in Manufacture. Sawmill waste consists of sawdust, slabs, edgings, trimmings, shavings, and of lumber too low in grade to be utilized. Additional waste occurs in seasoning, and a certain amount occurs in mill yards, wholesale yards, and retail yards. Sawdust, shavings, and other fine material make up 18 per cent of the total wood volume harvested, and this figure includes only that waste that occurs in primary manufacturing. The ordinary circular head saw in the sawmill cuts a ¼-inch saw kerf. This means that if the lumber is sawed into 1-inch boards, 20 per cent of the wood volume is converted into sawdust. If head saws are improperly aligned, they cut inaccurately. This results in a heavy loss of lumber in the planing mill where boards must be planed to uniform thickness. The planing mills usually remove about ⅛ inch from each surface planed.

Slabs and edgings are often reworked for lath, small-dimension stock, and other wood products. Inasmuch as slabs and edgings come from the outside of the log, they contain the highest grade of lumber. It is, therefore, often profitable to spend considerable effort in salvaging this material. Trimmings account for considerable waste in sawmills because softwood lumber is usually trimmed to even 2-foot lengths and hardwood lumber to 1-foot lengths. The waste that occurs at the sawmill totals 43 per cent and is distributed as follows: [1]

	Per cent
Saw kerf	10
Slabs	9
Edgings	8
Bark	8
Seasoning	4
Remanufacturing loss	3
Miscellaneous	1

Seasoning loss due to shrinkage in drying is unavoidable. Additional seasoning loss occurs, however, because of checking, splitting, warping, cupping, internal collapse, and other defects that may occur when lumber is improperly seasoned. Such defects may be of a minor nature, or they may be so severe as to render the entire board useless except for fuel.

One means of reducing waste and of upgrading low-quality lumber is to cut it into small-dimension sizes at the mill. Hardwood small-dimension products include turning squares, blanks for such products as gunstocks, and precut pieces for chairs and other furniture parts. This is normally a highly competitive field but does result in closer use of lumber at the mill.

Softwood small dimension may likewise be prepared in sizes suitable for a variety of uses. The trade name Milpac is given to softwood pieces cut to size, finished, and wrapped in paper for shipping. Such material is more convenient for household use than long lumber that must be cut before being taken to the basement workshop.

Waste in Lumber Yards. Fungi and insects will attack lumber in storage, in mills, concentration yards, and retail lumberyards if it is improperly piled and has a high moisture content. Green lumber sawed during warm weather is subject to sap stain unless it is treated with a stain-preventive solution and rapidly dried. Although sap stain, or blue stain, causes a degrade in the lumber, it does not weaken it for ordinary uses. Proper piling of lumber promotes rapid drying, which is one of the best ways to minimize staining during air seasoning and storage. If the lumber is piled out of doors, close piling should always be avoided (Fig. 41).

The powder-post beetle causes considerable damage to oak, ash, and hickory in lumberyards of the Northeast. Accurate estimates on the amount of damage are not available, but the over-all waste due to decay and insect damage to wood in service and in yards is estimated at approximately 10 per cent of annual lumber use. The losses are sufficiently heavy to justify necessary precautionary measures to minimize them. Losses in lumberyards in New York State are estimated at about $41,000 annually. The loss due to degrade in lumber resulting from insect work in the green logs is estimated at $63,200 annually. Termites are said to cause 40 million dollars worth of damage to wood in service throughout the United States.

Waste of Wood in the Construction Industry. Wood waste in the construction industry today is considerably less than it formerly was. This is due in part to the practice of trimming lumber to standard lengths used in construction. Further saving might be possible. Softwood is trimmed in 2-foot lengths, yet in house construction it is used in multiples of 16 inches, since studding and joists are normally set on 16-inch centers. Unnecessary use occurs when specifications for structures require heavier timbers than are actually needed to support the weights involved. The standard wall panel made up of 2 by 4 studdings with nailed wallboard or lath and plaster could be made equally stiff if constructed of studding 1½ by 3 inches with plywood glued on inside and exterior faces. A similar saving would be possible by using plywood as flooring glued to the tops and bottoms of 2 by 8 or even 2 by 6 joists. Gluing of plywood to the joist results in a far stiffer structure inasmuch as the glued structure spreads the weight over all joists instead of concentrating it on one or two, as is the case with normal nail construction.

Waste of Wood in Use. Wood-destroying insects and fungi attack untreated fence posts, poles, piling, ties, and other products that must be exposed to moisture or used in contact with the ground. Porches on houses are particularly subject to decay. The same is true of porch steps and other portions of the house that are exposed to moisture, especially if they are inadequately ventilated.

The use of insulation in house walls increases the decay hazard where moisture barriers have been placed on the outside wall. During winter, moisture from cooking, air conditioning, and other sources condenses on the moisture barrier. If the insulation becomes saturated and remains soggy throughout the summer period, it will cause decay in the studding, sheathing, and other members.

Faulty construction is responsible for most of the decay hazard in houses. Any type of construction which results in moisture collecting either in the form of rain water, snow melt, or condensed moisture vapor subjects wood to a decay hazard. This can be overcome by avoiding all types of leaks, by providing adequate ventilation, and by placing moisture barriers between the insulation and the interior wall surface.

Wood may give unsatisfactory service because of the use of green lumber, the wrong grade, improper species, or shoddy design and workmanship. The use of poor fastenings, poor finishing, and poor glue joints reduces the service life of furniture and wood in other uses.

Decay in wood that must be in contact with the soil, such as posts, poles, greenhouse flats and benches, and wood in vegetable-storage houses that must be kept moist can be minimized by treating with suitable preservatives. Many wood preservatives are now on the market. Manufacturers claim some are safe to use in milk houses, vegetable-storage bins, and greenhouse flats.

Stabilizing the Lumber Industry

Lumber production for the country as a whole has fluctuated widely in the past, and the fluctuations have been closely related to business and building cycles. Only in so far as our nation succeeds in controlling business cycles can fluctuation in lumber production be stabilized.

Even greater in significance than these major cycles, however, are fluctuations in production by regions, states, and individual companies (Fig. 36). Here depletion of standing timber supply has been a major factor. Some companies have been successful in escaping the "boom and bust" cycle that was the common pattern until recent years. Adjusting timber harvesting to growth on tributary lands is a fundamental requisite to stability. With a declining supply of raw material business volume may be maintained by increasing the intensity of processing and by expanding operations to include manufacture of pulp, veneer and plywood, wood chemicals, shingles, and other wood products.

The Urania Lumber Company of Louisiana has been outstandingly successful in adjusting timber cut to growth. The Weyerhaeuser Company and certain others have succeeded in diversifying their manufacturing operations so as to use all timber products from their lands. Such integrated operations help greatly to stabilize income. The Goodman and Crossett Lumber Companies with integrated pulp mills and by-products plants have maintained a good employment and business record despite depression and war. Good forestry and enlightened business integration have been strong stabilizing factors in these enterprises.

Seasons and Cycles in Light Construction. The building industry fluctuates widely in activity with seasons and with building cycles. Such fluctuation intensifies the business problems of the retail lumber dealer, the independent builder, and the prefabricator. Unless each maintains flexibility in organization and business volume, he will fail to take advantage of good times or will take heavy losses during slack periods. Prefabrication makes it possible to do much toward ironing out seasonal fluctuation by providing inside work during winter months, but the principle is more easily stated than effectuated. Prefabricated floors and panels are heavy and bulky, require large storage areas, and must be protected from weather until erected. It is far more economical to move such panels directly from the final point of assembly to the building site than in and out of storage.

Building cycles are especially hazardous to prefabricators. Business may virtually disappear for months at a time. Only the strongly financed can weather such inactive periods. The individual builders may lay off hands and thus survive a period of low activity with minimum loss. This is one reason that many small builders characterize the light-construction field today. Overextension of activity is fraught with heavy financial penalties.

These fluctuations are both hazards and opportunities. The alert, well-trained man who enters the business at the beginning of a building boom can become well established in a few years. If he then consolidates his gain, he may have a permanent business providing gratifying financial rewards.

Outlook for Lumber Production

In view of the fact that lumber is very useful and is easy to use, it seems likely that it will continue to be the major commodity from American forests. The future for lumber seems assured.

The same is by no means true of individual mills. Large mills have practically disappeared from the North; few are left in the South. The portable and moderate-sized stationary mills have taken their place. Many large western mills are almost certain to exhaust their timber supply in the relatively near future. How many can survive indefinitely will depend almost solely on their raw-material supply. National forest timber and sustained-yield forestry on company lands will be essential to permanence of large mills.

LITERATURE CITED

1. Brown, N. C. 1947. "Lumber: Manufacture, Conditioning, Grading, Distribution, and Use," John Wiley & Sons, Inc., New York. 344 pp.
2. Henderson, Hiram L. 1946. "The Air Seasoning and Kiln Drying of Wood," 3d ed., Albany, N.Y. Published privately. 332 pp.
3. Holbrook, Stewart H. 1938. "Holy Old Mackinaw," The Macmillan Company, New York. 278 pp.
4. Mathewson, J. S. 1930. "Air Seasoning of Wood," U.S. Department of Agriculture Technical Bulletin 174, Washington, D.C. 56 pp.
5. Steer, Henry B. 1948. "Lumber Production in the United States 1799–1946," U.S. Department of Agriculture, Miscellaneous Publication 699, Washington, D.C. 233 pp.
6. Thelen, Rolf. 1929. "Kiln Drying Handbook," U.S. Department of Agriculture Bulletin 1136, Washington, D.C. 96 pp.
7. U.S. Department of Commerce, Bureau of the Census. 1950. "Census of Manufactures: 1947," Vol. 1, "General Summary," Washington, D.C. 355 pp.
8. U.S. Department of Agriculture, Forest Service. 1946. "Cross-roads Conference for Sawmill Operators: How to Produce More and Better Lumber," Washington, D.C. 58 pp.

CHAPTER 9. *Wood, Its Nature, and Uses*

WOOD IS a material of such universal utility that, had nature not supplied it in great abundance, man would himself seek to invent such a commodity. Wood has high strength in relation to its weight. In tensile strength it is as strong, size for size, as steel. In compression it is twice as strong as steel on a weight basis. In stiffness it greatly exceeds steel on a weight basis. Across the grain, wood is relatively weak in tensile strength and compression, but along the grain it is very strong. Wood has low heat, sound, and electrical conductivity. It therefore makes warm walls, warm furniture, rooms relatively free from objectionable reverberations, and relatively soundproof walls. Wood is soft enough to be easily fastened and worked with simple hand and machine tools into a great variety of shapes and forms. Since it has intrinsic beauty, it is well suited for the manufacture of interior trim, furniture, and other commodities in which appearance is of great importance.

How a Tree Grows

Wood is made up of tiny cells that are formed in the two actively growing regions of the tree—the tip regions of root and stem, and the cambial region. Elongation of stems and roots occurs only at the tips. Here behind a cover of protecting tissue lies the apical meristem, or region of dividing cells, and here are formed the cells responsible for all increase in length of stem and root. The apical meristem gives rise to all cells that make up the primary stem tissue. Buds that develop into branches, leaves, or flowers are formed at the tip.

The apical meristem also gives rise to the cells that form the cambium a short distance back from the growing tip. The cambium also is a region of rapidly dividing cells. It forms between the primary xylem and phloem in the newly formed stem and gives rise to all secondary tissues found in the mature tree stem. The cells cut off on the inside of the cambium become woody, thereby giving strength to the main tree stem. Those cut off on the outside develop into the bark that protects the cambium and main stem and conducts fluids. Near the outer surface of the young stem is another bark-forming region called the cork cambium, which forms corky tissue. As the stem grows, the cork cambium is pushed outward, ruptures, and dies. New cork cambia are formed inside by young bark cells. This process is repeated continually as the tree grows.[1] The expansion of the main woody cylinder causes rupture of the outer bark, which eventually is sloughed off.

The wood on the tree stem is laid down each year in a sheath that extends from the root tips to the stem tips. This sheath completely surrounds the stem and once formed, becomes lignified, and develops into permanent tissue. Except during the year of its formation, the sheath does not itself increase in diameter or length. Each year a new sheath is formed on top of the old and thus the tree increases in girth. Each year also the actively growing stem tips form new cells that extend the length or height of the stem. Each new sheath is longer than its predecessor by the amount of total elongation that occurred during its year of formation.

Gross Structure of the Tree Stem

The mature tree stem from the center out contains a pith, a region of soft spongy cells having little lignification and therefore adding little to the strength of the tree stem. Outside the pith are the various sheaths of wood. When the stem is cut in cross section, we see these sheaths as rings, called growth rings. They extend from the pith to the cambial region that lies between the wood and the bark. Outside lies the bark, which may be made up of alternate layers of corky and fibrous tissue. Bark tissue in itself adds little to the strength of the tree stem.

If a small portion of the tree stem is broken, it tends to splinter, for the main elements of the stem are much longer than thick and are arranged parallel to the tree stem. This gives the tree stem its strength to resist bending and breaking as it sways in the breeze and bears its burden of leaves in summer or snow and ice in winter.

Between the pith and the bark there is often a zone of darkly colored wood and a zone of lightly colored wood. The dark wood is spoken of as heartwood and the light-colored wood as sapwood. This difference in color is due in part to the deposition of various chemical substances in the heartwood of the tree. The light-colored sapwood forms the main channel for movement of water and minerals up the tree trunk to the branches and leaves. Being highly permeable to water, sapwood on posts and other wood products exposed to attack by fungi decays readily. The dark-colored heartwood of most trees conducts little water, because the conducting vessels are either plugged by thin membranes, called tyloses, as in white oak, or impregnated with resin, gums, or other substances that impede water movement, as in longleaf pine. The heartwood functions chiefly as a supporting column for the crown. Generally it is considerably more decay-resistant than sapwood.

A cross section of the tree stem may also reveal a series of fine lines extending radially from the bark toward the pith. These are called rays. They are very prominent in oaks and inconspicuous in aspen and, therefore, help one to identify kinds of wood. The ray cells serve as a channel for movement of materials across the growth rings and as storage cells for reserve foods. Their function will be better understood after a description of the cellular structure of the tree stem.

Minute Structure of Wood

The Cellular Structure of a Coniferous Stem. When a very thin section of a woody stem such as white pine is prepared and examined under the microscope, the cellular make-up of the stem becomes evident (Fig. 44). With low power magnification the differences between springwood and summerwood that mark the annual rings of temperate-zone woods become evident. The cells formed in the spring are large and have large central openings; those formed in the late summer are flattened and have narrow openings in the center. A cross-section view will reveal here and there large channels lined by thin-walled cells. These are resin ducts which serve to transport the resin up, down, or across the tree stem. The major supporting elements of the woody stem are tracheids, which are small- to medium-sized, heavy-walled cells. When these cells are viewed in a section cut in a radial direction, it can be seen that they have small circular markings, known as bordered pits.

Running at right angles to the tracheids are the ray cells. These consist of ray tracheids and thin-walled ray cells. Since the thin-walled ray cells are weaker than tracheids, they are responsible for the wood's splitting along the line of a ray. Other types of cells in a pine stem are those that secrete resin. These are found both in the rays and in the vertical stem tissue.

The tracheids are no longer living cells nor do they perform any living function. Many of the thin-walled cells may also be dead but those that secrete resin are living. Other living cells enter into the storing and transformation of foods manufactured in the tree leaves.

The Cellular Structure of Hardwoods. The term "hardwood" is applied loosely to all forest trees that do not belong to the group known as conifers. A more appropriate name is porous woods, because these stems in contrast to those of the cone-bearing trees are made up of small wood fibers with very narrow cavities and of large thin-walled connected tubes that serve as the main conductive channels for movement of water and other materials up and down the tree stem. These show up as pores and are very conspicuous on a cross section of the wood. Their presence can

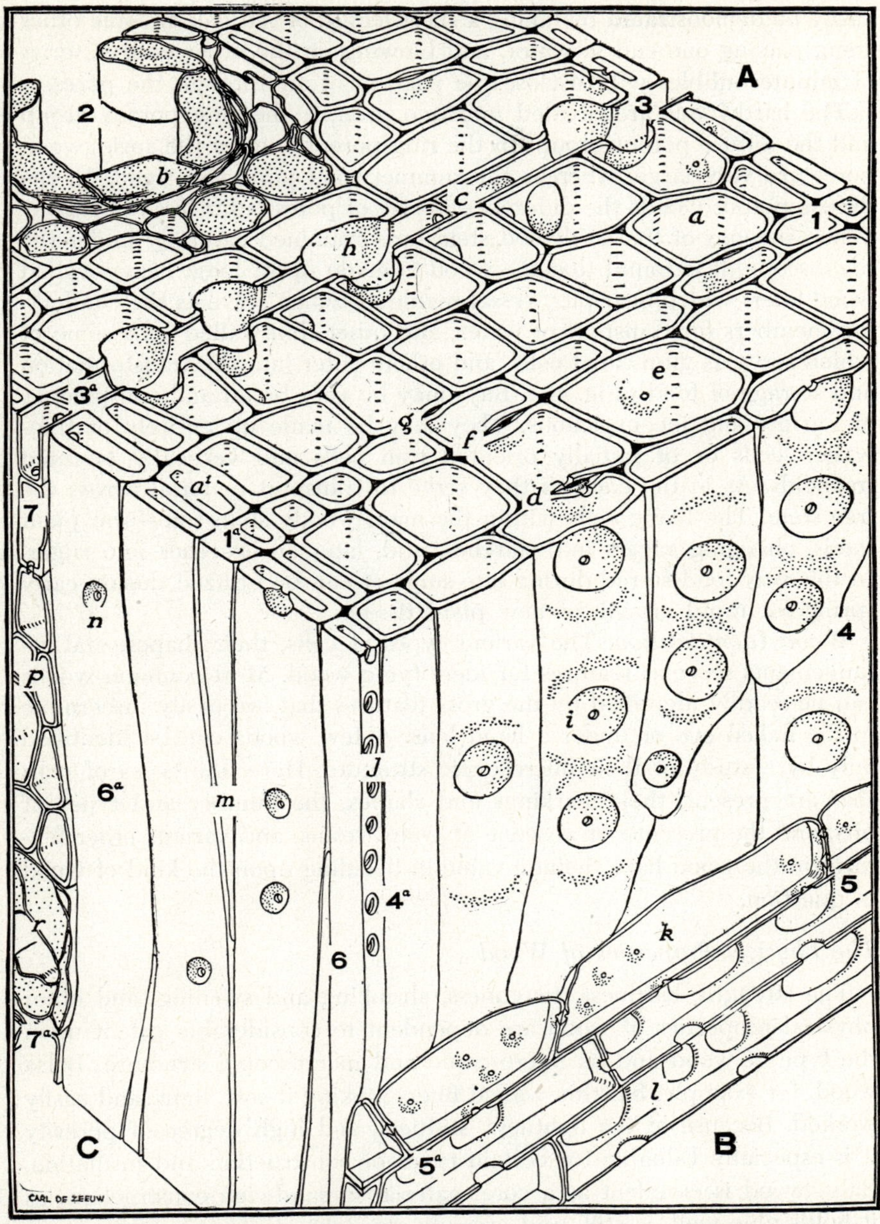

FIG. 44. Schematic diagram showing cellular structure of northern white pine wood. Face *A*, a cross section showing a resin canal 2, tracheids *a*. Face *B*, radial section. Face *C*, tangential section. (*From Brown, Panshin, and Forsaith, "Textbook of Wood Technology," Vol.* 1.)

easily be demonstrated by taking a short length of red oak or some other stem, placing one end in water, and blowing on the other end. A series of minute bubbles will disclose the passage of air through the pores.

The hardwoods are divided into two groups: the ring porous group and the diffuse porous group. In the ring porous species the springwood pores are very large, whereas the summerwood pores are small. In the diffuse porous woods the difference in size of pores is far less noticeable. When sections of the hardwood stems are examined carefully under the microscope, it is found that the wood is made up of long, heavy-walled wood fibers which bear pits, vessel segments which serve as the conducting members for transport of water, and other thin-walled cells some of which serve as gum-canal cells, and others enter into the transformation and storage of food (Fig. 45). Rays may be very broad and conspicuous or narrow and inconspicuous. They may be made up entirely of thin-walled cells or of radially oriented thin cells and vertically oriented tracheids. As in the conifers they serve to transport material across the tree stem. The living cells within the main woody stem transform plant foods, changing sugars into starches, and, later on, starches into sugar. In this way food stored during late summer can be utilized during early spring for rapid growth of new plant tissue.

Wood Identification. The various types of cells, their shapes, and arrangements serve as features for identifying wood. Most common woods can be readily identified by the gross features that are easily discernible to the naked eye or under a hand lens. A few woods can be identified only by a study of their microscopic structure. Here, the types of cells that are present, their markings and shapes, the number and types of pits, and the presence or absence of gums, resins, and various other features in the wood have definite value in deciding upon the kind of wood in question.

The Physical Properties of Wood

The strength, hardness, toughness, shrinking and swelling, and other physical properties of wood are dependent to considerable extent upon the type of wood and its microscopic and macroscopic structure. Balsa wood, for example, has thin-walled fibers making it soft, light, and easily worked. Because of the lightness, stiffness, and high degree of porosity it is especially valuable for certain types of construction and insulation. Balsa wood is excellent as a core material in sandwich construction, as it holds glue well, is stiff, and exceedingly light.

In contrast, hickory and oak are heavy. They have large pores, but they also have a high proportion of dense wood made up of heavy-walled fibers that make the wood hard and strong. Some kinds of wood are very straight-grained. White pine and redwood are good examples. Others

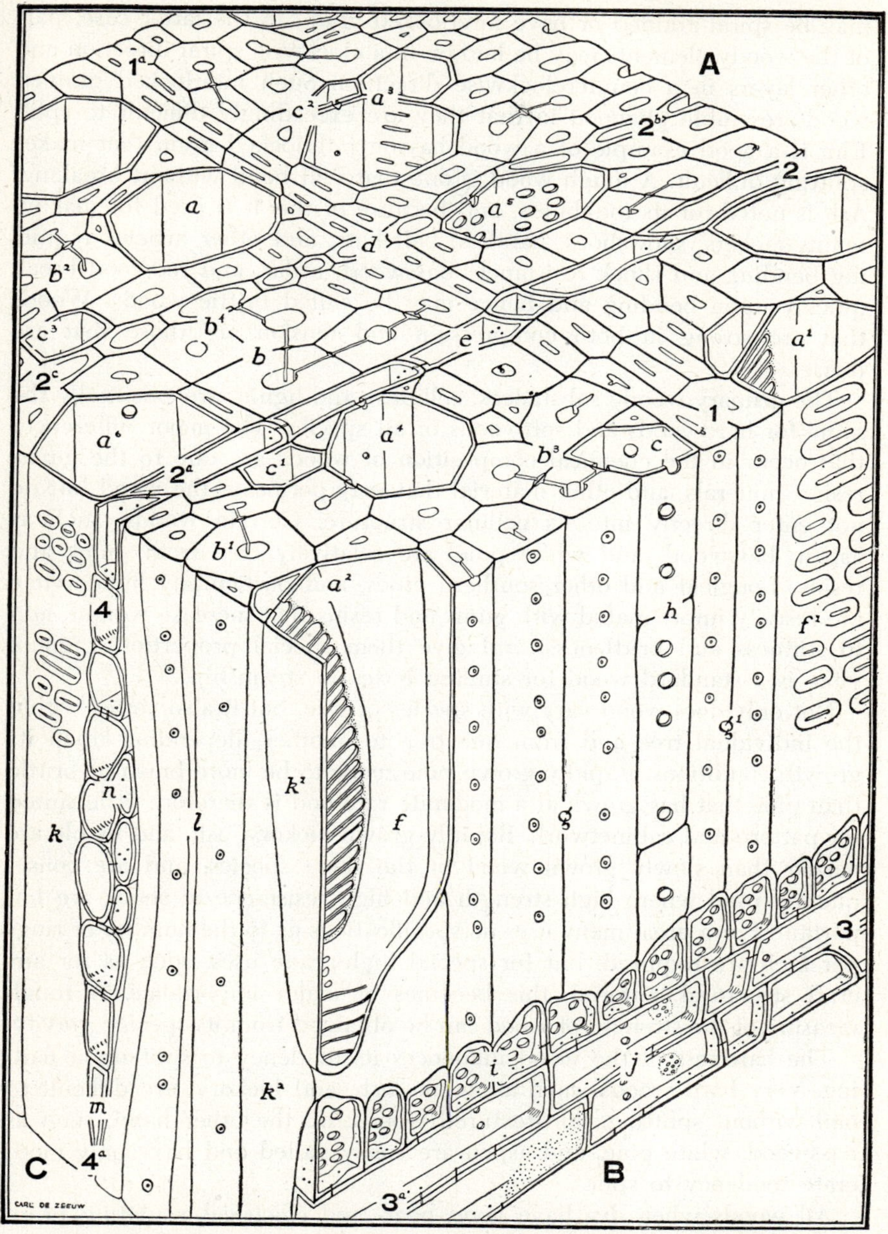

FIG. 45. Schematic drawing showing cellular structure of red gum. Face *A*, cross section showing vessels *a*, fiber tracheids *b*. Face *B*, radial section showing ray 3, and wood parenchyma *h*. Face *C*, showing ray cells 4 and fiber tracheid 1. (*From Brown, Panshin, and Forsaith, "Textbook of Wood Technology,"* Vol. 1.)

may be spiral-grained or have interlocked grain; in the latter case, part of the woody elements may be laid on in a clockwise spiral direction and other layers in a counterclockwise direction. Such interlocked grained woods resemble plywood in that they are exceedingly difficult to split. Elm is a good example of a wood having interlocked grain that makes splitting difficult. A tough wood is one that will bend without breaking. Ash is noted for its toughness, and because of this it is used for making tennis rackets, snowshoes, handles, ball bats, and other articles requiring bending and shock resistance. Basswood, cedar, and redwood break quickly upon bending and hence may be called brittle woods. Woods that are strong in both compression and tension are those that are dense.

The primary woody substances, cellulose and lignin, are essentially the same for hardwoods and softwoods of all species. The major differences that occur in the chemical composition of woods are due to the gums, resins, minerals, and other material that may permeate the wood but do not enter directly into its cellular structure. Certain woods, such as aspen, basswood, and white pine, are relatively free from gums and resins. Longleaf and other southern pines, and particularly lignumvitae are heavily impregnated with gums and resins that increase weight, add to stiffness and brittleness, and give them special properties. Lignum vitae is a standard wood for stuffing boxes in steamships.

Not only does wood vary with species of tree, but it also varies within the individual tree and from one tree to another, depending upon its growth conditions. Rapidly grown pine tends to be more brash or brittle than pine that has grown at a moderate rate and is therefore little suited for pattern and cabinetwork. Rapidly grown hickory, ash, and maple are harder than slowly grown wood of the same species, and are consequently used where high strength and high resistance to shock are important. For a great many uses no specifications as to the number of rings per inch are required, but for special high-grade uses, such as for aircraft structural material, this becomes of major importance. A rough measure of the strength of wood can be obtained from its specific gravity.

The hardness of the wood influences its tendency to split upon nailing. Very hard woods, such as maple, ash, and hickory, are difficult to nail without splitting. Soft-textured woods, on the other hand, such as basswood, white pine, and aspen are easily nailed and have only moderate tendency to split.[6]

All woods when dry have good heat- and electrical-insulating properties. Most of the gums and resins that are found in woods are themselves nonconductors of electricity. Some kinds of wood have a relatively high ash content. These when wet will conduct electricity. For this reason their moisture content can be estimated by means of an electrical moisture meter which measures the resistance between two electrodes when these are pushed into the wood. The open pores and thin-walled tracheids that are common in wood give it excellent heat-insulating qualities.

A property of wood that causes considerable difficulties in use is its tendency to shrink and swell with change in moisture content. Water occurs in wood as free water in the lumina of the cells and as absorbed moisture within the cell walls. The free water can be evaporated from the wood without causing any pronounced shrinkage. The point at which all moisture has been withdrawn from the cellular spaces and yet cell walls are saturated is called the fiber saturation point. This occurs at about 25 to 30 per cent moisture content. As the wood dries below the fiber saturation point, shrinkage occurs and continues to occur until the wood is thoroughly dried. Shrinkage across the grain exceeds many times that along the grain. Wood for commercial use is dried to about 6 per cent equilibrium moisture content. Equilibrium moisture content is that at which the wood neither gains nor loses moisture when surrounded by air of a given relative humidity and temperature. Wood that is to be used for furniture must be dried to about a 6 per cent equilibrium content; that to be used for exterior walls, to about 12 per cent. The tendency of wood to shrink or swell can be largely prevented by impregnating the wood with certain chemicals that will hold it in swollen condition. If chemicals having the capacity to swell wood are then polymerized, they form a plastic substance with the wood that maintains the wood at the swollen dimension.

Uses of Wood

Design of Wooden Structures. Joints require major attention by the wood designer, for it is here that failures are most common. The strongest wood joints are those that are held by means of glue. The use of modern synthetic adhesives has made it possible to fabricate a large number of articles from wood which have essentially the same strength in all directions and which can be formed and bolted without any weak joints.

The stiffness of a beam varies directly with the square of its depth. Therefore, wooden members that are to support floors and heavy structures are made deep in comparison with their breadth or thickness. In designing wooden structures, it is important to bear in mind the type of loading that will be involved. Structures are strongest if the load is along the lengthwise dimension of the beam rather than at right angles to it. Other simple engineering principles apply to all structural design. Trusses should be built up of triangular sections so that each section will be rigid. Use should be made of those forms and shapes that give the desired degree of stiffness or flexibility in all designs of wooden structures.

Small as well as large wooden structures need careful design. Good design takes advantage of the highly useful properties of wood and offsets its weaknesses. Furniture and many other household articles will give unsatisfactory service if designed with weak joints or improperly finished. Glue joints that loosen in time are especially troublesome. If modern design, based on an understanding of wood properties, is combined with modern machining, gluing, and finishing, wooden articles can give pleasure and satisfaction during long periods of use.

Veneer. Veneer is wood cut in sheets of thicknesses from $\frac{1}{100}$ to $\frac{1}{4}$ inch. Veneer $\frac{1}{100}$ inch in thickness is used for aircraft plywood. The thick veneers, $\frac{1}{8}$ to $\frac{1}{4}$ inch, are used for containers or stamped articles. Shoe shanks are stamped from veneer about $\frac{1}{4}$ inch in thickness. Face veneer for furniture is $\frac{1}{28}$ inch thick.

Fine veneers are used in furniture and office panels; common veneers in the manufacture of many small items that run into big-time business. Items such as medical splints, tongue depressors, fruit containers, ice-cream spoons, veneer baskets, shipping containers, all come wholly or partially from veneer. The industry is characterized by small and large operations requiring varying degrees of technical skill depending upon the quality of the product being sliced or sawed.

Veneer is manufactured by three separate processes and is called rotary veneer, sliced veneer, and sawed veneer. Rotary veneer is made from large logs cut into appropriate lengths corresponding to the width of the veneer desired. The bolts are soaked and steamed, then placed on a huge lathe (Fig. 46). The veneer bolt turns against a heavy knife that extends the full width of the veneer bolt.[4] The first few sheets that come off extend only part way around the bolt, but as the bolt becomes truly cylindrical, the veneer is unwound in a continuous sheet. It is clipped into appropriate lengths and defective sections are discarded. After it passes through a dryer, it is ready for manufacture into plywood or other products. When the veneer is used for making baskets, tongue depressors, and novelties, the processing is carried to completion before the veneer is dried.[2]

Sliced veneer is preferred to rotary cut for most exposed furniture surfaces. The sawed flitch or crotch is steamed, fastened in a heavy clamp, and moved in an oblique direction past a huge, sharp knife that cuts off the veneer in thin slices the length of the flitch being cut. Sliced veneer is kept in the order in which the sheets are removed to facilitate matching of grain. It is dried and shipped to furniture or other manufacturers. Great skill is required in matching veneer to obtain the most beautiful patterns of grain.

FIG. 46. Cutting knot-free Douglas-fir veneer. (*A.F.P.I. photo.*)

Sawed veneer is cut from flitches by thin circular or band saws that remove the veneer in thicknesses of 1/8 to 1/4 inch. Even when band saws are used, the saw kerf wastes a considerable amount of the wood. Sawed-veneer sheets also are kept in order to facilitate pattern matching. Veneer sawing is not common today. Huge veneer lathes and slicers permit manufacture of veneer much more cheaply than it can be done by sawing.

Plywood. Plywood consists of thin layers of wood glued together so that the grain of each successive layer is at an angle to the layer next to it. The finished product is strong in all directions, changes little in dimensions, and is superior to solid wood for many construction and allied purposes. The plywood industry is characterized by large corporations, large independent operators, and by highly technical problems of production. At present it is faced with the problem of maintaining production with a decreasing supply of large clear logs. Adjustments are being made so that small medium-grade logs may be used.

Veneer for plywood manufacture is usually rotary-cut unless the plywood is to be used for special furniture stock. The veneers are first separated into core stock and face stock. Core stock can be of less uniform texture and can include certain defects, provided they do not result in complete absence of woody material. Where knotholes exist, they must be patched (Fig. 47). Knots in face stock are also removed and small

FIG. 47. Patching knotholes in veneer. In a single operation this machine stamps out a circular hole surrounding the knot and inserts a patch. (*A.F.P.I. photo.*)

plugs inserted either by hand or by modern patching machines. Core stock may be of one kind of wood and face stock of another.

Plywood is constructed of three, five, seven, or any other odd-numbered layers of veneer.[4] The two surface layers have the grain running in the same direction, but each successive layer of veneer is placed at right angles to the preceding. Cross-banded layers are coated top and bottom with glue. The laid-up panels are pressed at room or slightly elevated temperatures. When hot presses are used, one or two panels are pressed between steam-heated cauls (Fig. 48).

Many adhesives used in modern plywood require heat for curing. Adhesives may be of the quick-setting, synthetic resins such as the phenol formaldehydes, urea formaldehydes, melamines, and resorcinols, or they may be made of animal or vegetable glues of the cold-setting type. After the adhesives have cured, the plywood panels are taken from the presses, the edges are trimmed square, and the surfaces sanded. The plywood is then ready to package for shipping to the consumer.

Laminated Wood. Laminated wood consists of successive laminae glued together with the grain parallel. The Second World War with its shortage of metallic structural material brought the industry into prominence. Laminated beams and rafters coupled with timber connectors permitted the construction of the world's largest clear floor space structures (blimp hangers). Such beams also prove highly desirable for overhead craneways.

Laminated timbers may be small or of huge size. The small timbers are used to make skis, aircraft propellers, and other products requiring high strength and uniformity. Large timbers are laminated for boat keels and skegs, beams, arches, and other purposes. Large timbers, particularly if curved, are built up of 1-inch lumber. This is planed to high tolerance, and end joints scarfed to increase the strength. Since it is difficult to apply heat uniformly to such massive timbers, cold-setting adhesives are ordinarily used. The individual boards are laid up in a jig and arranged so that pressure may be applied uniformly throughout the length of the glued-up member. This may be done with various types of clamps and levers.

The problems in this industry are many, including high cost of production, waste in manufacture, and high wood requirements.[4] Many research and production problems still face the industry. Lamination does, however, permit fabrication from boards of timbers with greater strength ratios than solid wood members. It also permits wood to remain in the large-construction field despite scarcity of large-size timber.

FIG. 48. Huge hot presses are used to cure synthetic resin-bonded plywood. (*Pluswood, Inc. Photo by Chas. Meissner & Assoc. A.F.P.I.*)

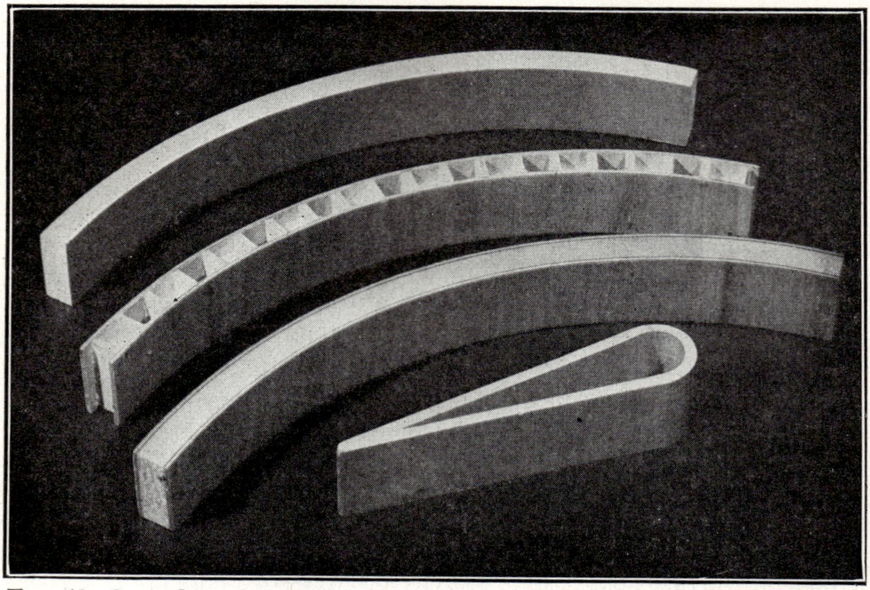

FIG. 49. Curved sandwich construction of balsa wood and plywood. Especially useful where light weight must be combined with stiffness. (*A.F.P.I. photo.*)

Sandwich and Box Construction. Special types of structures used in aircraft and for other purposes where high strength and stiffness must be combined with light weight are sandwich construction, box beams, and laminated shapes.[2] Sandwich construction as applied in the mosquito bomber consisted of inner and outer shells of 5-ply birch plywood about ⅛ inch in thickness separated by a ½-inch thickness of balsa wood. The entire 11-layer assembly was laid up using waterproof adhesive and was pressure-molded at one time. Where metal fastenings were to be anchored, maple was substituted for balsa wood. The fuselage was very easy to repair with simple equipment. This same principle is used in manufacturing flush doors and other products (Fig. 49).

Box beams and shapes are hollow structures with solid or laminated wood for the two main stress-bearing members, but these are connected by glued-on plywood faces that cause the two members to act as one and greatly stiffen the beam. The hollow, glued-up mast for small sailboats is an example of a box beam using solid wood instead of plywood for connecting members. Glued, laminated shapes and forms may be of box construction, or they may be arranged in T's, angles, I beam, and other forms. They may actually be built-up trusses with glued-on plywood faces. Wide possibilities exist for using wood in such novel ways for aircraft and other purposes. The potentialities have been only partially explored.

Wood Treating

When wood is to be exposed to destruction by insects, marine borers, fungi, and fire, wood treating to prolong its service life has long been practiced. Its importance increases as wood becomes scarce and as costs of replacing wood in service mount. The quantities treated by products for the year 1949 are shown in Table 2.

TABLE 2. WOOD TREATED IN 1949*

Product and unit	*Quantity treated, million*
Crossties	40.0
Timber and lumber, bd. ft.	509.6
Switch ties, bd. ft.	130.4
Piles, lin. ft.	16.3
Poles	6.0

* 1949 Wood Preserving Statistics, *1950 Wood Preserving News*, 28:138–143.

Creosote, either full strength or in solution with coal tar, petroleum, or other materials, was used for 95 per cent of all wood treated.

For treating ties, piling, posts, and poles, coal tar creosote is cheap and effective. This may be applied as a hot dip or diluted with solvents and applied with a brush. Results with these methods are only partially successful, for the treated layer is thin, and insects, fungi, or marine organisms will enter through cracks, checks, and abrasions. Far more effective protection is provided by placing the timbers in a huge cylinder where they may be treated with preservative under pressure. Piling, which requires heaviest treatment, is first heated and evacuated. The hot preservative is admitted to the cylinder without releasing the vacuum. This is called the full-cell process. Ties, posts, and poles which require light treatment are placed first under air pressure in the cylinder. Then, hot preservative is pumped in under higher pressure. When all the pressure is released, the excess preservative is forced out of the cells and cavities by the entrapped air. This is called the empty-cell process. Other materials used in 1948 included chromated zinc chloride, Wolman salts, and pentachlorophenol. The same treating techniques may be used with these as with creosote.

Commercial wood-treating plants operating in 1948 totaled 236. Of these 171 were pressure plants, 51 nonpressure, and 14 a combination of the two.[5]

Wood may also be treated with dyes or other chemicals to modify its appearance and properties. Of considerable importance is the treatment of wood with chemicals that reduce inflammability. Wood cannot be made fireproof, *i.e.*, treated so that it will not oxidize when heated to high temperature, but it can be treated so that it will not support combustion. Effective fire-retardant chemicals are diammonium phosphate, monoammonium phosphate, and phosphoric acid. Other chemicals that will stop combustion are aluminum sulfate, ammonium bromide, ammonium chloride, boric acid, and monobasic zinc phosphate. Still others may reduce either flaming or glowing.[6]

Wood-processing Industries

Families in Maine that sell gasoline and wooden novelties to tourists in summer may spend much of the winter making these novelties in a home workshop. Such a family business uses a minimum of raw material.

In a central New York village famous for its early history is a small sawmill and wood-turning establishment, which furnishes employment for about 30 workers of the community. About 500,000 board feet of beech and maple logs are used each year to make banister rungs, chair rungs, and various other furniture parts.

A company in eastern Maine, employing about 75 workers, manufactures ice-cream spoons and shoe shanks. It uses about 20,000 cords a year of selected paper birch veneer bolts that are cut into veneer. Spoons and shanks are stamped from the veneer, dried, and polished in a rotating drum with soapstone.

A company in western Pennsylvania makes wooden toys and brush handles. In 1941 it employed about 800 workers and used 5 million board feet of maple and beech lumber. Various specially designed machines turned out different shapes of handles. Machining must be to high tolerance. The company also manufactures educational toys.

At Grand Rapids, Mich., are huge factories that manufacture all kinds of furniture from wood and other materials.

These are examples of wood-processing industries. Other wood processors include those manufacturing boxes, crating, cars, boats, musical instruments, trunks, baskets, fixtures, and burial caskets.[3] They vary in complexity from the man who nails a few boards together to make a loading pallet to the manufacture of piano movements and hand-carved shoe lasts.

The 1947 U.S. Census classified wood processing under 25 categories. A few of the more important are shown in Table 3. This gives the number of establishments, men employed, and value added by manufacture. The industries are not classified strictly on the basis of wood used. Together, however, the wood-processing industries employ more workers and create greater values than sawmills and planing mills.

Furniture manufacture exceeds all other wood-processing industries in value of products. The industry is centered in the eastern United States. Important producing states with percentage of total output attributed to each are: North Carolina, 12.0; New York, 10.5; Illinois, 9.0; Indiana, 8.5; Virginia, 7.0; Pennsylvania, 6.5; California, 6.0; and Michigan, 5.5.[4]

TABLE 3. STATISTICS ON SOME IMPORTANT WOOD-PROCESSING INDUSTRIES FOR THE YEAR 1947*

Type of industry	Number of establishments	Number of men employed	Value added by manufacture, thousands of dollars
Veneer mills	156	10,461	36,858
Plywood plants	162	26,499	146,087
Wooden containers	1,516	69,104	233,453
Wood house furniture	2,251	132,709	492,738
Wood office furniture	115	6,601	26,839
Screens, shades, and blinds	1,058	25,973	116,396
Miscellaneous wood products	2,349	65,514	260,517

* U.S. Department of Commerce. 1950. "Census of Manufactures: 1947," Vol. I, General Summary, Washington, D.C. 355 pp.

The wood-processing industry is primarily a small-business field. Only a few companies are sufficiently large to have their securities listed on the great national exchanges. Many are individually or family-owned. Because they are small, they offer, as a rule, both the advantages and limitations of small industry. When well organized, adequately financed, and skillfully managed, such small industries may be highly profitable. Each industry has its own special problems of procuring raw material, preparing it for processing, processing, and selling the final product.

Broadly speaking, wood goes through the following operations as it is manufactured into commodities: sawing, seasoning, resawing, machining, gluing or fastening with metal fastenings, finishing, and packaging. These together with other operations that might be considered as departments in a manufacturing plant will be discussed individually. Most require technical control by an engineer, chemist, or wood technologist for best results.

Conditioning Wood for Manufacture. Wood to be fabricated must first be properly conditioned. If the wood is to be bent or cut into veneer, it must be soaked and steamed. Soaking in chemicals that have mild plasticizing action facilitates bending. Many wood products, particularly those made from veneer or turned on a lathe, may be completely processed before drying occurs. Other products such as shoe lasts and bowling pins are rough-turned before seasoning. Each industrial plant is likely to own and operate its own dry kiln observing the precautions mentioned earlier.

Machining of Wood. Wood-cutting operations include ripping and crosscutting with ordinary saws and with special hollow-ground, planer-type saws that cut smooth surfaces; planing the surface with jointers and planers; molding the wood by the use of variously shaped, high-speed cutterheads; turning wood on various types of lathes; boring and routing wood by the use of bits and router tools; and carving of wood either by hand or by use of mechanical routers.

Each operation has its own technical problems. Rough lumber is resawed, ripped, and crosscut very much as these operations are carried out in the sawmill. Various types of table and swing saws may be used, including those that move to and fro in a straight line. To cut wood along curved lines, narrow-bladed band saws are used. For very intricate work the jig saw may be used.

Jointing, planing, molding, turning, and routing require careful workmanship and technical control to avoid splitting, roughened grain, and overheating of wood or cutter. Special attention must be given to the shape of the cutting knife, angle at which the knife cuts the wood, depth of cut, speed of cut, especially for rotating tools, and the amount of wood removed by each tooth of the saw or each sweep of a cutting knife. All of these factors are controlled in a machine by the method of sharpening the knives, the angle at which they are set, the speed of the cutterhead, the number of cutting edges, the rate of feed, and the depth of cut in one operation (Fig. 50).

Little detailed research has gone into the problems involved in the machining of wood. This subject has just recently attracted the attention of forest-products research institutions. Laboratories are now being organized to study in detail the effects of varying the several factors in machine operation on the quality of job done with wood of different species and wood conditioned by various methods. Obviously, this is a field of operation in which a wood technologist has much to contribute. His services are needed not only by the industries that operate wood-cutting machines but also by those that design and manufacture the machines.

Fastening of Wood. The weakest parts of products manufactured of wood are likely to be the joints. This is especially true of furniture but is true of wood used in trusses, beams, and general construction. Joint weakness is accentuated by the weakness of wood across the grain. Various mechanical devices have been constructed to facilitate and improve

Fig. 50. Planing and shaping lumber. Weyerhaeuser Timber Mills. (*A.F.P.I. photo.*)

wood fastening: wooden pegs or dowels; nails of various sorts, including rosin-coated, cement-coated, and roughened-galvanized nails; corrugated metal fastenings; wood staples; screws, including lag screws; bolts; rods; and various forms of timber connectors. The effectiveness of all of these mechanical fastenings varies with the way in which they are used and with the wood in which they are used.

Modern timber connectors are just coming into widespread use in the United States. This is a field where considerable additional promotion needs to be done by those who understand the effectiveness of these connectors and how they should be used.

Gluing. Gluing is one of the most satisfactory means of fastening two pieces of wood together. A good glue joint is as strong as the wood itself. It avoids the removing or crushing of woody substance needed to achieve maximum strength, such as is inevitably involved in the use of metal or wooden fastenings. Modern glues and adhesives are made from a great variety of raw material and may be used for widely varying purposes. Some are virtually waterproof and can be used on wood that is to be subject to repeated wetting and drying; others are moisture-resistant; some have little or no resistance to moisture. Glues also have other properties that influence their effectiveness when used with wood. Some are subject to crazing or other forms of breakdown after prolonged use. For effective use of modern glues the factory should have a glue expert who understands the properties of modern adhesives and the conditions under which they must be used for best results. A number of wood technologists have been able to save many times their salary through reducing rejects caused by faulty gluing.

Sanding and Finishing. Sanding is normally done after the final cutting and machining have been completed. It is a relatively simple operation that requires but ordinary care for good results. It should be well done because it is very difficult to cover up a rough surface by paints and varnishes.

Final finishing of the fabricated wooden product after manufacture and assembly involves applying various seals, stains, lacquers, varnishes, paints or other coatings to the wood surface. Many of these must be sanded or rubbed between coats and buffed after the application of the final coating. Many modern varnishes are making use of new plastic materials containing various solvents as thinners. These require careful technical control for best results.

Modern wood finishing may be said to be of two types: the penetrating oil seals, stains, varnishes, and resins; and those finishes applied to the surface of the wood such as shellac, varnish, lacquer, paint, and related products. The penetrating seals and particularly the penetration of wood with chemicals that form a hard resin upon curing give a surface of maximum resistance to marring. Even when such a surface is marred, the mar can usually be removed by light sanding and buffing. The penetrating oil seals have a similar advantage in that once a mar or a soiled spot occurs, it can be removed by rubbing the spot with steel wool moistened with the sealing material.

Few of the modern finishes applied to the surface of wood are as resistant to wear as the wood substance itself. This is especially important in flooring. Varnishes are readily worn away. Wood penetrated with an oil seal or with a chemical substance that will polymerize provides a very long-wearing surface. Knowledge of wood and of the conditions of service to which the wood is to be subjected is necessary for applying the most effective finish to the wood. The paint-shop foreman of former days can no longer cope effectively with the many new products that are now available for wood finishing.

Research and Quality Control. Every large manufacturing plant must have its inspectors and testers to maintain quality of product. Oftentimes such a plant has a special research staff whose responsibility it is to test the products of the company and of its competitors and to determine means by which products can be made more cheaply and better. Wood technologists are employed in practically all research organizations of the wood-processing industries, but only the larger companies feel that they can afford a research organization.

Packaging. Packaging the finished product is the work of an important division in many plants. The packaging industry for the United States as a whole does an estimated 5 billion-dollar business yearly. Experts on packaging, therefore, have a wide field in which to exercise their skills. Modern packaging involves not only mechanical protection but protection against moisture and fumes that might mar or destroy the product during transit. During the war period packaging received tremendous impetus. Early in the Second World War a high percentage of motor parts, delicate instruments, and other products were delivered to the war zones too badly damaged to be serviceable. A vast research and training program, which involved special designing of packages for every product, was then undertaken. Special care was given to metal products that must be protected against moisture and salt water. The result was a packaging technique that made it possible for packages to be immersed in salt water or to stand unprotected for two or three months on the shores of tropical islands without damage to the contents.

The industry has shown a tendency to return to prewar methods in recent years, primarily because these are cheaper. Experts believe that many of the war-introduced techniques should become a permanent part of the operations of the packaging industry. This industry needs specialists, not only in the use of wood, but in the use of moisture barriers of various types and in the use of hygroscopic materials that will absorb moisture existing inside the package at the time it is made up.

Plant Layout and Design. Small wood-fabricating plants as a group might profit much from adopting modern plant layout and design in their operations. A typical small wood-processing plant may have its machines arranged on the machining floor more or less at random. Products to be machined are transported in boxes mounted on dollies from one machine to another. This causes a maximum of handling and of disruption to normal flow of operations. Men who have specialized in layout of woodworking plants have been able to perform important services in improving flow of materials and output of factories. This is a field of study about which every man planning to enter the wood-processing field would do well to inform himself.

Management, Sales, and Administration. The wood-processing industries and those servicing them require sales management, production management, and general administration. Such positions are ordinarily filled by men with long experience in the wood-processing industries. Managers are now seeking younger men with technical background who understand how to apply modern technology and administrative techniques so as to increase output, improve quality, and lower costs.

Waste in Wood Processing

Waste occurs in both primary and secondary wood processing (Table 4).

TABLE 4. MANUFACTURING WASTE IN RELATION TO TOTAL COMMODITY DRAIN, BY PRINCIPAL FOREST PRODUCTS, 1944*

Product	Total commodity drain, million cu. ft.	Waste as a per cent of drain		Actual product as a per cent of drain
		Used for fuel	Not used at all	
Hewed crossties	363.5	4	66	30
Cooperage stock	173.8	23	49	28
Fence posts	219.9	1	9	90
Veneer	391.7	30	36	34
Round mine timbers	225.0	2	9	89
Shingles	71.7	18	45	37
Other	515.8	17	20	63
Total	1,961.4	14	33	53

* U.S. Department of Agriculture, Forest Service. 1947. Wood Waste in the United States, Report 4, from "A Reappraisal of the Forest Situation," Washington, D.C.

Wood Turning. The wood-turning industry has very heavy waste. Most of the material that is turned is first sawed square or split and seasoned before it is turned. Losses in turning of squared blanks often run as high as 75 per cent of the wood material. The loss is particularly heavy in turning such products as bowling pins, shoe lasts, spools, spindles, bobbins, shuttles, gunstocks, handles, ball bats, and related products. The amount of manufacturing waste that occurs in various industries as listed by the U.S. Forest Service is given in Table 4.[7]

Veneer. Veneer manufacture entails a number of losses. Short unusable lengths of veneer are produced by the rotary process in truing up the veneer bolt before continuous sheets can be taken off. The central core of a veneer bolt is also lost from use as veneer, since it is too small for continued turning and oftentimes contains knots that would make the veneer of little value. Veneer cores are sometimes sawed into small-dimension lumber, and they are frequently used for fuel. Veneer scraps are produced by trimming to remove defects. Veneer bolts are cut long to permit edge trimming to avoid cracks caused by end checking. These scraps together with those from rounding up the bolt are used in part to prepare patches to plug knotholes, but a high percentage of this material is lost. Veneer waste is commonly burned for fuel to operate the plant. Veneer loss also occurs during seasoning when the veneer may crack and check.

Secondary Wood-processing Waste. The amount of waste that occurs in secondary wood processing varies widely from product to product. In products such as ladder rungs, turned furniture, and handles, the waste runs as high as 75 per cent. In furniture manufacture it may run even higher. A high degree of waste is inevitable in products that must be turned, molded, and shaped by various means. Waste also occurs as a result of cutting out defects and of buying improper lengths of lumber for the product to be made. Total waste is estimated at 463 million cubic feet. All is probably used for fuel or other purposes.

A special form of waste is due to overspecification by lumber buyers. Much high-grade lumber is used for purposes for which low-grade would do just as well. This lowers the value of poorer grades and they become a drug on the market. Cutting around knots and other defects in lumber is expensive after the lumber reaches the stage of final manufacture. Greater waste almost inevitably occurs here than would occur if this cutting were done in a small-dimension plant that would have markets for a wide variety of shapes and sizes. Elimination of this type of waste, which is really not waste in the usual sense but simply the use of a high-grade product where a low-grade product would do just as well, awaits the development of an intensive market for small-dimension material. The organization of such a market is difficult and would require a considerable amount of capital, a large inventory, and a large sales organization.

Reduction of waste and the discovery of profitable uses for waste are also technical jobs in the wood-processing industries. The manufacture of wood flour from factory waste is but one example of converting waste to good use. Opportunities also exist to collect waste products from one manufacturer or process to be used by another.

Waste in other products is discussed in Chap. 11.

LITERATURE CITED

1. Brown, H. P., A. J. Panshin, and C. C. Forsaith. 1949. "Textbook of Wood Technology," Vol. 1, McGraw-Hill Book Company, Inc., New York. 652 pp.
2. Brown, Nelson C. 1950. "Forest Products," John Wiley & Sons, Inc., New York. 399 pp.
3. Hoyle, Raymond J., and John R. Stillinger. 1949. "Wood-using Industries of New York," New York State College of Forestry, Syracuse, N.Y. 133 pp.
4. Panshin, A. J., E. S. Harrar, W. J. Baker, and P. B. Proctor. 1950. "Forest Products—Their Sources, Production, and Utilization," McGraw-Hill Book Company, Inc., New York. 549 pp.

5. Steer, Henry B. 1949. Wood Preservation Statistics 1948, "Proceedings, American Wood-preservers' Association," 45:407–450, Washington, D.C.
6. U.S. Department of Agriculture, Forest Products Laboratory. 1940. "Wood Handbook," Washington, D.C. 326 pp.
7. U.S. Department of Agriculture, Forest Service. 1947. Wood Waste in the United States, Report 4, from "A Reappraisal of the Forest Situation," Washington, D.C. 45 pp.

CHAPTER 10. *Wood Chemistry, Paper, and Plastics*

WOOD CHEMISTRY embraces the study of cellulose and lignin and other wood constituents, their molecular structure and properties, their chemical behavior, their breakdown by microorganisms, and their use in industry. Wood chemistry really begins with photosynthesis, that is, with what goes on in the living chloroplast when carbon dioxide and water are combined to form sugar intermediates. These initial simple substances are transported throughout the living plant and converted into a host of additional materials found in living trees. The study of the chemistry of wood also includes the processes that take place naturally when fungi, insects, and other organisms break down cellulose and lignin and convert wood back into carbon dioxide and water from which it was originally synthesized. Principles of wood chemistry are involved in the free combustion of wood and in destructive distillation of wood.

A great many important industries depend upon wood chemistry. These include the wood-chemical industry which is concerned primarily with wood distillation, the naval-stores industry, and the various extractive industries which prepare such materials as tannins, waxes, oils, pharmaceuticals, essential oils, and other products. The pulp and paper industry is concerned with separating the fibers of wood by mechanical or chemical means and forming sheets of paper and paperboard from them. The plastics industry is concerned with various cellulose and lignin derivatives that may be chemically dissolved and reprecipitated to form filaments, sheets, and molded products. The explosives industry is concerned with nitrocellulose. Others use lignin derivatives such as vanillin.

Various modified wood products depend upon impregnating wood with plasticizing and polymerizing agents. When soaked in certain plasticizing agents, wood can be twisted, bent, and to some extent molded. It will remain permanently in the molded or bent shape if the plasticizing agent is polymerized after shaping. Wood impregnated with phenol formaldehyde resins can form Compreg, a product in which the wood is compressed greatly and therefore densified. Ethyl alcohol and fodder yeast can be made from the sugars formed when cellulose is hydrolyzed.

Furfural can also be obtained. This can be prepared from waste liquors as a by-product of acid pulping or by the hydrolysis of wood directly, though the latter has not been economically feasible.

Composition of Wood and Occurrence of Components

Wood is a complex organic tissue, delicately patterned both anatomically and chemically. Chemically it contains two types of components: those that are insoluble in neutral solvents, or nonvolatile [8] with steam, or both; and those that can be removed by such solvents or steam.[7] The latter are spoken of as extractives, or extraneous substances, and include tannins, waxes, oils, resins, gums, sugars, pigments, and others. The former, the wood substance, or cell-wall substance, is made up of lignin and the polysaccharide system termed holocellulose. The latter is made up of cellulose and the hemicelluloses together with minor amounts of pectic materials. Wood contains about 20 to 30 per cent lignin. Temperate-zone hardwoods contain less lignin than the softwoods. About 50 to 60 per cent of the cell wall of wood is composed of cellulose, and about 20 to 30 per cent is comprised of hemicelluloses. In wood the individual fibers are held together with an amorphous material in the middle lamellae which is primarily lignin together with a certain proportion of hemicellulose and perhaps some cellulose. The lignin in the middle lamellae not only cements the cells together but adds rigidity and strength to the wood. The primary and secondary cell walls contain a relatively small amount of hemicellulose and lignin and a large amount of cellulose, though, actually, both lignin and hemicellulose are found together with cellulose in all portions of cell walls throughout the woody stem.

Since both cellulose and lignin are inert to ordinary chemicals, wood is not easily corroded in the air or in salt water. Cellulose and lignin both, however, can be hydrolyzed to yield simple sugars that will support the life of fungi and various other wood-destroying organisms. For this reason wood decays. Neither pure cellulose nor pure lignin contains ash. Wood ash contains a fairly high percentage of potassium and calcium. The ash content of wood is due largely to the deposition of chemicals in the tree stem from the water that is transported from the roots. A small part of the chemicals enters into the live tissue, in the rays, cambium, and bark. Consequently, wood ash contains all of those elements that are essential for plant growth, excepting the carbon, hydrogen, oxygen, nitrogen, and sulfur that are lost during combustion.

The Chemical Nature of Cellulose

Cellulose is the major component of the cell wall of woody and other plants. Cellulose is a chemical compound of high molecular weight made up of units of anhydroglucose, that is, glucose from which a molecule of water has been subtracted. The empirical formula for cellulose is written as $(C_6H_{10}O_5)_n$, where *n* equals the number of glucose anhydride units combined in the molecule. Cellulose can be quantitatively hydrolyzed to yield glucose. Native cotton cellulose has more than 3,500 anhydroglucose units per molecule, purified cotton linter 1,000 to 3,000, wood pulps 600 to 1,000, and commercially regenerated cellulose, such as rayon, 200 to 600 units. [4, 8] The cellulose molecule is large enough to be classed with colloids.

Cellulose molecules are linear or threadlike in form. Most of them are organized parallel to each other and more or less parallel to the fiber axis. In a highly organized parallel condition, these molecules form microcrystals, whose existence can be proved by X rays. It is the linear nature of the cellulose molecules, together with their tightly packed parallel organizations, which gives cellulose fibers their strength.

Cellulose from all plants has the same chemical composition. Cellulose is closely related to starch chemically and has the same empirical formula. When treated under controlled hydrolysis, however, starch breaks down into maltose and cellulose into cellobiose. These are identical in composition but differ in stereochemical configuration. Cellulose can be arbitrarily divided into three groups—alpha, beta, and gamma cellulose. Alpha cellulose, which contains the longest molecular chains, remains undissolved when treated with 17.5 per cent aqueous sodium hydroxide; beta cellulose dissolves but may be precipitated by acidification of the alkaline solution; and gamma cellulose, which is made up of the shortest molecular chains, remains in solution. Cotton has about 98 per cent alpha cellulose. Wood pulp has a much lower alpha cellulose content and is higher in amounts of beta and gamma cellulose.

Plant cell walls contain other substances known as hemicelluloses. These have been loosely defined as those portions of the cell wall that may be extracted by dilute alkalies or by mild hydrolysis. However, portions of the hemicelluloses are quite resistant to chemical action and remain in the purified wood pulp. The hemicelluloses are made up of long chains of sugar units quite like the cellulose, but the nature of the simple sugars may differ from cellulose. Hemicelluloses when hydrolyzed may yield glucose, galactose or mannose (all six-carbon hexoses) and xylose or arabinose (five-carbon pentoses). Hemicelluloses which are made up mainly of pentose sugar units are known as pentosans. These upon hydrolysis yield pentoses that cannot be converted directly into alcohol by most yeast. However, certain yeasts have been discovered that convert pentose sugars into proteins and fat usable as fodder for animals. The pentose units can also be converted to furfural.

It is possible by appropriate treatment of wood with chlorine and 95 per cent alcohol containing 3 per cent monoethanolamine or by other chemical means to dissolve away the lignin and leave the total polysaccharide fraction. This product is called holocellulose and is the nonlignin portion of the cell wall or the total cellulose plus hemicelluoses. Woody plants contain 70 to 80 per cent holocellulose.

Cellulose is separated from most of the other constituents of wood by the industrial chemical pulping processes, the sulfate, the sulfite, and soda processes.

Properties and Derivatives of Cellulose. Cellulose has high affinity for water. Even after cellulose has been dried in a vacuum oven over phosphorus pentoxide, it still retains 0.35 per cent of water. It is, therefore, one of the most powerful dehydrating agents. Dry, fibrous materials are good electrical insulators, but it is almost impossible to keep woody material dry. The electrical conductivity of wood increases over 100,000 fold in the range from 7 to 30 per cent moisture.

Wood pulp or cellulose fiber may be mechanically "hydrated." This is one of the actions that takes place when wood pulp is beaten, and it is purely mechanical, not chemical. The fibers hold more water mechanically because of the increased area of the finely broken fibers. The pulp tends to become gelatinous. The glassine grades of paper pulp are made by "hydrating" the pulp in the beater. Since such pulp gives up its moisture very slowly, it can be run only at a slow rate on a paper machine. It is said to have a low freeness. After prolonged beating the fibrous nature of wood pulp almost completely disappears, and the whole becomes a gelatinous mass, which, upon drying, becomes a horny, amorphous substance.

Cellulose can be dispersed in a cuprammonium solution known as Schweizer's solution.[4] From such a solution the cellulose can be recoagulated in a dilute acid bath. Rayon manufactured by the cuprammonium process is known as Bemberg rayon. The viscose process of manufacturing artificial silk is more widely used today. In this method cellulose is combined with caustic soda and the alkali cellulose is reacted with carbon disulfide to give a product, cellulose xanthate, soluble in dilute sodium hydroxide. From this cellulose xanthate, cellulose can be regenerated by acidification. If the viscose solution is forced through a spinneret and the cellulose regenerated in the acid bath, a thread or filament is formed. If the viscose solution is extruded through a long slit, it forms a sheet, which, upon drying, becomes cellophane.

Cellulose treated with a mixture of nitric and sulfuric acids forms nitrocellulose, which has high explosive properties.[8] Since nitrocellulose was originally made entirely from cotton, it was spoken of as guncotton. Now it is also made from wood cellulose. Nitrocellulose containing less nitrogen than guncotton is used for lacquers and plastics. Cellulose forms esters with acetic acid known as cellulose acetate, which, being thermoplastic, may be softened by heat. Various plastic materials are made of cellulose acetate, including rayon, films for photographic work, and various other products. Lacquers are made of mixed esters of cellulose acetate and butyrate. Ethyl and methyl ethers can also be made from cellulose. Plasticized ethyl cellulose may be used as an insulating coating for wire and may be extruded in strips for the manufacturing of materials for furniture seats. It is flexible, tough, dimensionally stable, has low moisture absorption, can be easily fabricated, and takes printing.

Lignin

Lignin is more difficult to separate from woody cell walls than is cellulose. The chemistry of lignin is still imperfectly understood. Whether the material after it is extracted from the cell wall by various chemical methods is the same that occurs in the cell is not known, but it is generally conceded that certain modifications occur during the extracting process. For these reasons and others the molecular weight and structure of lignin remain unknown. When lignin is chemically broken down by various methods, it yields a number of compounds that are believed to be present in modified form in the lignin molecule. Among these are vanillin, cyclohexylpropane, acetoguaiacone (derivative of material found in lignumvitae), β-vanilloyl ethyl alcohol and acetyl vanilloyl; hydroxybenzoic acid and isohemipinic acid.[2] Vanillin is now manufactured from lignosulfonic acid formed during the sulfite cooking process of making wood pulp.

The structure of lignin is known to be very complex. It differs from the polysaccharides of the cell wall in that it is partly aromatic in nature. Lignin in itself has plastic properties and may ultimately prove to be a valuable agent in the wood-plastics field. It will react with certain other plasticizing and polymerizing agents to form new plastic materials. Lignin in common with coal and other products can be hydrogenated, that is, treated with hydrogen under elevated temperatures and pressures to yield various petroleum-type derivatives. These include products resembling gasolines and lubricating oils.

Making Paper Pulp from Wood

The largest wood-chemical industry by far is the pulp and paper industry. It is important because of the wide and growing use of paper and paper products and because pulp preparation is the first step in the manufacture of cellulose-based plastics.

The Chinese made paper in the year A.D. 105, using bark from the mulberry tree and old linen and fish nets. The Arabs learned papermaking from the Chinese and improved and perfected the process. It was not until early in the eleventh century that paper was first made in

Europe (in Spain) and not until 1690 that the first mill was established in the United States. During most of this period, paper was made from linen rags, and because of the scarcity of raw material the production of paper did not keep pace with the demand.

Paper was not made successfully from wood until the middle of the nineteenth century. Today, although limited amounts are made from straw and rags, practically all paper is made from wood.

Scope of the Pulp and Paper Industry. The two important major operations in making paper products are pulp production and paper and paperboard manufacture. A paper mill may obtain pulp from the company's own pulp mills or it may purchase pulp from domestic or foreign pulp companies. Paper mills with integrated pulp mills have operated at much greater advantage in recent years than nonintegrated mills.

Closely related to paper manufacture is the manufacture of paperboards for shipping containers, paper insulating felts, paper building boards, and paper-base plastics. Converting plants for making boxboard, corrugated board, paper cups and plates, building boards, and other similar products are sometimes operated in conjunction with the paper mill. Some of the operations which may be carried out in the converting plant include laminating, saturating, coating, printing, and fabricating.

Pulp companies sometimes operate by-product chemical plants to convert pulping waste liquors into useful products, such as lignin, alcohol, vanillin, and fodder yeasts.

Pulp is manufactured by treatment of wood or other fibrous raw material in such a way that the fibers are separated into a pulpy mass. This can be done by either chemical or mechanical means or by a combination of chemical and mechanical means. A number of different pulping processes in use [5] are described below. The amount of pulp produced by each major process is shown in Fig. 51.

In all processes except for pulps made into coarse grades of building boards, clean, bark-free wood is required. Most of the bark removal is done at the pulp mill.

Groundwood. The making of groundwood is a purely mechanical operation. The wood billets are forced against the face of a large grindstone at a pressure of about 40 pounds per square inch. The axis of the billet is parallel to the axis of the stone, because the longest fibered pulp is obtained in this way. The face of the stone is "dressed" or roughened so that it tears the fibers loose from the wood. Water is played on the stone to remove the groundwood as a pulp or "slush" and to maintain the temperature of grinding at the proper point (about 160°F.). The pulp is screened to remove knots and lumps of wood, thickened by removing water, and then sent to a paper mill for making into paper.

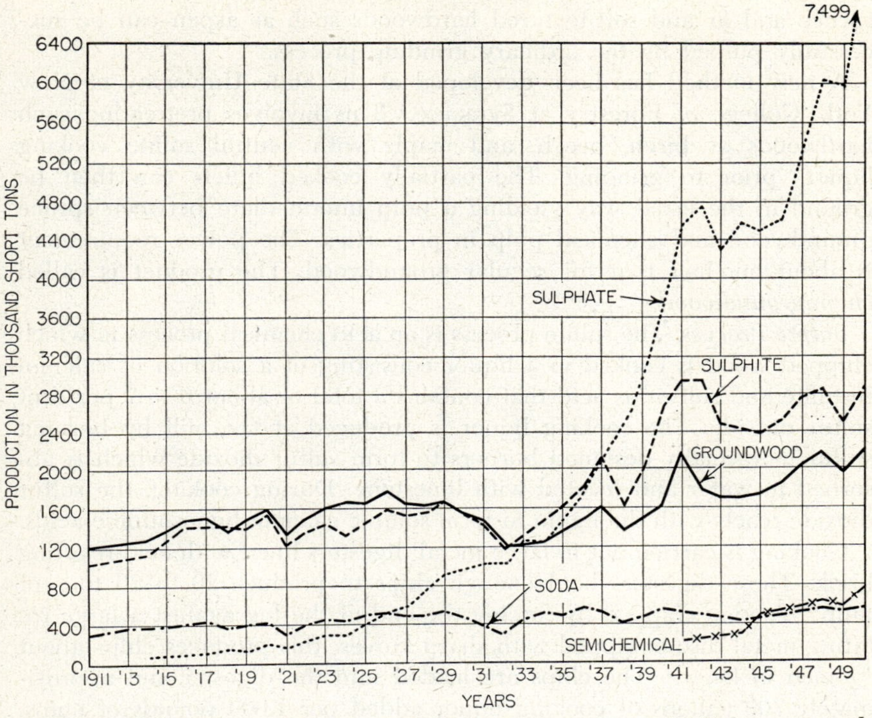

FIG. 51. Wood-pulp production in United States 1911 to 1947 by principal processes.

Groundwood is a low-grade, low-cost pulp, which is suitable only for the cheaper grades of paper and paperboard. The pulp consists largely of broken fibers, fiber fragments, and fiber bundles. Groundwood, which has low strength, must generally be mixed with a stronger pulp. Newsprint, for example, is made from 85 per cent groundwood and 15 per cent stronger pulp. Papers made from groundwood are not permanent. The chief advantage of groundwood is low cost which is derived from the high yield (92 per cent) of fiber from the original wood. The groundwood process is growing in importance because of its increased use in printing papers. Recent improvements in making "bleached" groundwood have added to the number of grades of paper in which this pulp can be used. The production of groundwood in the United States in 1949 amounted to 1,962,141 tons. The price was \$85 to \$100 per ton in 1951.

The principal item of manufacturing cost is for power. The power requirement for grinding spruce and fir is between 50 and 60 horsepower days per ton (895 to 1,074 kilowatt hours). Only softwoods such as spruce and fir and soft-textured hardwoods such as aspen can be successfully pulped by the ordinary grinding process.

A new method has been developed at the State University of New York, College of Forestry at Syracuse.[3] This involves pretreating such hardwoods as birch, beech, and maple with neutral sulfite cooking liquors prior to grinding. The partially cooked billets can then be ground in the usual way yielding a pulp intermediate between spruce groundwood and chemical pulp in properties. The power requirement is about one-half that for regular groundwood. The product is called *chemigroundwood.*

Sulfite Process. The sulfite process is an acid chemical process in which chipped wood is cooked in a liquor consisting of a solution of calcium bisulfite and sulfurous acid that contains a total of about 6 to 7 per cent sulfur dioxide. The cooking liquor is produced at the mill by burning sulfur in specially designed burners to form sulfur dioxide which is absorbed in water and reacted with limestone. During cooking, the sulfur dioxide reacts with the lignin to form soluble calcium lignosulfonic acids.

Cooking is carried out in large metal digesters lined with acid-resisting brick. These digesters hold enough chips to produce 20 to 30 tons of pulp. Wood is chipped by forcing the end of the log against a large rotating metal disc equipped with sharp knives; this produces chips about ¾ inch in length. The chips are loaded into the digester and approximately 700 gallons of cooking liquor added per 1,000 pounds of chips. Then the digester cover is bolted on and live steam introduced. The temperature is raised to about 260 to 300°F. and held there for a period of four to fifteen hours. The time of cooking depends upon the species of wood used, the concentration of cooking liquor, and the grade of pulp desired. During cooking, the pressure continues to rise and the digester must be "relieved" at intervals to maintain a pressure of about 75 to 100 pounds per square inch. The schedule of relieving the digester is one of the most important factors in the control of the cook, since it determines the concentration of sulfur dioxide in solution and controls the temperature.

As soon as the cook is completed, the digester charge is blown into a large tank where the pulp is washed with large quantities of water to remove the dissolved lignin and spent liquor. The yield of pulp is about 45 to 50 per cent of the original weight of wood. Most of this is pure cellulose. The waste liquor, which contains most of the lignin and the remaining components of the wood, is usually run into a stream. This waste of raw material and pollution of water has encouraged research efforts toward use of waste sulfite liquor. Recent work has shown that the waste can be recovered and reused when ammonia or magnesium are used as the base of the liquor in place of calcium. Various useful products, such as alcohol, fodder yeast, and lignin derivatives, can be recovered from the waste liquor by appropriate techniques.

Sulfite is a high-quality pulp. In the unbleached grades it is used in newsprint, catalogue, glassine, paperboard, and wrapping paper. After bleaching with hypochlorite, it is suitable for white bond and ledger papers. In 1949, production was 2,537,084 tons. The price in 1951 was \$132 to \$150 per ton for the unbleached grade and \$135 to \$165 per ton for the bleached grade. The species used for the sulfite process are the long-fibered coniferous woods such as spruce, hemlock, and balsam. Because of their high resin content the pines are not suitable, but in recent years some hardwood sulfite pulp has been produced.

Soda Process. The soda process is an alkaline chemical process, using as the cooking chemical sodium hydroxide, which is made at the mill by reacting sodium carbonate with lime. Unlike the sulfite process, this process permits the recovery and reuse of the waste liquors. These are concentrated by evaporation and then burned in furnaces where the dissolved organic matter in the waste liquor (mostly lignin) is burned and the alkali recovered as sodium carbonate.

The cooking operation is carried out in unlined steel digesters, which are usually smaller than sulfite digesters. The cooking liquor is a 5 per cent solution of sodium hydroxide. Enough of this is added to produce a weight of chemicals about 20 per cent of the weight of the wood. Cooking is done at a temperature of about 340°F. for a period of four to six hours. The yield of soda pulp after bleaching is about 43 per cent of the original wood weight.

Soda pulp is the only pulp which has not increased in production through the years. The process is used only for the pulping of hardwoods, and is the least important of the major pulping processes, but, even so, 491,822 tons of pulp were produced in 1949. The pulp is low in strength and is used in combination with stronger pulps in the manufacture of book and magazine papers. The price of bleached soda pulp was about \$140 per ton in the fall of 1949.

Sulfate Process. The cooking liquor in the sulfate process consists of a 5 per cent solution of alkali of which about 75 per cent is sodium hydroxide and 25 per cent is sodium sulfide. Enough liquor is added to the digester to equal 20 per cent of the weight of wood. Cooking temperature is about the same as the soda process (340°F.). Cooking times vary from one and a half to seven hours depending upon whether a coarse grade or a bleachable grade of pulp is desired. The yield of pulp varies from 45 to 55 per cent of the weight of the wood.

The chemicals are recovered from the waste liquor by evaporating most of the water, adding sodium sulfate to replace lost alkali, and burning the cake that contains inflammable lignin and saponified resins. The heat produced is sufficient to evaporate the water and to supply steam for the digesters.

The sulfate process produces a strong pulp in a short cooking cycle. The process is well suited to the pulping of pine and is widely used in the southern states. The unbleached grades, or "kraft" pulp, are used in wrapping papers, corrugated board, fiber boxes, and paper bags. The pulp is rather difficult to bleach but improved bleaching processes have made available white grades suitable for writing and other high-grade papers. Almost any species can be pulped by the sulfate process.

The sulfate has been the fastest growing pulping process and is now the largest. Total production of sulfate pulp in the United States in 1948 was 5,968,557 tons. The price in 1951 was $135 to $200 per ton for the bleached grades. Southern kraft pulp (southern yellow pine) is generally about $10 per ton cheaper than northern (spruce and jack pine) and imported kraft pulps (spruce and Scotch pine).

Semichemical Process. The semichemical process is a combination chemical and mechanical process. The cooking chemical may be any of the conventional cooking chemicals, although generally a mixture of sodium bicarbonate and sodium sulfite is used. Hardwood chips are furnished to a digester and cooking liquor added at a concentration of about ½ pound per gallon. The temperature of cooking is about 250 to 340°F., and the time varies from two to five hours. After cooking, the chips are discharged to a refiner consisting of one or two rotary metal discs which reduce the chips to a fibrous state (Fig. 52). The yield of usable pulp is 70 to 90 per cent of the original wood weight.

Semichemical is a relatively coarse pulp most of which is used in paperboards. No semichemical pulp was sold on the market in 1950.

Although most semichemical pulps being manufactured in 1951 were of low grade, the process is suitable for manufacturing high-grade pulps when conditions of cooking, refining, and bleaching are properly adjusted. Bleaching may reduce the yield by 10 to 15 per cent, giving a final yield approximating that of the sulfite process. The high-grade semichemical pulps are suited for use in magazines and glassine and for related uses. Refining the partially cooked chips consumes considerable power. At least two passes through the attrition mill are required to produce high-grade semichemical pulp. This method is still being developed and may ultimately attain importance commensurate with other methods now in use.

Whole Wood Fiber. There are other special pulping processes which are used to reduce wood to the fibrous state with little or no chemical. One of the most important of these is the Masonite process, in which wood chips are heated for a very short time in a special "gun" with steam at a pressure of 1,000 pounds per square inch. The chips are then dis-

FIG. 52. A disc refiner for wood pulp and chips.

charged through a quick-opening valve where the explosive action of the entrapped steam reduces the softened chips to a fibrous state. The fibers can then be formed into a loose, porous insulating material or pressed into hard board.

Another whole-fiber pulp is made by the Asplund process. Wood chips are forced into a cylindrical reaction chamber by means of a screw feed. Chemicals and steam are added and after retention up to 20 minutes, the softened chips are fed into a disc refiner. The yield is 90 per cent. The pulp is used for coarse building boards.

Technology of Paper Manufacture

Paper manufacture is a separate operation from pulp manufacture. Pulp received at the paper mill is mixed with water and subjected to beating or refining in special beater or refining machines. During beating the pulp is further subjected to a rubbing or crushing action which "hydrates" and fibrillates the individual fibers. Without this treatment it would be impossible to make strong paper (Fig. 53).

To the pulp in the beater the various materials used in making paper are added. Most papers must be sized so that they will not absorb water or ink. Sizing is accomplished by adding a solution of rosin soap to the

Fig. 53. Sampling pulp from the beater.

pulp followed by a solution of alum which precipitates the rosin on the fibers. Clay may be added to improve the smoothness and increase the opacity, starch to increase strength, and dyestuffs to color paper. If the paper must retain its strength when wet, urea formaldehyde or melamine formaldehyde resins may be introduced.

The thoroughly beaten pulp mixture is diluted to form a suspension of about 1 per cent fiber and pumped to the paper machine. The modern paper machine is one of the largest and most expensive individual machines used in any industry. There are two types, the Fourdrinier for making paper and paperboard and the cylinder for making multi-ply paperboard. The Fourdrinier machine consists of a horizontal, endless wire sieve on which the dilute fiber suspension is flowed (Fig. 54). The water drains through the wire and the fibers are felted and interlocked with each other on top of the wire. The wet fibrous mat is then picked up by a special felt and carried first through a series of presses which compact the paper and press out the water and then over a series of steam-heated dryers which lower the moisture content of the paper to 5 to 7 per cent (Fig. 55). At the end of the machine the paper is passed through a series of highly polished metal rolls, called a calender stack, where the paper is given a final polish. Since all these operations are carried out in continuous sequence from the time the fiber suspension is

FIG. 54. Drawing off a sample of white water from the paper machine.

FIG. 55. Measuring the temperature of a fast-moving sheet on the paper-drying rolls.

first flowed onto the wire, close synchronization of the various paper-machine sections is necessary. Fourdrinier machines operate at speeds of 200 to 2,200 feet per minute. A paper mill may operate anywhere from one to ten machines.

Cylinder machines differ from Fourdrinier machines in that the paper is built up in plies instead of being laid down in a single layer. A cylinder machine consists of a series of vats in each of which is suspended a hollow wire-covered cylinder. The stock is pumped into the vats and a thin layer picked up on the cylinder. The thin layer is then transferred to an endless felt which picks up succeeding layers until a multiple layer of the desired thickness is formed. This paper is then pressed and dried in the same manner as paper made on a Fourdrinier machine. The particular advantage of cylinder machines is that different stock can be used in the various plies.

After leaving the paper machine, the paper may be given one or more special treatments in a converting plant. It may be run through a supercalender, be embossed, surface-sized with glue or starch, or coated with pigment and adhesive.

Technology of Paper-Plastics Combinations

Closely related to the pulp and paper industry is the plastics industry. Paper, in the generic sense, is combined with plastics materials by three principal methods. The first involves the treatment of processed paper in sheet form by saturating or coating resins or both. In the saturation method the paper is brought below the level of the resin bath in a suitable trough and the resin-laden paper then passed between two pressure rolls in order to control the amount of resin in the paper sheet. Both sulfite and sulfate papers are used, the former with "elastic" resins and the latter with fusible thermosetting plastics. Papers treated with elastic resins (*e.g.*, the rubbers) are subsequently coated by conventional knife coaters and used for the manufacture of many commercial objects. Papers treated with fusible, thermosetting resins are generally plied up under heat and pressure to form infusible, hard, paper-resin panels.

The second method is to add a water-soluble resin or emulsion to the paper stock during the beating or refining operations. Thermoplastic resins are used to make flexible board and cover stock, such as artificial leather products. Thermosetting resins are used for paper packaging, electrical moldings, laminates, cold and hot pulp moldings, and the like.

The third method is use of filler pulps and wood flours in thermosetting resin coatings and moldings. Wet or semidry fibers are incorporated into the resin material prior to the fabrication of the plastic. The filler tends to toughen the final molding.

Importance of the Pulp and Paper Industry

The paper industry group is the eleventh largest in the country in value added by manufacture.[6] The importance of the paper industry in the national welfare is even greater, because of the essential part which paper plays in every phase of modern life.

The paper industry has shown remarkable growth, particularly in view of the ancient origin of the industry. The widespread use and increasing consumption can be attributed to the low cost and versatility of paper. No low-cost substitute exists for paper as a wrapping, packaging, printing, and writing material.

Paper consumption moves up and down with the general business cycle, but the long-term trend in paper consumption has been steadily upward. In 1948, the total consumption of paper in the United States was over 26 million tons.[1] This amounted to a per capita consumption of 357 pounds, an increase of 14 pounds over 1947 and an increase of 300 pounds over 1899 (Table 5). Increase in per capita consumption coupled with a rising population spells long-term growth in paper consumption.

TABLE 5. UNITED STATES PRODUCTION OF PAPER AND PAPERBOARD BY GRADES, IN TONS [1]

Grades	1939	1948
Writing paper	594,594	908,890
Other fine paper	128,508	232,812
Book paper	1,534,591	2,338,671
Groundwood printing and specialty papers	567,559	808,110
Newsprint paper	954,259	875,760
Tissue paper	645,223	1,187,739
Wrapping paper and special industrial paper	2,238,993	3,461,121
Building paper and insulation	659,090	1,348,321
All other papers	185,342	108,614
Total paper	7,448,148	11,193,156
Total paperboard and building board	6,025,494	10,773,407
Total paper and paperboard	13,509,642	21,921,757

In 1948 the primary pulp and paper industry employed 198,000 workers in 37 states. Products made of paper and paperboard provided employment to 250,000 additional workers. In addition to mill workers 300,000 woods workers received part or all of their income from pulpwood operation. Moreover, in 1947, 715,000 workers were directly employed in printing and publishing establishments.[6] Pulp production was at an all-time high of 13 million tons and paper production was 22 million tons. Capital investment in the industry stood at approximately 3.5 billion dollars and the value of the products turned out was estimated at about 4 billion.

The paper industry requires heavy investment and large-scale operations to function efficiently. A single, modern paper machine costs in the neighborhood of 1 million dollars, and the construction of a modern pulp and paper mill with a single paper machine costs 20 to 30 million dollars. The paper industry ranks highest among all industries in the capital invested per worker—$18,000 to $19,000. The industry is normally highly competitive.

Raw-material Supply. The basic raw material of the paper industry is wood. In 1948 slightly over 21,000,000 cords of wood were consumed by the paper industry. Of the total wood consumed, 2,244,000 cords, mostly spruce, were obtained from Canada. The paper mills in New York State consume annually about 700,000 cords of pulpwood, nearly half of which is imported from Canada. The average price paid for pulpwood in 1948 was $18.80 per cord; the price of spruce was about $23 per cord and that of yellow pine about $11 per cord. The species used for pulp production vary according to the section. In the southern states nearly 90 per cent of the pulpwood is southern yellow pine. In the North, the principal species are spruce, fir, eastern hemlock, and pine. In the western states, pulpwood is chiefly spruce, true fir, and western hemlock. Hardwood species are less desirable than the softwoods, but the use of hardwoods is increasing because of the diminishing supplies of softwoods. In the North the principal hardwoods used are aspen, birch, beech, and maple. In the South the gums and oaks are used. Pulpwood accounts for about 14 per cent of the forest drain.

Spruce and fir pulpwood in the Northeast and Lake States has been sharply depleted so that paper mills in these states must draw heavily on Canada and Scandinavia for wood and pulp. The United States imports about 83 per cent of the newsprint consumed, mostly from Canada.

For the United States as a whole, ample reserves of woods suitable for pulp exist in the forest. The industry is obliged, however, to process less spruce and more hemlock, fir, pine, and hardwoods. Integration of sawmills and pulp mills, a common practice in Scandinavia, is under way to some extent in the United States.

Domestic pulp consumption for paper in 1948 was 14,331,505 tons, of which about 2,200,000 were imported. Over 7,500,000 tons of waste paper was reprocessed in 1948. Consumption of waste paper has shown a significant decline since the war in contrast to the increasing consumption of virgin pulp. Other fibrous materials used amounted to 1,451,000 tons.[1]

In order to operate effectively, pulp mills must be near their source of wood. Paper mills, on the other hand, are sometimes located near their markets instead of near their source of raw material. Paperboard mills which use waste paper as a large part of their raw material are often located in or near large metropolitan areas in order to be near their supply of waste paper. An essential requirement in the location of both pulp mills and paper mills is an adequate supply of good water. The amount of process fresh water used by a paper mill each day varies from 432,000 to 50,000,000 gallons, depending upon the size of the mill and the type of paper being made. A sulfite or sulfate pulp mill requires in the neighborhood of 92,000 gallons of water daily for each ton of pulp produced, which in the case of a 400-ton mill would result in a daily consumption of about 40,000,000 gallons of water. Chemical supply, labor supply, and good transportation facilities are also important.

World Trade. The paper industry is an important factor in world trade. In 1948, imports of pulp, pulpwood, and newsprint amounted to 753 million dollars. The value of paper exports was only 137 million dollars.

Distribution of Mills and Working Conditions

Paper mills are distributed in fairly well-defined regions of the country. The principal ones are the Lake States, Ohio, eastern Pennsylvania, New York, Massachusetts, Maine, the southern states, and the West Coast states. Concentration facilitates business and professional association. Most mills are located close to their source of wood supply in small communities which are usually chiefly dependent upon the paper and pulp mill for their livelihood. Geographical regions are often associated with the production of particular grades of paper. Paper mills in the South make principally kraft papers, mills in Massachusetts are noted for their rag-content writing papers, mills in Michigan make a high percentage of all book papers, and the mills in New York State make principally groundwood specialties and catalogue papers. The South now accounts for approximately 50 per cent of the wood pulp produced in the United States.

The paper industry is progressive. Although it was one of the last of the major industries to adopt modern scientific methods, once started, it made rapid progress. The industry employs a large number of well-trained chemists and engineers. The paper industry finances a graduate school at Lawrence University, known as the Institute of Paper Chemistry, where men are specially trained for the paper industry. Other schools training men for this industry are the State University of New York, College of Forestry, Massachusetts Institute of Technology, University of Maine, Western Michigan College, and McGill University in Canada, and still other schools have one or more courses in papermaking.

Cellulose and Plastics Industries

Industries that manufacture rayon, cellulose plastics, and cellulose derivatives are closely related to the pulp and paper industries. Much of the alpha cellulose used for dissolving purposes is produced by companies that also manufacture papers, and closely allied techniques are used.

The plastics industries use wood in other forms also. Wood flour and finely ground charcoal are used as extenders in various plastics. Wood fiber or pulp may similarly be used, though the plastic mixture may require preforming because fiber interferes with plastic flow. Lignin is used to a limited extent as an extender and also as a plasticizing agent. Many companies in the wood-fiber and lignin-plastic fields entered this as a side line to paper or pulp-product manufacture.

Other plastics manufacturers use wood flour or wood carbon in ways differing widely from the techniques of pulp and paper manufacture. Small companies may manufacture plastic novelties, machine parts, or household articles. Rubber and many chemical manufactures differ radically from pulp and paper manufacture; yet their technicians may be members of the Technical Association of the Pulp and Paper Industry (TAPPI). TAPPI has a plastics committee, meeting annually, which grew from a small group in 1946 to a membership in excess of 300 by 1950. It serves, therefore, as a strong liaison body between papermakers and plastics manufacturers.

The plastics industry has a bright outlook, as the products, output, and companies involved are expanding rapidly. The industry has places for both large and small companies.

LITERATURE CITED

1. American Paper and Pulp Association. 1949. "The Statistics of Paper," 2d ed., New York. 63 pp.
2. Brauns, F. E. 1948. The Proven Chemistry of Lignin, "Nature of the Chemical Components of Wood," pp. 108–132, TAPPI Monograph Series 6, Technical Association of the Pulp and Paper Industry, New York.
3. Libby, C. Earl, and Frederic W. O'Neil. 1950. "The Manufacture of Chemigroundwood Pulp from Hardwoods," New York State College of Forestry, Syracuse, N.Y. 42 pp.
4. Plunguian, Mark. 1943. "Cellulose Chemistry," Chemical Publishing Company, Inc., Brooklyn, N.Y. 96 pp.
5. Stephenson, J. Newell, ed. 1950. "Pulp and Paper Manufacture," Vol. 1, "Preparation and Treatment of Wood Pulp," McGraw-Hill Book Company, Inc., New York. 1043 pp.
6. U.S. Department of Commerce, Bureau of the Census. 1950. "Census of Manufactures: 1947," Vol. 1, "General Summary," Washington, D.C. 355 pp.
7. West, Clarence J., ed. 1948. "Nature of the Chemical Components of Wood," TAPPI Monograph Series 6, Technical Association of the Pulp and Paper Industry, New York. 234 pp.
8. Wise, Louis E. 1944. "Wood Chemistry," Reinhold Publishing Corporation, New York. 900 pp.

CHAPTER 11. *Secondary Forest Products*

LUMBER AND OTHER sawed timbers are the major product of American forests. Second in importance is pulpwood, which constitutes a drain on the forest of one quarter the volume of sawtimber. Many additional round products are harvested, including poles, piling, posts, mine timbers, chemical wood, and fuel. Together with pulpwood these products require almost as much wood as is cut for sawtimber. Accurate information is not available, since no nationwide records are made of posts, fuel wood, and other products cut for local consumption.

Products of Wood

Crossties, Cooperage, Shingles, and Excelsior. Crossties require about 1.8 billion board feet of wood annually. Despite considerable research both in the United States and Europe, no satisfactory substitute for the wooden crosstie has been developed. During the period 1929 to 1943, 129 ties were replaced annually per mile of track, which contains about 2,000 to 3,000 ties.[8] Estimated annual requirement for the period 1950 to 1955 is 52 million ties per year, somewhat under 1942 usage on a mileage basis. The life of untreated ties is 5 to 8 years, that of creosoted ties 20 to 30 years.[1] About 60 per cent of the crossties used in 1945 were sawed; the remainder were hand-hewn. Oak has long been a favorite wood for crossties and accounts for almost 40 per cent of those in use, but southern pine and Douglas-fir are also widely used. Locally many other species are accepted, such as hemlock, larch, lodgepole pine, gum, hickory, beech, and ash. The heavy crosstie is 7 inches by 9 inches by 8 feet 5 inches and contains 45 board feet. The crosstie market provides a good outlet for small-sized logs of medium quality, provided the wood will hold spikes, is strong enough to withstand the impact of heavily loaded trains passing over the rails, and may be successfully treated with preservative for decay prevention.

The cooperage industry consumes about 775 million board feet annually. Slack cooperage is manufactured of southern pines, western pines, gum, and other species. The barrel or keg consists of staves, head, and hoops. Staves may be cut by a special cylindrical saw, but this is wasteful, as considerable wood is consumed as saw kerf. Larger mills steam the stave bolts and slice the staves from quartered bolts with a heavy knife in a stave cutter.[1] The staves are seasoned, jointed so as to fit closely, and beveled and grooved for the head. Heading stock is cut on special saws, seasoned, surfaced, jointed, laid up, and turned into individual barrelheads. Hoops are mostly of metal but may be made of elm, hickory, or ash. Barrels are assembled at place of use on special forms.

Tight cooperage is manufactured by similar processes. High quality material—white oak is preferred—is used for staves and heading. All machining must be carefully done and final assembly must be accurate to make the finished barrel watertight. White oak suitable for tight cooperage brings prices in excess of those for veneer logs.

Shingles are sawed from carefully selected bolts trimmed to exact length and quartered. The shingle bolt carriage operates so as to cut tapered shingles, the thick end cut from first one, then the other, end of the bolt.[1] Shingles are cut green, trimmed, bailed, and seasoned, and bales are re-pressed and tied. The shingles are then ready for shipping. The preferred shingle wood is western redcedar, but northern white cedar, cypress, and redwood also make durable shingles. Edge-grain western redcedar shingles may last 30 to 40 years. When used as siding on houses, they may have an even longer life. Stained shingles used as siding on the author's residence show little evidence of deterioration after 35 years of use.

The volume of wood used in shingles is about 400 to 500 million board feet annually.[8]

A product of some importance is excelsior, which is used largely for packing and upholstery. Excelsior is made from 36- to 56-inch bolts of aspen, cottonwood, yellow pines, basswood, and yellow-poplar.[5] The wood is dried and shredded by special machines. Only sound, straight-grained wood is suitable for excelsior. Three main grades are manufactured. A cord of 56-inch wood produces about 1 ton of excelsior.

Pulpwood. Most pulpwood is cut by wood-pulp manufacturers operating their own crews and camps or by contractors. Pulpwood is not an active article of trade that is cut and stocked by dealers to be sold upon demand to pulp companies. Farmers and independent cutters may cut and deliver pulpwood to a mill, but rarely do they have more than one mill close enough to provide a ready market. Pulpwood specifications are determined by the mill or mills purchasing wood in a given locality. In the Lake States, wood is commonly cut in 100-inch lengths and is purchased by the double cord. In New York and New England it is cut to

4-foot lengths and sold by the standard cord. In Pennsylvania the 50-inch cord is used and in the South the 60-inch cord.

Pulp companies prefer to buy the wood ready peeled. Hand peeling during spring and early summer gives the cleanest wood. River-driven wood in Maine is usually floated with the bark on. By the time the wood reaches the mill, the bark has already been rubbed off or can be easily detached in a drum or other type of mechanical debarker. Chemical treatment to loosen bark is coming into use in Canada and in parts of the United States. This has the advantage of permitting partial seasoning before felling. Other phases of pulpwood logging have been covered in Chap. 7. Pulpwood requirements are estimated at 1,660 million cubic feet annually, of which 5,784 billion board feet are cut from sawtimber-sized trees.[8]

POLES AND PILING. Poles are used to carry telephone, telegraph, and electric wires, and for limited other uses, such as supports for flood lights and clotheslines and as building material for log cabins. Piling is used for docks, wharves, abutments, piers, for bridge foundations and trestlework, and to prepare foundations for buildings.[1] Specifications for piling usually require that a line from the center of the butt to the center of the top stay within the pile for its entire length. Comparable straightness is desired in poles (Fig. 56).

FIG. 56. Our country uses about 38 million linear feet of piling annually for dock and other construction. (*A.F.P.I. photo.*)

Straight, reasonably soft, cylindrical stems are desired for poles. Western redcedar, northern white cedar, and Atlantic white cedar are favored because of high durability, but southern pine is also extensively used. Most poles are treated with preservatives to above the ground line. It is here that decay hazard is most acute. Piling is often treated for its entire length. Although the same species used for poles may be used for piling, harder and stronger wood is acceptable and may be preferred. All piling must withstand the shock of the pile driver.

Poles are set with the large end in the ground; piles are driven small end first. Both poles and piles are sold by length; 30, 40, 50, 60, and 70 feet are commonly used. Both poles and piling must be peeled. Together they represent an annual drain on our forests estimated to be 452 million board feet annually.[8]

MINE TIMBERS. Round timbers are extensively used in mines to support the roof of drifts and stopes. Heavy timbers are used to carry the weight; small timbers, called lagging, to deck the roofs and line the walls. Flattened timbers are used for mine ties, and special double-pointed 18-inch pieces, called sprags, to lock wheels of mine cars to stop them. Mine timbers are specified as to size and species according to availability and the use intended. Timbers should be strong and stiff and crack to give warning before breaking. Durability is desirable in drifts and shafts but is of minor importance in places requiring protection for only a few weeks while the vein is being worked. Where durability is important, treated timbers may be used. The annual requirement of mine timbers is estimated at 220 million cubic feet of which 312 million board feet are cut from sawtimber-sized trees.[8]

POSTS. Posts are cut of available woods in lengths of 5 to 8 or more feet. Strong and durable woods are preferred, such as cedar, locust, cypress, oak, and tamarack. Eastern redcedar, northern white cedar, Atlantic white cedar, and western redcedar make durable posts. If treated, fence posts can be of any wood that takes treatment well. Posts are used in the round and split in halves and quarters, and most are set in holes. In some eastern states the posts are sharpened and driven in spring just after the frost leaves the ground. Untreated posts of nondurable woods last 3 to 7 years under eastern service conditions; the durability of treated posts is 12 to 20 years, depending on the effectiveness of treatment. About 480 million cubic feet of wood is cut annually for posts, of which 492 million board feet is from sawtimber-sized trees.[8]

FUEL WOOD. Second only to lumber is the wood used for fuel in the United States. The total cut is estimated to be 1,836 million cubic feet of which 3,066 million board feet come from sawtimber-sized trees,[8] but because wood is heavy, bulky, and inconvenient to use as fuel, it is rapidly being displaced by coal, oil, and natural gas. Mechanized handling and manufacture are essential to low cost and competition with other fuels. The heating value of a cord of dry maple wood is equivalent to about that of a ton of coal. As a luxury for fireplaces fuel wood will have a market for years to come. It offers enterprising foresters a chance to develop local markets for low-quality trees that should be removed to favor growth of better trees. Sawmill waste is commonly used for fuel, especially slabs and edgings. Tops and limbs may be used but are laborious to work up.

SHAVINGS AND SAWDUST. Shavings and sawdust are used for stock bedding, fuel, and packing. They may be burned directly or made into briquettes in special machines. Special sawdust burners have been devised for use in western states where sawdust is plentiful. In the dairy region of the Northeast, sawdust and shavings find a ready local market for cattle bedding. They are superior to straw for this purpose because of their high absorptive capacity.

Sawdust, shavings, and chips may be applied directly to fields as a soil amendment. They loosen clay soils, help hold moisture in sands, and break down into humus that improves both sands and clays. Usually nitrogen must be applied with fine wood to support the growth of the organisms that cause decay. Portable chippers are now in operation that can chip limb wood and light branches for application to soil.

Both sawdust and shavings are used for packing articles for shipping where less resilience is needed than that provided by excelsior. If bark-free, they may also be used for manufacture of wood flour and for hydrolysis to make alcohol. The latter use is dependent upon large quantities of waste available at one locality and at a low price. Even so, wood hydrolysis is still to be proved commercially.

Extractives and Derivatives. NAVAL STORES. Turpentine and rosin are obtained by distilling the resinous exudate from the tappings of longleaf and slash pine in the Southeast. The field operation consists of chipping through bark of living trees, hanging resin cups, and delivering crude gum to the still. The industry is restricted to the southern pine region and characterized by many small distilling operations.

Tapping trees for naval-stores production is a relatively slow and laborious process. A fresh streak must be made weekly. Every three or four weeks the gum is "dipped" from the cups in which it is collected and taken to the still for refining. During the Second World War, resin yields were increased by substituting bark chipping for deep chipping and by applying sulfuric acid to the freshly cut streak (Figs. 57 and 58). This keeps the resin canals open and prolongs the flow of resin. Turpentine distillation must be carefully controlled to make sure that the product is uniform and taken off only between proper boiling-point levels. If burning occurs, the quality of the product deteriorates rapidly,

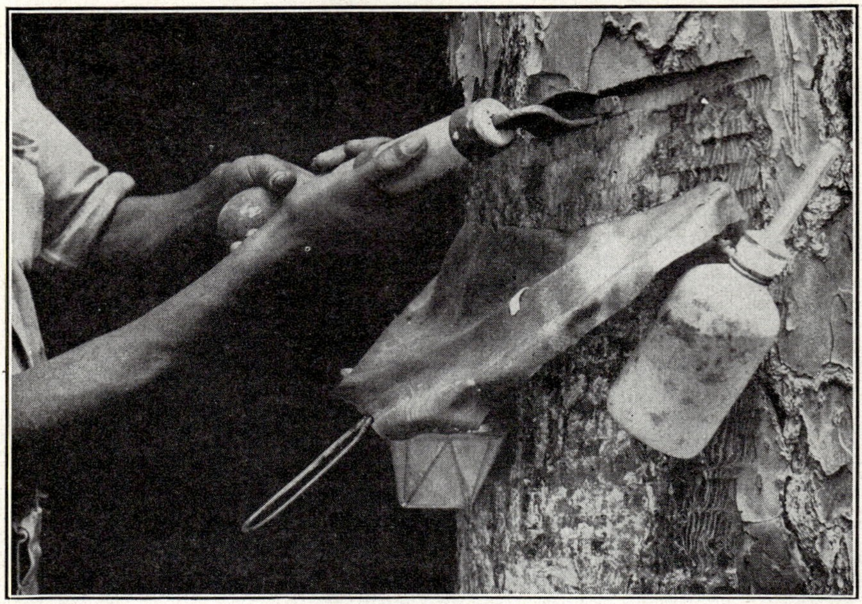

FIG. 57. Bark chipping for resin flow. The resin is collected in the cup beneath the apron. (*U.S. Forest Service photo.*)

FIG. 58. Spraying the freshly chipped streak with acid to stimulate flow of resin. Acid-treated streaks produce almost twice as much resin flow as untreated streaks. (*U.S. Forest Service photo.*)

and the gum should be kept as free as possible of dirt, water, and other foreign material.

Turpentine, rosin, and pine oil are extracted with hot gasoline from chips of the stump wood of longleaf and slash pine. This process is called "steam and solvent extraction." The operations are on a large scale with elaborate equipment and trained men. Continual search is made for new products and ways to reduce costs. The products are made to exact specifications. Although turpentine has long been the most important solvent used in paint manufacture, it has been losing ground of late to other organic solvents. Rosin is the most effective material for sizing in paper manufacture, and is widely used in paint, varnish, and soap manufacture (Fig. 59).

Turpentine and rosin may also be recovered from the sulfate pulping process. This method is used considerably in Europe where gum rosin is difficult to harvest because of low yield from native pines. In the pulping process the wood is chipped and pulped, and the saponified resins and oils which rise to the top of the digester are skimmed off. These skimmings are acidified to yield "tall oil" having about 45 per cent resin acids. These, in turn, are refined by distillation, extraction, or by special

FIG. 59. Selective cupping for naval stores, Olustee Experimental Forest. The first 10-year cycle is now in its fourth year. After 10 years of cupping, the worked trees will be harvested and a second cycle begun on the trees marked with an *X*. Seed trees unworked will still remain. (*U.S. Forest Service photo.*)

chemical treatment to yield rosin, fatty acids, and sterols. Output of naval stores in the United States for the year 1947–1948 was 294,028 barrels of turpentine and 828,128 drums of rosin. The total income of the industry was 39 million dollars.[11]

OTHER WOOD EXTRACTIVES. Other extractive substances in wood are tannins, various phenolic constituents, and various dyes and coloring matters.[12] The phenolic and other resin compounds in wood may be used to a limited extent in identifying species.

Tannins occur in a great number of woody and herbaceous plants. The chief sources of tannins used in the United States are quebracho imported from South America, chestnut and hemlock from the United States, mangrove from Borneo, and wattle from South Africa.[3] Shortage of tannin materials during the war caused the United States considerable concern, and special efforts were made to harvest sumac leaves, scrub oaks, and other products for tannin. The barks of red spruce and Douglas-fir can also be used as sources of tannin. The tannins vary considerably in their nature and also according to the method of extracting them from the wood. Usually tannins are obtained by steeping the chipped wood or bark in hot water.

Various phenolic substances are found in wood, but these normally occur in low concentrations, 0.5 per cent or less. The phenolics, nevertheless, have considerable influence on the durability of wood, for they are toxic to wood-destroying organisms. Tannins serve in like manner to discourage wood-destroying organisms.

Woods yield a number of natural coloring materials or dye substances. These belong to four major chemical groups: the quinones, hydroquinones, benzopyrans, and chromones.[6] Maclurin is a pale yellow dye material found in mulberry and in Osage orange. Walnut hulls contain a material, juglone, that produces a dark dye for wool and silk. Various tropical woods have been used as sources of dye materials, among them the greenheart woods of the Bignoniaceae, Caesalpiniaceae, and the Lauraceae, *Haematoxylon*, sumac, and others. The chemistry of natural dyestuffs is complicated.

WOOD DISTILLATION. When wood is heated with a limited amount of air, it breaks down into a number of products, among them methyl alcohol, acetic acid, various phenols which are found in wood tar, and other products. Charcoal is the end product from the wood itself. The gases formed upon heating wood are combustible and may be converted into liquid fuels.[4, 13] Wood gas was used extensively by the Germans, Swedes, and other nations during the Second World War when they had limited amounts of petroleum fuels. It remains a potential source of fuel wherever petroleum and coal are unavailable.

The chief product of wood distillation, charcoal, must be made of wood. Charcoal is used for domestic cooking and for broiling in restaurants and dining cars. It is often used on shipboard for cooking and is the main cooking fuel of the tropics. Charcoal has limited use in the steel industry for manufacturing certain grades of high-carbon steel. Because it produces a reducing flame, charcoal is used for heating soldering irons used by tinsmiths. For the same reason charcoal blankets—fine charcoal treated with a fire retardant—have been used by the steel industry to cover the molten metal while it is being poured. Activated carbon is made by treating fine charcoal with superheated steam. This has far greater adsorptive capacity than ordinary charcoal and hence is widely used in gas masks, in city water purification, and for other purposes where high adsorptive properties are required. Charcoal is also used for curing tobacco.

Wood-chemical plants can use top and limb wood unsuited for lumber or pulp manufacture and can use heavy slabs from sawmills.

The wood-distillation industry achieved substantial prosperity between 1910 and 1920 but declined rapidly between 1920 and 1940. A sharp but short-lived revival occurred from 1940 to 1945 and an almost equally sharp decline followed. Many plants of the Northeast closed and others had ceased to recover products other than charcoal by 1950. A market for charcoal seems likely to continue indefinitely, although the industry at present is one of waning importance. Coke now replaces charcoal in most steel furnaces, and synthetic methanol, acetic acid, and acetone are made more cheaply and efficiently from natural gas and petroleum than from wood. However, if new and cheaper methods of harvesting and distilling can be discovered, the industry may again become important.

Very little research has been done on wood pyrolysis. The original wood-chemical industry is a relatively simple one and has never been sufficiently profitable so that companies could accumulate financial reserves for research to keep the industry abreast of modern developments. Had as much effort been spent on studying the by-products from wood distillation as has been spent on the by-products of coking of coal, it is probable that wood pyrolysis would be a flourishing industry in our country today.[4]

Other Forest Products

Maple Products. Maple products bring in a substantial income to farmers and others who engage in tapping the trees and refining the sap. The capital required to refine sap was about $2,000 in 1950, and a minimum-size operation is one of about 500 pails.

Production of maple syrup in the United States attained an all-time high of 6,612,000 gallons in 1860. This level was approximated in 1918, when production was 6,564,000 gallons. Since this time production has dropped to 2,500,000 gallons for the 10-year period 1933 to 1942 and 2,250,000 gallons for the 10-year period 1938 to 1947.[2] The price of maple syrup in recent years has averaged from $1.18 per gallon in 1933 to $5.10 per gallon in 1947. Retail prices were as high as $6 to $8 per gallon during the period 1945 to 1950. Vermont led all states in 1949 with output equivalent to 4,627,000 pounds of sugar. It was followed closely by New York with 4,332,000 pounds. Other leading states were Ohio, Michigan, Pennsylvania, Wisconsin, Massachusetts, New Hampshire, Maryland, and Maine.[14] Production is carried out mainly by farmers as an early spring activity before cultivation for field crops begins. Even in communities where "sugaring" is common practice, only a small percentage of suitable trees are tapped.

The average sugar concentration of sap is about 2 per cent. This means that 43 gallons of sap are required to produce 1 gallon of syrup weighing 11 pounds. In a normal season about ⅓ gallon of syrup may be realized for each pail hung. An ideal sugar bush should have no more than 50 well-spaced trees per acre. The crowns should be large and have full access to sunlight. It is desirable to have some trees of all ages so that young trees can replace the old ones as they become defective. Trees 10 inches in diameter are large enough to tap. These support one bucket; trees 16 to 20 inches, two buckets; trees 21 to 25, three buckets; trees 26 inches and over, four buckets (Fig. 60). A special sap house and evaporator are essential if good-quality syrup is to be produced.

The main cost of manufacture is for labor in tapping trees, in collecting sap, and in tending the evaporator. Off-season work includes cleaning equipment and harvesting fuel. A cord of dry fuel is required for each 20 to 25 gallons of syrup produced.

Maple sugar is produced by continued boiling of the sap until crystallization will occur upon cooling. Special care is required in handling the sap to make sure that fermentation does not occur in the pails, pipe lines, or in the sap storage tanks, and evaporation must be vigilantly watched to avoid burning. Considerable skill is required to produce the best quality syrup.

Maple syrup is strictly a luxury commodity selling for four or five times the price of cane-sugar syrups. Pure maple syrup is used to flavor other syrups and to flavor confections, ice cream, and other products, and lower grades of syrup are used in curing tobacco.

Christmas Trees and Greenery. It is estimated that approximately 28 million Christmas trees are used annually in the United States. Over 21 million of these are cut in the United States and more than 5 million are imported.[7] The northeastern and Middle Atlantic states provide about

6.5 million trees, the Lake States over 5 million, the southern states about 3 million, and the Pacific Coast and Northwest over 6 million trees. About 1.5 million trees are cut annually from established plantations and the remainder from natural stands. The most common sizes of Christmas trees are 5 to 7 feet, though they range from 2 to 20 feet or more.

The value of the industry in the United States is from 20 to 50 million dollars, depending upon whether wholesale or retail prices are considered. In some states Christmas-tree cutting begins as early as October. At least one company in Minnesota harvests trees almost the year around,

FIG. 60. Sap pail on a sugar maple tree, Paul Smith Experimental Forest. Maple sugar and maple syrup represent an important forest industry in the Northeast. Note how the spile is inserted in the tree so that it slopes downward and how the pail is attached by a short hook to the spile. Pails are frequently covered to shelter them from rain and snow and to prevent undue accumulation of dirt. Note the three holes at the left of the pail where spiles had been inserted in other years.

placing the trees in cold storage as soon as possible after cutting. One man can cut and yard about 200 trees a day.

Plantations for Christmas trees are often pruned to produce well-formed trees. Trees are normally set about 3 by 3 or 4 by 4 feet so that from 2,000 to 4,000 trees per acre are possible. Many growers have found Christmas-tree plantations a very profitable form of land use, yielding returns of $10 or more per acre annually. Under sustained-yield production a grower could count on no more than 100 trees per acre annually. Favored species in the United States are balsam fir, Douglas-fir, black spruce, redcedar, and white spruce.

Holly for Christmas wreaths, mistletoe, laurel, and other evergreen material are valuable for decorative purposes. The exploitation of Christmas greenery other than Christmas trees is almost exclusively in the hands of people who understand how to harvest and fabricate the material for sale but who are not particularly skilled in perpetuating the supply of growing material. The state of Maryland has undertaken to control the harvesting of holly.

Various decorative materials are harvested from the forest at seasons of the year other than Christmas. Wood ferns are extensively used in some parts of the country with all types of floral decorations. Ground pine and other club mosses are used both for Christmas wreaths and to form chains or ropes for weddings. Evergreen branches are extensively used for decorations throughout the winter period, and pussy willow, red osier dogwood and other tree branches in early spring. Flowering dogwood is especially sought in late spring when the floral bracts are in full display but its collection in many states is prohibited. Redbud, cherry, and peach boughs are also used during the season they are in flower.

Tree Seed. The harvesting, extracting, and storage of tree seed is becoming an increasingly important business in the United States. Vast quantities of seed are used by state and Federal nurseries engaged in tree-planting programs and by private nurseries as well. As seed for forest planting must be of high quality and be collected from parent trees adapted to the climate in which the tree seedlings are to be set out, great care is needed in seed-collecting operations. The U.S. Department of Agriculture makes it a policy to use only seed of known origin in all plantations on national forests and on other plantings the department sponsors. Suppliers must provide adequate evidence to verify place and year of collection before seeds are purchased by public nurseries. Local seed is to be used whenever available. Local seed is defined as that collected from an area similar in climatic influences and within 100 miles of the planting site and at an elevation differing no more than 1,000 feet from that of the planting site. When local seed is not available, seed is to be obtained from regions having as nearly as possible the same climate and same altitude and latitude.[10]

Private forest planting has increased very rapidly during the last decade under the stimulus of the "tree-farm" program. Opportunities, therefore, are gradually increasing for private tree-seed dealers. A tree-seed collector must know his territory, he must have contacts with nurserymen who will buy seed, and he must be in a position to obtain rapidly a considerable field force to engage in seed-collecting activities. Seed collecting is a highly seasonal business and must be done at exactly the right time—after the seed is ripe, but before it has been dispersed.

Seed of conifers is obtained by collecting the cones and extracting the seed from them. Extraction may be carried on in a heated building, but best results are obtained through the use of a carefully controlled cone kiln. After extraction, the seed must be dewinged and cleaned. These are delicate operations which must be gently performed to avoid injuring the seed. Most conifer seed keeps best if dried to appropriate moisture content and stored in sealed containers at low temperature.

Seeds borne in fleshy fruits must be extracted by maceration in water and the washing away of the pulp. Walnuts can be run through a corn husker. Final cleaning can be done by screening and fanning, using commercial fanning mills equipped with suitable screens.

Fruits and Nuts. Many wild fruits and nuts enjoy good local and even regional markets. Black walnuts, hickory nuts, butternuts, pecans, hazelnuts, piñon pine, and various other edible nuts are collected and sold locally for human food. Beechnuts, acorns, and other nutritious fruits and nuts serve as food for wildlife and are often used by domestic animals, particularly swine. Locally, they form an important resource of the forest.

The forest also produces a variety of berries and fruits useful for human food. Many have a good local or regional market. Among these are blueberries, cranberries, persimmons, wild raspberries, strawberries, blackberries, and cherries. Little attempt is made to cultivate or control the harvest of products other than cranberries and blueberries. Cultivating these crops is a highly technical agricultural pursuit that requires great attention to detail for success.

Medicinal Herbs and Pharmaceuticals. In the past the forest has furnished a number of plants used for pharmaceutical purposes. Among these have been wild ginseng, sarsaparilla, yellowroot, and a host of others. The medicinal drugs from these products are now produced, however, almost entirely by drug companies through synthetic processes or from cultivated plants. Only a limited market exists for wild medicinal herbs.

Oil of wintergreen and other flavoring extracts may be obtained from the twigs and leaves of various forest plants. Oil of wintergreen is now made almost exclusively from black birch twigs by steam distillation. Various other waxes, oils, and flavoring materials are obtained from the forest and enter into commercial use. Among them is the gum of the balsam, which is used as a cementing material in the optical industry and for the mounting of permanent microscope slides.

Tropical-forest Products. Cinchona bark produces the alkaloid quinine and closely related products useful in the prevention and cure of malaria and other fevers.

Rubber, chicle, dyestuffs, and tanning materials are still imported into the United States from tropical and subtropical regions, and restricted imports during the war period caused considerable concern. Efforts have been made to grow cork oak in California, cinchona in Costa Rica, and dyestuffs in Puerto Rico.

The synthetic rubber industry has developed sufficiently today so that our economy is not likely to be seriously disrupted should the source of natural rubber be cut off. Natural rubber, however, is still used for a great variety of products for which it is better suited than modern synthetic rubbers. Attempts to produce natural rubber from guayule and Koksaghys were successful during the Second World War but highly expensive.

Waste in Secondary Products. Little usable waste occurs when trees are cut for round products such as poles, piling, posts, pulpwood, fuel, chemical wood and excelsior. However, crooked, limby trees are not well suited for economical manufacture of any of these products.

Hand hewing of crossties is especially wasteful, averaging 4.3 cubic feet per tie. The material cut away might be cut into usable lumber in many cases. As much as 83.2 million cubic feet of wood is wasted annually in tie cutting.[9] The cooperage industry creates waste estimated at 67.1 million cubic feet annually and the shingle industry 19.3 million cubic feet.

LITERATURE CITED

1. Brown, Nelson C. 1950. "Forest Products," John Wiley & Sons, Inc., New York. 399 pp.
2. Cope, J. A. 1949. "Maple Sirup and Sugar," Cornell Extension Bulletin 397, Ithaca, N.Y. 32 pp.
3. Flinn, Edwin S. 1948. Wood Tannin, "Nature of the Chemical Components of Wood," pp. 146–166, TAPPI Monograph Series 6, Technical Association of the Pulp and Paper Industry, New York.
4. Gillet, Alfred. 1949. "The Future of Wood Pyrolysis," pp. 45–51, Fourth

Meeting of the Technical Committee on Wood Chemistry, Food and Agriculture Organization of the United Nations, Brussels, Belgium.

5. Panshin, A. J., E. S. Harrar, W. J. Baker, and P. B. Proctor. 1950. "Forest Products—Their Sources, Production, and Utilization," McGraw-Hill Book Company, Inc., New York. 549 pp.
6. Shriner, Ralph. 1948. A Review of Natural Coloring Matters in Wood, "Nature of the Chemical Components of Wood," TAPPI Monograph Series 6, pp. 182–219, Technical Association of the Pulp and Paper Industry, New York.
7. Sowder, Arthur M. 1949. Christmas Trees—The Industry, "Trees, The Yearbook of Agriculture," pp. 248–251, U.S. Department of Agriculture, Washington, D.C.
8. U.S. Department of Agriculture, Forest Service. 1946. Potential Requirements for Timber Products in the United States, Report 2, from "A Reappraisal of the Forest Situation," Washington, D.C. 70 pp.
9. U.S. Department of Agriculture, Forest Service. 1947. Wood Waste in the United States, Report 4, from "A Reappraisal of the Forest Situation," Washington, D.C. 45 pp.
10. U.S. Department of Agriculture, Forest Service. 1948. "Woody-plant Seed Manual," Miscellaneous Publication 654, Washington, D.C. 416 pp.
11. Ward, Jay. 1949. Naval Stores: The Industry, "Trees, The Yearbook of Agriculture," pp. 286–291, U.S. Department of Agriculture, Washington, D.C.
12. West, Clarence J., ed. 1948. "Nature of the Chemical Components of Wood," TAPPI Monograph Series 6, Technical Association of the Pulp and Paper Industry, New York. 234 pp.
13. Wise, Louis E. 1944. "Wood Chemistry," Reinhold Publishing Corporation, New York. 900 pp.
14. Woods, John B. 1949. Our New Maple Sugar Crop, *American Forests,* 55(6):21, 30.

CHAPTER 12. *Wildlife and Range Management*

SINCE COLONIAL DAYS the American public has cherished the right to hunt, fish, and trap. Perhaps this more than any one feature of modern-day life gives us a feeling of kinship with the pioneers who settled our land. Physical hardihood is required to tramp all day with rod or gun. Skill is needed to stalk a game animal, bring down a bird on the wing, or cast a fly so that it floats naturally across the center of a small pool. The ability to find and take fish and game develops a certain self-assurance in the individual.

Rich and poor men alike enjoy hunting and fishing. Success depends far more on skill, patience, and physical vigor than on expensive equipment. These sports, therefore, evoke a feeling of equality among all devotees. Even the humblest citizen may find himself the center of admiration if he excels with rod or gun.

For several decades the only measures taken to promote the welfare of wildlife were legislative acts to regulate hunting, fishing, and trapping. Decade by decade these became more restrictive while the hunters and fishermen increased in numbers and in dissatisfaction over the scarcity of game. Thus the stage was set for scientific wildlife management.

What Is Meant by Wildlife and Wildlife Management

Wildlife includes all animals that make their homes in forest, field, and wasteland and that obtain the necessities of life from the natural environment without substantial aid from man at any time during their life cycle. Wildlife does not include domesticated animals that must be fed by man or cared for during at least a part of their lives. Nor does it include man's pests the house mouse and brown rat that gain their livelihood from the food man grows to feed himself and his domestic animals. Wildlife is a much broader term than game, as it includes such varied creatures as hawks, ospreys, songbirds, shrews, red squirrels, the fur-bearing animals, porcupines, and woodchucks.

Fish may or may not be included as wildlife, but the wildlife manager is often called upon to deal with fish as well as with land animals. The term wildlife has a further connotation. Animals kept in a zoo are not wildlife but caged animals; neither are penned foxes on a fur farm strictly wildlife. To be wildlife the animals must be free to move about in their natural environment and to obtain from their environment all their life requirements. If they are dependent upon man for one or more of their life requirements, they become to that extent domesticated. Likewise, if the animals are maintained within an enclosure and not free to move about, they become semidomesticated.

Wildlife management covers those activities undertaken to increase the value of wildlife to mankind. These activities are similar in purpose, but radically different in technique, from those the farmer engages in to care for herds and flocks. The farmer directly controls his breeding stock, feeds his herds, and confines his animals so that they can be easily cared for. The wildlife manager accomplishes his ends primarily by indirect means. He devotes his main efforts to the sustained production and control of game, fish, fur bearers, and other wild vertebrate animals in their natural environment. Wildlife management serves the interests of fishermen and hunters, of cameramen and recreationists, of trappers and fur dealers, of those who profit directly from the sale of hunting rights, sporting goods, cameras, and binoculars, and of those engaged in the tourist trade. Wildlife management is of concern to forest and range-land managers because of the interrelationships, both beneficial and harmful, between the wild animals and the commercial resources of forest and range.

Values of Wildlife. It is very difficult to assign direct monetary values to wildlife. Opinions vary widely, from those who declare that the monetary value of wildlife is negligible to those who maintain that wildlife performs priceless services to mankind without which we would soon be overwhelmed by destructive insect pests. We find that wherever man lives, he affects the wildlife that inhabit the land. He takes land for his cultivated crops, his factories, his homes, his cities, his highways and railways. His rapidly moving vehicles destroy wild creatures that get in his path. Where mankind presses the heaviest on the soil, in such regions as India and China, he also presses the heaviest on the wildlife. In no part of the world, however, has man's effect on wildlife been so destructive as to eliminate man himself from the environment. It is where man first destroyed the productivity of the soil that man and wild creatures both have been driven off the land.

Professor R. T. King defines wildlife values under the following seven headings: *

* From special, mimeographed lecture notes of R. T. King.

1. COMMERCIAL VALUES. Income derived from the sale of wild animals or their products or from direct and controlled use of wild animals and their progeny. This results in either the destruction of the animals or their transformation from a wild state to a domestic or semidomestic state, *e.g.*, commercial fishes and all their products, furs, fur and game farming, game as food, domestication.

2. RECREATIONAL VALUES. Moneys expended in the pursuit of wildlife in connection with sports and hobbies such as hunting, fishing, hiking, touring, camping (to the extent that these last three are based on the attractions of wildlife), collecting (nonscientific), and photography (as a hobby). This includes sums spent for equipment, supplies, wearing apparel, license fees, transportation, provisions, board and lodging, guide service, etc. It may or may not result in the destruction of the animals.

3. BIOLOGICAL VALUES. The worth of the services rendered to man by wild animals, *e.g.*, insect and rodent control; soil formation, enrichment, tillage, and aeration; water conservation, sanitation, suppression of diseases, recovery and conversion of materials not otherwise recoverable, pollination, etc. Some of these services could be performed by man in the absence of animals but only by increasing operating costs; for others man is totally dependent on wild animals.

4. SOCIAL VALUES. Values accruing to the community as a result of the presence of wild animals, *e.g.*, increased opportunity for wholesome and economical outdoor recreation, hobbies, and adventure; utilization of leisure time; enhancement of interest in surroundings and required activities; increased real-estate values; income from otherwise idle lands; increased farm income; offsetting carrying charges incurred in other connections; creation of marketing possibilities for minor products; alleviation of monotony; and increased physical and mental health.

5. ESTHETIC VALUES. Values of objects and places possessing beauty, affording inspiration, contributing to the arts through music, poetry, literature, and painting, and possessing historical and patriotic significance. (In these last-mentioned respects native species of wild animals in their natural surroundings are similar to sites of physiographic, political, military, and biographical interest.) These values are largely purely personal but are, nevertheless, of vital concern to practically everyone spending any time in the out-of-doors. They are also the values that induce a goodly number to become interested in our natural resources and active in their conservation.

6. SCIENTIFIC VALUES. Values realized through the use of wild animals and their populations as a means for investigating certain fundamental and widespread natural phenomena that may affect man's interests either directly or indirectly, *e.g.*, population behavior, diseases and their spread, varying virulence of pathogens, certain aspects of nutrition and reproduc-

tion, ecological relations, population dispersal, phenological phenomena, speciation, social organization, etc. These values are of particular interest to the ecologist, pathologist, epidemiologist, and sociologist, and, indirectly, to all who benefit by their worth.

7. NEGATIVE VALUES. Value of property destroyed or damaged by wild animals and costs incurred in efforts to limit or reduce wild-animal populations and to prevent their access to or use of crops, stored products, and structures. This includes losses to growing and stored crops, such as hay, grains, fruits, vegetables, to livestock on farms, to grasses and livestock on the range, to standing, stored, and manufactured timber products, to processed and stored foods, and to manufactured goods and articles; damages to buildings, rights of way, dikes, levees, irrigation and drainage ditches, terraces, and other structures; losses from diseases harbored or transmitted by wild animals; and costs of protecting, repelling, poisoning, trapping, and other means of control.

The sums of all these several values is the total value to mankind of wildlife and is the justification for wildlife conservation. That some of these values run into substantial sums is evident from the fact that the Federal government collected in 1948 over $11,000,000 through an 11 per cent tax on sporting arms and ammunition. The total collected from 1938 to 1948 was $48,175,000.[4]

Attitudes toward Wildlife. Just as men differ in their appraisal of the value of wildlife, so also do they differ in their attitude toward wildlife. Some have no interest in wildlife nor do they feel that wildlife in any way affects their welfare. Others oppose any use of wildlife by mankind. Many people interested in wildlife are concerned with restricted use, that is, fish to be taken by the rod and game to be taken by the gun. Conservationists recognize and appreciate all of the values of wildlife, but they may or may not oppose direct efforts by man to manage wildlife in his own interests.

The wildlife manager is concerned with all values of wildlife and with the proper balance between wildlife and other resources present in its natural environment. Those who would burn the forest to provide food for deer take a very narrow view indeed of wildlife management. Those who would maintain pure stands of spruce over extensive areas where hardwood and other mixed species tend naturally to grow likewise have a limited understanding and appreciation of the importance of multiple use of forest land. Very little land in our country produces its highest over-all returns as deer pasture or as a pulpwood forest. Balanced and integrated use of the wildlife, forest, soil, water, and recreational resources of the land ordinarily yields the highest long-term return for the benefit of mankind.

Measures to Conserve Wildlife. When the pioneers first landed on the shores of North America, they found abundant wildlife, which they took with little thought for the future. Gradually, as our country became settled, wildlife diminished. Both sportsmen and others attributed this decline in game to the toll of hunters. To check depletion, legislatures introduced closed seasons, prohibited market hunting, and established bag limits in the hope that this would protect and restore wildlife, but these measures proved only partially effective.

Certain species were removed from the list of game animals. First were the songbirds and other wild creatures of benefit to mankind but of little use as game. Species approaching extinction were completely protected. Special regulations were introduced restricting the methods that might be used in taking wildlife. Baiting game and the use of live decoys in the taking of waterfowl were prohibited. Finally, the restrictions have limited the hours of the day as well as the days of the week on which wildlife might be taken.

Men sought to control predators. Sportsmen have burned up a good deal of ammunition in shooting crows, hawks, owls, and other predatory birds and animals. Men's understanding of what constituted a harmful predator or "vermin" varied widely. Some sought to destroy all hawks, owls, and eagles; some, all fish-eating birds, such as herons, pelicans, and kingfishers; some, the carnivorous animals, such as bears, wolves, and coyotes; others, fur-bearing animals, such as foxes, weasels, minks, and skunks; and still others, rodents, such as woodchucks, porcupines, and squirrels. All these were believed to destroy game or some things valuable to man. Unregulated and unplanned destruction of predators has in the opinion of many biologists caused at least as much harm as good.[5] They feel that the predators perform a useful function in eliminating the weak and diseased animals and in harvesting the natural increase above that which the environment can support.

Wildlife refuges were established to conserve wildlife. The original idea was to have a few large refuges which could serve as sanctuary and provide a reserve stock of wildlife to replenish depleted territory. Such measures have been most effective when the refuges are relatively small in size, many in number, and interspersed with land used for other purposes where the creatures can find foods that do not occur in the refuge. Modern refuge programs embrace sound management of land, including forests, range, and croplands, for the purpose of protecting and increasing native wildlife. Surplus animals are removed by regulated hunting or live trapping to restock other areas.

Artificial replenishment of wildlife has also been attempted. This has been widely practiced in the stocking of trout streams and in the rearing of pheasants and other wildlife for release on natural ranges. Foreign species have also been introduced. These measures have provided sport for many individuals who might otherwise have been unable to take game.

All above-named measures have been included in the wildlife managers' techniques. Wildlife management encompasses all these and more. Productivity of wildlife depends upon the breeding habits of the animals and the productivity of their environment. Man cannot manipulate animal habits, but he can manipulate the environment.

Behavior of Species and Populations. Species of animals differ widely in their breeding habits, feeding habits, cruising radius, and gregariousness. Differences in these characteristics cause the variation found in range and distribution of species, densities tolerated, hunting pressures withstood, and the influence animals have individually and collectively on their environment. Some animals bear only one or two young annually; others may bear up to 10 or more. The ruffed grouse commonly lays 8 to 14 eggs in a nest; the loon only 2. The white-footed mouse may bear 4 to 8 litters of 6 to 8 mice each during a year. The cottontail rabbit may produce 3 to 4 litters of 4 to 5 young each in a year. The bear produces 1 to 2 young every other year.

The distribution of animals in a given area naturally will depend upon how selective they are in their feeding. Animals that eat a wide variety of foods may find their life requirements within a relatively narrow area and consequently abundant, while those selective in feeding will find them sparse. The ruffed grouse, being quite selective in its feeding, never occurs in dense flocks. It is nongregarious, and up to 5 acres are required to support a single pair of birds. Other birds, such as wild ducks, geese, blackbirds, and starlings, can be seen in flocks.

Cruising radius is the distance an animal or bird will travel to meet its daily life requirements. All factors essential for its welfare must be found within its daily cruising radius, else the animal will leave the area. It is for this reason that large areas devoted to a single crop, whether this be wheat, potatoes, or spruce, tend to produce a small amount of wildlife, whereas those that are highly diversified with woodland interspersed with pasture, clean-tilled crops, and waste areas tend to support dense wildlife populations.

Animal cruising radii for a few species have been determined as follows:

Ring-necked pheasant	2½	miles
Ruffed grouse	⅝	"
Bobwhite quail	¼	"
Cottontail rabbit	¼	"
White-tailed deer	7	"

Some animals tend to exhibit group behavior that is not readily discernible in the individual animals. For instance, there must be limits of density of population in both directions. The animals must be sufficiently numerous to find mates but not so dense as to overstock their range. A population must be of a certain minimum size and occupy certain minimum range if it is to survive. This is particularly true of such animals as the Rocky Mountain sheep, that feed in bands and that require fairly large areas to maintain themselves. Some animals have special tolerances for others of different species and for various factors of the environment. Others are limited in their tolerance and will drive out populations of other animals or leave themselves. The population should have a proper sex and age-class distribution if it is to remain vigorous and healthy. Some populations can readily be transplanted to a new location, whereas others can establish themselves in a new environment only with difficulty. All these considerations influence what the game manager can do successfully to promote the welfare of wildlife.

Range Essentials. A wildlife range must provide all the essentials for the life processes of wildlife. It must contain food, coverts, and water sufficient for all species concerned, for both sexes, for all age classes of the various species, and for all activities and life functions of these species. For example, it is not sufficient simply to have plenty of young sprouts as food for deer. There must also be well-distributed water and particularly water available during the period when fawns are born so that the does need not travel far from the newborn fawns to find food and water (Fig. 61).

All essentials of the wildlife range must likewise be well distributed over the range and with the essentials sufficiently close to one another so that an animal can meet its life requirement within its cruising radius. This naturally varies between adults and young and between nursing females and mature males.

Factors that tend to restrict wildlife population are rigorous winters, unseasonable cold or rainy weather—especially during the period when young are born—prolonged drought, failures of nut, fruit, and forage, accidents, disease outbreaks, predators, parasites, pressure of domestic animals, land clearing for cultivation, swamp drainage, forest fires, and hunting. Man has probably done just as much, if not more, to restrict wildlife through his appropriation of wildlife range for his crops, domestic animals, roads, dwellings, and industries and by the starting of forest fires and other acts of carelessness, than by all his hunting and trapping. Wildlife is favored by abundant food, water, coverts, and such special requirements as salt for herbivores, gravel for gallinaceous birds, and hibernating places. Not only must all necessities for life be present in the environment, but they must be within reach of the animal during

FIG. 61. Deer frequent openings in the forest to feed on tree sprouts, shrubs, and herbs. Huntington Forest, N.Y.

his normal daily travel. Interspersion of food, water, and cover is essential to abundant wildlife population, just as interspersion of human food, water, raw materials, power and markets are essential to dense human population [7] (Fig. 62).

Developing a Wildlife Management Program. Management of an area for wildlife, like management for timber, requires an inventory of resources. This means determining the type and populations of wildlife present and the condition of the animals. It is equally important to survey the food and cover available to support them. Next, measures are considered to improve the environment for use by wildlife. Such measures might include logging, if the area is composed primarily of mature timber; planting, if there is an excess of open land; creation of coverts through encouraging growth of shrubs and leaving large tops on the ground; and developing water supply.

If he is to avoid shortsighted measures, the wildlife manager must have a certain philosophy. All wildlife must be included within his sphere, not just game alone; he is managing environment primarily; and he must keep in mind the interests of all concerned with wildlife and the land supporting it. Man's purpose is served best if many types of animals are present where they can be seen, enjoyed, and studied. Animals, like timber, can and should be managed on a sustained-yield basis. Wildlife management can yield a certain income to the landowner which may help to offset carrying charges. In turn, other forest products can be harvested from lands managed primarily for game, thereby offsetting in part the cost of managing the game. Wildlife management cannot be divorced from administration of the land on which it occurs. It is but one of many products of the land of value to the landowner and the community. Finally, to develop the art of wildlife management high technical training is required.

PUBLIC POLICIES AND PROGRAMS. Migratory birds, including waterfowl, are under Federal jurisdiction. All other wildlife is considered the property of the state in which it is located, because wildlife can freely move from one man's land to another's. The owner of the land does not have the right to take wildlife as he sees fit on his own property; he must conform to the game laws established in his state. The only exception is that Indians may take fish and game at any time on reservations. Wildlife on national forests is likewise considered the property of the state

FIG. 62. A fox den, Huntington Wildlife Forest Preserve, N.Y.

in which the forest is located. The courts have ruled, however, that the U.S. Forest Service does have the right to control wildlife on the national forests where excessive populations jeopardize the forest.

The Forest Service has sought to bring about control of overpopulation through cooperative action with the states rather than by direct measures. Cooperation with states in controlled hunting programs is also carried out by the Bureau of Land Management and Fish and Wildlife Service.

The Federal government has jurisdiction over the wildlife on national parks and national monuments. Here the government attempts to maintain an abundant and vigorous population of wild animals.

FISH AND WILDLIFE SERVICE. The Fish and Wildlife Service of the Department of the Interior is the government's expert agency for management of fish and wildlife resources. This Service is responsible for the management of some 282 national wildlife refuges aggregating over 18 million acres. These refuges include 196 migratory-waterfowl refuges, totaling 3,122,000 acres; 48 colonial nongame-birds refuges, totaling 91,212 acres; 16 for big game, totaling 10,652,000 acres; and 21 general refuges, totaling 4,221,000 acres. A number of the reserves are located on land administered by other agencies, but 10,987,000 acres are under the sole or primary jurisdiction of the Fish and Wildlife Service.*

Other major responsibilities of the Fish and Wildlife Service include research in wildlife management, administration of Federal aid to states for the protection and management of wildlife, maintenance of inland fisheries, conservation of salmon, expert service on river-basin development plans, marine-fishery research, utilization of fish resources, control of predators and rodents, administration of Alaskan fisheries, administration of Federal statutes for the protection of wild fowl, cooperation with the American republics in the enforcement of the migratory bird treaties and other wildlife work, research on whaling, and various international fishery problems.[2, 4]

In addition to its regular activities, the Fish and Wildlife Service serves as an expert agency for other government land-managing agencies such as the Bureau of Land Management, Indian Service, National Park Service, the U.S. Forest Service, and the Soil Conservation Service.

STATE CONSERVATION DEPARTMENTS. The major activities in game management are centered in the state conservation departments. These issue licenses, protect game against illegal hunting and fishing, operate hatcheries, advise legislatures on seasons and bag limits, manage state refuges and public fishing streams and shooting grounds, and conduct research in game production. The latter activity is often shared with or delegated

* Personal communication from Arthur A. Davis, 1950.

entirely to the state agricultural colleges. By far the largest percentage of wildlife specialists are employed by the states. It is estimated that the states employ approximately 4,000 men engaged as game protectors and a total of some 400 biologists. The biologists are commonly stationed at the headquarters of the state fish and game departments where they assist in administration and planning and carry on research. The availability of substantial funds under the Pittman-Robertson Act has greatly stimulated the development of measures deemed useful in the protection and propagation of wildlife. During the fiscal year ending 1950, the Federal government distributed over 11.6 million dollars to the states from this fund, and the states were obliged to match this with an additional 3.9 million dollars.[2]

Many of the more progressive fish and game departments in the several states publish a monthly magazine to acquaint their citizens with the activities of the department. As a consequence, the public generally is becoming much better informed on wildlife management. In a number of states the fish and game departments have been placed under civil service. This improves employment opportunities and the outlook for trained biologists.

The states have need for men who can write about wildlife and who can deliver talks before schools, sportsmen's clubs, and other groups interested in wildlife. There is an especially attractive field for biologists with a flair for popular writing.

OTHER AGENCIES. Forest and other wild-land owners have implied responsibility for wildlife. The owner should cooperate with public agencies to permit public use of his land when not in conflict with private use and should carry out operations so that they favor wildlife. Care in forest management, distributing operations over a wide area and over a number of years, and avoiding needless destruction of wildlife dens, coverts, and breeding areas are helpful.

The farmer has responsibility, because he owns a substantial area of land that produces wildlife. More than other landowners he may find that wildlife interferes with his activities. Game requires food, and the farmer feels that he has certain rights to the game he has fed. He has the right to control hunting on his land. Farmers have planted waste areas to game food and have taken other measures to promote the welfare of game. It is only fair that they receive some consideration from the public and the sportsman for their contribution to the welfare of wildlife.[5, 7]

Sportsmen's clubs have a major responsibility in wildlife management. It is their members, more than any other group, who determine public policy and who contribute funds for wildlife research and wildlife management. It is they, also, who lease land for special hunting privileges.

Sportsmen's clubs have been wont in the past to rely more on hunches and expressions of opinion by old hunters and trappers than upon biologic facts. The day for such behavior is rapidly passing as the modern game manager gains his place in the councils of such organizations. The sportsmen's clubs are ideal organizations for the dissemination of information on wildlife and for enlisting the interests and energies of people in taking needed steps to improve wildlife habitats. Through sportsmen's magazines an excellent opportunity is available to interest a wide group of individuals in scientific game management. It is the responsibility of sportsmen's organizations to be sure that their actions are based on facts rather than on unverified opinions.

The general public also has a responsibility in wildlife management. Its responsibility is somewhat broader than that of sportsmen's clubs or landowners, because wildlife has values that go beyond those realized by the landowner and by the sportsman. It is to the American public as a whole that we must look for appropriate legislation to prevent indiscriminate destruction of valuable wild creatures and for review of proposals made by landowners and sportsmen. The public also will have the responsibility for providing some of the funds for carrying on wildlife work. The wildlife manager in turn has a responsibility to the public to provide it accurate, unbiased information on wildlife questions and particularly to help the public to appreciate the values inherent in the wildlife resource. Only as he serves the public well and is appreciated by them, can the wildlife manager expect to see his profession grow to full fruition.

The Job of the Wildlife Manager. The typical wildlife manager is a state employee having responsibility for the protection and well-being of wildlife over an area of some 30,000 acres or more. His responsibility carries little authority. He must enforce laws passed by a legislature that may or may not pay heed to his recommendations. Since he has administrative control over little if any land, he cannot direct that trees be planted, grains be left unharvested, dams be erected, or even that hunting be excluded, except on refuges. His job is still thought of as that of a protector rather than that of a manager.

How then does he get results? The imaginative protector is one who knows his landowners and encourages them to plant game food and cover plants with aid from Pittman-Robertson funds, who addresses sportsmen's clubs on wildlife requirements, who enlists the interests of school children in studying wild creatures and improving nesting areas, and who promotes good sportsmanship among the public. He is, in fact, more a teacher than a warden.

WORKING CONDITIONS. Working conditions for the wildlife manager are somewhat more demanding than those for the forest manager. The wildlife manager must be alert to prevent trespass, especially during hunting seasons and during those periods of the year that are critical for wildlife. This means that he must often be in the field during stormy weather of winter and during the period when young are born in the spring and summer. As wildlife do not observe an eight-hour day, the wildlife manager must often be in the field before dawn and work until after dark. This is particularly true of the wildlife research man. The field is certainly not one for the man who has limited energy and who shuns physical hardship. To be successful, he must enjoy working with wildlife and be willing to find his compensation in the understanding and intimate association with wild creatures that his job permits.

A wildlife manager should be a good plant and animal ecologist. He should understand the management of the resources the land produces other than wildlife. The most efficient habitat management often can be effected through a well-coordinated land-use program. This requires that he also be proficient in public relations. The fact that 300,000 to 500,000 people annually visit refuges places additional emphasis on his capacity to maintain good relations with the public. His work is as versatile and challenging as that of a national forest ranger or supervisor.

He should be a good photographer as his work is likely to require the taking of pictures under extremely unfavorable conditions. He must be, above all, a man of great patience and of keen insight and understanding. He must be willing, if need be, to sit for hours at a time in a blind, studying the behavior of animals and collecting information that can be used in their more effective management. He must know the life requirements and life habits of the various species of wildlife he seeks to influence. He must be prepared to study these animals both as active creatures under natural conditions and as laboratory animals under semiconfinement, and he must be prepared to perform autopsies to determine parasitism, disease, development of embryos, and food habits. For all these reasons the job of wildlife manager represents a true profession, one requiring intimate study and lifelong scholarship for success. Men in wildlife work may rise to important administrative posts in the Fish and Wildlife Service, Forest Service, Soil Conservation Service, and in state conservation departments.

Men can gain both inspiration and appreciation of the subject from the biographies of the great biologists who have contributed much to our understanding of wild creatures. Such men as Henri Fabre, J. H. Comstock, and W. M. Wheeler, though primarily entomologists, used methods of study similar to those employed by all biologists. Charles Darwin, David Starr Jordan, C. Hart Merriam, Vernon Bailey, William L. McAtee, and Aldo Leopold have all made unique contributions.

Range Management

Range management is the protection and use of grasslands and other lands furnishing forage for the production of livestock, usable water, and other valuable resources. It requires proper use of ranges now in good condition and the application of proper methods to restore depleted and eroded ranges. Integration and coordination of the several uses are also involved.

Range management is a major concern of men charged with the responsibility for regulating grazing on western public lands, including forests, except coastal forests of the Pacific Northwest. It is a concern of foresters who work in the southern United States, for here also considerable numbers of domestic animals, such as cattle, swine, and horses, feed on forest land. Since the southern range differs radically from the western range, it presents its own peculiar problems.

Importance of Range Management. Grazing occurs on some 1,112,000,000 acres of land in the continental United States.[6] This includes farm pasture, cropland grazed after harvest, and forested and wooded range.

The major range problems occur on the grazing lands of the West, which cover some 728 million acres or nearly 40 per cent of the total land area of the continental United States. A line running from about the center of North Dakota down to the mouth of the Rio Grande roughly separates the range territory from the eastern states, where little open range is available. Livestock growing is one of the major enterprises in these western states, and the livestock industry is closely integrated with other farming activities. It does not, however, represent the greatest dollar production from the land, being surpassed by timber production and by crops grown on both irrigated and dry-land farms. By 1900 the industry had reached a peak of some 21 million animal units, and it attained the same peak in 1920. One animal unit is defined as one beef animal or one horse. The number in 1936 was 17 million animal units. Beef cattle from the western range are commonly marketed in the fall of the year. Some go directly to slaughter, but a higher percentage are sent to feeding lots in Iowa, Illinois, and Kansas where they are fattened for market. The use of the western range, therefore, influences agriculture in the West and in the Middle West and affects food prices throughout the country.[10]

The Condition of the Western Range. Grazing on public lands was carried out for decades with no general control other than that ranchers themselves enforced. In 1905 the Forest Service began to regulate grazing on national forests, and in 1934 the Grazing Service began to administer the grazing districts.

Serious overuse existed before controls were attempted and this situation could not immediately be corrected. As a result range deterioration has become widespread. It accounts in large part for the decline in livestock numbers noted above. Lack of precise appraisal of early conditions and of extensive recent surveys makes it impossible to give a true picture of range depletion resulting from domestic animals. The material in this section is based chiefly on the Forest Service report "The Western Range," which was prepared in 1935 when the western states were in the grip of a dry cycle and at a time when the livestock industry was off balance from the general economic depression.

Regardless of the type of ownership, climate, and topography, the Forest Service estimates the range to be 52 per cent depleted from its virgin condition. Depletion goes through the following cycle: First, overuse of the most palatable species leads to their decline in vigor. Gradually these plants die out and weeds or worthless shrubby species take their place. As the more palatable species of grass are killed out, sheet erosion begins and leads oftentimes to rill erosion and gullying.

The most productive land is the valley bottoms, which are often park-like lands surrounded by forests. Once the grass is badly overgrazed, gullying begins, lowering the water table, and this, in turn, hastens further decline in vigor of vegetation.

One of the most disheartening results of overgrazing is the fate that overtakes mountain meadows. These moist, lush grasslands serve as reservoirs holding back water from snow melt and winter rains and feeding it crystal clear to mountain brooks. They are a delight to the eye and a symbol of plenty to man and beast alike. Mountain meadows are particularly vulnerable to overgrazing, which cuts up the turf, packs the soil, and accelerates runoff and erosion. This leads to deep gullies that drain away the moisture, thus exposing the grass to drought (Fig. 63). Eventually these rich meadows may lose their character entirely and be invaded by trees or woody scrub growth.

Natural meadow restoration is possible where grazing is excluded, as in national parks. Willows, aspen, and alders invade the eroded meadow; these in turn provide food for beaver that dam the gullies. Gradually the flooded land fills in, and after the beaver colony moves out to find other feed, the grasses may take over. This natural process sometimes moves with gratifying rapidity, though, more commonly, centuries rather than decades are required.

Estimates show no less than 589 million acres of range land to be more or less seriously eroded. As a consequence, soil productivity has been reduced and watershed services impaired. Tremendous quantities of silt have clogged major western streams. Reservoirs and irrigation systems are slowly being choked in many states.

FIG. 63. An active gully eats its way across a heavily grazed mountain meadow. Sequoia National Forest, Calif. (*U.S. Forest Service photo.*)

The types and conditions of the major range divisions are shown in Table 6 below.[8, 9, 10] It should be kept in mind that rainfall must be higher in the South than in the North to produce the same degree of vegetative thrift.

TABLE 6. TYPES AND CONDITIONS OF RANGE LANDS

Range type	Area, thousand acres	Average rainfall, inches	Area required per animal unit of grazing capacity per month, acres	Percentage depletion
Tall grass	18,513	20–40	¾–1½	21
Short grass	198,092	13	2½–10	49
Pacific bunch grass	42,534	10–20	4–6	51
Semidesert grass	89,274	15–18	6–7	55
Sagebrush grass	96,528	10–20	9–15	67
Southern-desert shrub	26,986	5–10		62
Salt-desert shrub	40,858	5–10	18–20	71
Piñon-juniper	75,728	6–15	8–10	60
Woodland-chaparral	13,406	10–20	2–3	50
Open forest	126,367	15–30+	6–10	33

Range Ownership and Use. Ownership of western range land in 1935 was estimated to be as follows: [10]

	Area, million acres
National forests	88
Grazing districts	128
Indian lands	48
Other Federal lands	23
State and county	66
Private	375
Total	728

Private lands make up 52 per cent of the total, Federal lands 39 per cent, and state and county lands 9 per cent. These several ownerships, aside from Indian lands, are highly interspersed. But even more significant, operation of a high proportion of ranch units requires use of both public and private land. A fairly typical operator might have a home ranch with irrigated hay lands and range for late fall and winter use. In spring and fall his animals might graze on public lands administered by the Bureau of Land Management and in midsummer on national forests. Successful operation is thus contingent on integrated use of public and private land. Although such use patterns have existed for decades, still the individual rancher may have no long-term leases or other rights that assure him use of public land. It is only natural for him to defend his use of public land, which he comes to consider his right. Deprived of such use, his ranch might prove to be submarginal.

Officers charged with administering public land have a responsibility to the ranchers. They also have other resources to protect. On the grazing districts, forage is the chief resource. To produce this in abundance, overuse must be avoided and, above all, the soil must be protected from erosion. The national forests occupy high lands that receive most of the snow and rainfall and are therefore the source of water for irrigation and municipal use. Also they produce timber, game animals, and afford much recreation. All these uses as well as the soil must be protected.

With such conflicts in ownership and use it was inevitable that controversy should arise.

Controversy over Range Control. Some 34,000 settlers and stockmen graze animals on national forest land under permits which specify the number of animals and the part of the forest on which they may be grazed during a stated period (Fig. 64). In 1934 Congress passed the Taylor Grazing Act "to prevent overgrazing and soil deterioration," which set up in the Department of the Interior the Grazing Service to police the public range lands.

The western livestock growers are well organized. The National Wool Growers Association represents sheepmen and the American National

FIG. 64. Cattle grazing, Deschutes National Forest, Ore. (*U.S. Forest Service photo.*)

Cattlemen's Association the cattlemen. Though collectively the people in the range-livestock industry total only about 800,000 or 5 per cent of the population of the region, they exert much political influence.[3] In some states they control state policy, and they have been powerful enough to challenge the Forest Service and Grazing Service in their management of public lands. By lobbying against appropriations, they temporarily succeeded in destroying the effectiveness of the latter, but grazing work has been reorganized under the Bureau of Land Management. Pressure on these bureaus has brought counterpressure from conservationists interested in soil, irrigation water, recreation, and wildlife.

Controversy between stockmen as represented by their associations and the U.S. Departments of Agriculture and the Interior led to investigation by a Congressional subcommittee. Their report in 1948 recommended: [1]

1. That the Forest Act be amended to provide that grazing, recreation, and wildlife be made basic uses of national forest lands.
2. That advisory boards on the national forest be given legal status.
3. That the present policy of transfer cuts be discontinued and that any cuts that may be necessary be made for protection purposes only.
4. That the Forest Service undertake a vigorous program of range improvement including: (*a*) water development, (*b*) reseeding, (*c*) fencing, and (*d*) rodent and poisonous weed control.

5. That the Forest Service undertake a policy that will enable the permittees to participate in and contribute to a greatly extended range-improvement program.

Some of these measures will help considerably to ease the burden of the stockmen. Water development, reseeding, and fencing will distribute animals and restore overgrazed areas, provided the animals are removed long enough for reseeded range to recover.

What livestock interests seek is to establish vested rights in the public range lands that can be transferred when their ranches are sold. Without this their livelihood remains subject to control by public officers. Ranches that can be sold with grazing privileges on either national forests or public grazing districts bring a higher price as a result of the privilege, even though no guarantee of its continuation is extended.

Stockmen claim that privately owned range lands are better managed than public lands. Cases can be cited to support the claim. However, ranch owners with limited personal resources cannot be expected to make the investment in fencing, water development, reseeding, and other measures needed for region-wide range restoration. Most small owners are better served by having this work done for them by public agencies. Many willingly cooperate to this end, realizing that it means a better living for them in the years ahead. Large owners, on the other hand, can, and many do, install improvement measures.

It would be unfair to attribute all western range depletion to the livestock industry. Deer, elk, antelope, and even mountain sheep when overabundant locally deplete their range. The result where hunting pressure is light is herd starvation during critical months.

RANCH MANAGEMENT. Much of the controversy that has arisen between the western stockmen and the Forest Service is traceable to the latter's being concerned with the condition of the range rather than with the condition of the rancher. Range management cannot be divorced from ranch management. The approach the Soil Conservation Service has used to the erosion problem is being used with good effect in approaching the range problem.

Certain ranches in the West, like many farms in the eastern states, are too low in productivity for the owner to compete successfully in modern agricultural enterprise. Some of the small ranches should be retired from production just as marginal eastern farm lands have been retired to forest. Grouping two or more small ranch units together to form one that is suitable for continuous operation may help in other cases. A great deal of ranch consolidation has gone on, as Clawson points out.[3] Progress is slow because of many difficulties, especially the necessity for adjusting the ranch for a balanced operation. Balance is needed among spring, summer, fall, and winter range and winter feed from hay land.

Ranches also must be adjusted in size to the labor supply. The smallest ranch is one that can be operated successfully by one man, except for extra help for such seasonal jobs as haying. Where winter feeding is necessary, this means about 200 cattle; where year-round grazing is available, 400. The average cattle ranch in 1926 had 185 head and the average sheep ranch about 850 head.

This leads to a third difficulty, the capital required to operate a ranch. In 1929 the average size of ranches was 3,042 acres, the value of land and buildings $34,641, and the value of products sold annually $7,358. This contrasts with an average product sale of only $2,000 for all farms during the same year. The average ranch income was, however, considerably above the median, which was in the neighborhood of $3,000. By 1950 median income was double that of 1929.[3]

Fluctuating prices of livestock are a serious difficulty, especially when coupled with weather cycles. Beef-cattle prices at the farm show the following fluctuations: [3]

Year	*Price per pound*
1910	$.05
1917	.10
1922	.06
1929	.10
1933	.04
1937	.07
1940	.08
1942	.10
1945	.11
1948	.23

Weather cycles are of greatest importance. Good years tend to come together and to be followed by a series of bad years. Throughout the northern range region, rainfall varies over cycles of about 16 years in duration; in the central region cycles are of about 8 to 12 years' duration, and for the southern region, 5 years'.[3] This tendency for wet years to be grouped together causes great temptation for ranchers to expand operations during good years. If they kept their herds adjusted constantly to the forage available during the driest years, such as the 1930's, production of livestock in many sections of the range area would be less than half what it could be and actually was by allowing herds to build up in good years. Income of ranchers has accordingly benefited, because the high prices of the 1940's coincided with heavy rainfall on the range. Had ranchers not expanded operations during this period, they would have missed a lifetime opportunity to accumulate savings, and meat supply during wartime would have been scarce indeed. The difficulty with accepting this pattern of range management lies in the impossibility at present of predicting wet and dry cycles and the slowness of ranchers in curtailing operations until nature or prices force them to do so.

Other difficulties that confront the rancher seeking to maintain a profitable operation are fluctuations in meat consumption, variations between relative profitability of sheep and cattle, transportation costs, poisonous and injurious plants, diseases of range animals, severe winters, irregular supply of credit, isolated living, and lack of outside employment opportunities during slack periods on the ranch.

Intelligent action is being taken by many ranchers to improve the breed of livestock, to improve breeding stock on individual ranches, to develop the range by establishing watering tanks, reseeding, fencing, and other measures, and by adopting ranch management practices that provide efficient operations (Figs. 65 and 66).

Public agencies are lending a helping hand through research by the Forest Service and agricultural experiment stations, through demonstration and education by the Agriculture Extension Service and the Soil Conservation Service, and through improved management practices by the Forest Service and Bureau of Land Management. The approach to range improvement through better ranch management by the Soil Conservation Service and Bureau of Land Management offers real promise of long-term gains. A control system that rewards the man who improves his range and penalizes the one who does not would hasten improvement of the public range.

The western livestock rancher has a complicated and difficult job. To earn his livelihood, he must cope successfully with fluctuating livestock prices, unfavorable climate, rough topography, and depleted range cover. He must use large areas of land including public lands over which he has little if any direct control. He needs and deserves understanding consideration from public officers and other range users whose activities affect his own. A huge task lies before all to extend rapidly efforts to restore the range lands to full productivity. It is a task calling for high skill in applying improvement measures and adjusting uses.

THE RANGE MANAGER'S JOB. The successful range manager must understand the ranching business. He must be in a position to work out adjustments in public-land use with a minimum of hardship and with maximum benefits. The absence of long-term commitments to range users and particularly the tendency, which was widespread in earlier years, to reduce the grazing permits of large ranch operators to favor the medium- and small-sized operators simply added to the difficulty and increased the pressure on the range.

FIG. 65. Water hole, Ouray Grazing District, Colo. (*Bureau of Land Management photo.*)

FIG. 66. Crested wheat grass reseeded on wind-blown bottom land, Utah. (*Bureau of Land Management photo.*)

Range managers perform a wide variety of tasks. Each spring they determine when the range is ready for grazing. The range managers travel over the range, examining the forage and its stage in development, correlate this with other phenological information, and so decide upon the opening dates for admission of grazing animals. Foothill range opens up far earlier than mountain range and the low range much earlier than the high. South-facing slopes are usually ready before north-facing slopes.

The range manager determines whether the range is in good, fair, or poor condition and whether the range is improving or deteriorating from year to year. To perform these tasks, he must be able to recognize the various species of forage plants at all stages in their development. He must know their relative palatability, their requirements for moisture and light, and their nutritional values in various stages. Above all, he must know how to recognize the signs of range deterioration in its incipient as well as advanced stages.

Once the condition of the range is thoroughly appraised, it is usually possible to decide upon measures needed to maintain it in good condition or to improve it if it is in poor condition. Reseeding is one of these measures. This is most effective if the soil can first be cultivated. Oftentimes this may require the removal of sagebrush and other vegetation that will crowd out new seedlings before they could get well established. Seed may be sown by ordinary grain drills after the soil is prepared (Fig. 66). The land must be protected after reseeding until the new seedlings are in a position to take possession of the soil. Badly deteriorated ranges grown up with sagebrush and various other shrubs and weeds, including poisonous plants, must often be given a drastic treatment before they can be rehabilitated. Mountain meadows require special measures. These lands are often used as sources of hay for winter feed. If overgrazed, they gully, lowering the water table and thus reducing the capacity of the meadow to produce hay. Clogging the gullies with vegetation and debris and reducing the stocking can help to restore such meadows for valuable grazing in the future.

Proper distribution of livestock is very important in good range management (Fig. 67). Distribution can be effected in part by distributing watering holes, by distributing salt widely over the range, and by riding to drive the animals from overgrazed to underused areas (Fig. 68). Herding of sheep has proved very effective as a means of getting uniform use of large grazing areas. As a consequence, sheep range in many sections of the West is in better condition than cattle range, even though the sheep, if improperly herded, are far more destructive than cattle.

Rotation grazing is another technique that is sometimes used to rehabilitate the range. This may be done both seasonally and from year to year. If the range has been only moderately overgrazed, deferred graz-

FIG. 67. Severe erosion caused by overgrazing. Navaho Indian Reservation. (*Bureau of Land Management photo.*)

FIG. 68. Proper distribution of salt and water helps to obtain full range use with minimum overuse. (*Bureau of Land Management photo.*)

ing, rotation grazing, and reducing the stocking are often the only steps that are required to start the range on the mend. Where areas have been intensively overgrazed for a long period of time, mere exclusion of animals will not result in prompt range restoration. Soil rebuilding, reseeding, and other measures are required in these cases.

The range manager, therefore, must know intimately the stock rancher and stock ranch management as well as the range plants and their interrelationships. To acquire this information requires intensive study in college as well as a great deal of actual field study.

LITERATURE CITED

1. Barrett Committee Report. 1948. "Report of Hearings before the Committee on Public Lands and the Subcommittee on Public Lands of the Committee on Public Lands, House of Representatives, on Forest Service Policy and Public Lands Policy," Report 2456, 80th Congress, 2d Session, Washington, D.C. 15 pp.
2. Chapman, Oscar L. 1950. "Annual Report of the Secretary of the Interior" (fiscal year ended June 30, 1950), U.S. Government Printing Office, Washington, D.C. 411 pp.
3. Clawson, Marion. 1950. "The Western Range Livestock Industry," McGraw-Hill Book Company, Inc., New York. 401 pp.
4. Day, Albert M. 1948. Fish and Wildlife Service, "Annual Report of the Secretary of the Interior" (fiscal year ended June 30, 1948), pp. 285–315, U.S. Government Printing Office, Washington, D.C.
5. Leopold, Aldo. 1933. "Game Management," Charles Scribner's Sons, New York. 481 pp.
6. Stoddart, Laurence A., and Arthur D. Smith. 1943. "Range Management," McGraw-Hill Book Company, Inc., New York. 547 pp.
7. Trippensee, Reuben Edwin. 1948. "Wildlife Management," McGraw-Hill Book Company, Inc., New York. 479 pp.
8. U.S. Department of Agriculture. 1941. "Climate and Man, The Yearbook of Agriculture," Washington, D.C. 1248 pp.
9. U.S. Department of Agriculture. 1948. "Grass, The Yearbook of Agriculture," Washington, D.C. 892 pp.
10. U.S. Department of Agriculture, Forest Service. 1936. "The Western Range," Senate Document 199, 74th Congress, 2d Session, Washington, D.C. 620 pp.

CHAPTER 13. *Forests and Recreation*

A LEADING PROFESSOR of public administration, who himself attained subcabinet rank in two departments of our Federal government, has expressed his opinion of our country's forests. He declares that forests are worth what we pay for their care because of their value for wildlife, soil, and watershed protection, and for recreation, especially the latter; the timber we harvest from them represents pure profit. This statement comes not from a recreation enthusiast but from a man whose job it has been to decide how public money can be spent to yield greatest national benefits.

Almost every human being has a desire to visit out-of-the-way places, to learn the ways of life of creatures other than man, and to experience some of the wonders of nature. Some gratify their urge for exploration by reading and creative art, some by scientific enquiry, and many by travel and outdoor living.

The urge to escape the restrictions of organized society are easily gratified in a pioneer land. As land becomes settled, signs bearing such captions as "No Trespassing," "Private Property," "Keep Out" frustrate the city dweller seeking a Sunday outing. The demand goes up for parks and public playgrounds.

A public park formally laid out with neat signs and uniformed park attendants to ensure that people stay off the grass, pick no flowers, and play no games may afford some aesthetic pleasure, but it offers little relaxation to restore mind and body. Nor does the overcrowded picnic area differ much from an overcrowded restaurant.

The purpose of recreation planning in forest and other natural settings is to provide the visiting public with the necessities for enjoying forests and parks with a minimum of restrictions (Fig. 69).

Values of Recreation

Recreation has been defined as the pleasurable and constructive use of leisure time. It involves far more than escape from boredom. It satisfies physical and mental health needs for relaxation from strain through

FIG. 69. Park planning should be developed in detail on the drawing board and with models before construction starts on the ground.

the devotion of the body and mind to constructive activity differing from the tasks of routine breadwinning.

The physical and mental health of people as well as their love of country is improved by recreation. Since health and general well-being are the concerns of society as a whole, recreation becomes the concern of government.

The desire of people for recreation gives it commercial value. Expenditures for recreation rank it among the biggest industries of the nation. They include costs of motion pictures, radios, clubs, resorts, sporting equipment, books, and magazines. The American Automobile Association released an estimate that 10 billion dollars was spent for *recreation travel* in the United States in 1946. The New York State Department of Commerce has determined that over 400 million dollars is spent on vacations in New York State every year.

Outdoor recreation involves big expenditure and covers a variety of activities. The automobile industry provides means of travel, highways and parkways lead to places of interest, hotels and resorts provide eating and sleeping accommodations, and the acquisition, development, and maintenance of park lands provide definite facilities. All these and many others are of major importance in the recreation story.

Use of Forests for Recreation

Forests, because they occupy large areas of land, offer a degree of seclusion, and shelter many interesting wild creatures, are especially popular with recreationists. Unless excluded, the public will naturally use both public and private forest land for hiking, picnicking, hunting, and collecting. Public foresters recognize this use and encourage it. The national forests were destinations for 27,368,000 visits by recreationists in 1950.

State forests served likewise as recreational areas open to public use, and the same is true of many private forest holdings. To regulate use so that it brings maximum pleasure to forest visitors and causes insignificant damage to the forest, various facilities are erected and maintained. The forester thereby becomes in some degree a park manager. He must be guided by the same principles that guide park managers if he is to perform his services to recreationists cheaply and effectively. Gradually there has arisen a profession of park management that embraces planning, design, construction, operation, and maintenance of parks. It draws on the fields of regional planning, landscape designing, civil engineering, forestry, human behavior, and public administration.

Park management may be divided into planning and administration. The first has to do with developing land for use; the second is concerned with operation and maintenance. Actually, these two divisions merge many times as it is necessary for the planner to consider operation, program, and maintenance if he is to plan well. The man concerned with administration is interested to see that development is such as will enable him to administer the area in the best possible manner.

The forest manager will need to avail himself of the knowledge and services of park managers to the extent that recreational use is developed on his property. Some forests may have virtually no recreation use. Others, such as the Adirondack Forest Preserve and the Cook County Forest Preserve, are used only for recreational, wildlife, and watershed purposes. On most national and state forests, recreation is considered a major use. The broad principles of park management therefore have a place in forestry.

The Growth of Park Systems in the United States

The one thing that was present in great abundance in early America was land. Only with growth of urban centers did there develop need for parks, which were of two types—the formal park and the playground. Gradually these were extended to include bathing beaches, picnic areas, and hiking and bridle paths.

With the discovery in our country of great natural wonders, including hot springs, waterfalls, geyser basins, big trees, and rugged mountain scenery, came the demand for national parks. The thought of such areas falling into the hands of commercial exploiters was distasteful to high-minded citizens, who desired that this wonderland heritage be preserved for public enjoyment. Areas of historical or lesser scenic value were set aside for national monuments and state parks. Still later has come the demand for public fishing and shooting grounds. By 1950 parks and recreation areas were administered by schools, villages, counties, cities, states, and the Federal government.

Planning Park Systems

The parks and park systems described above have grown up largely as a result of public interest and demand. No attempt to study in detail the recreational needs of large populations was made prior to 1933. Even those that have been made since that time have not reached national scope. Park development has been largely fortuitous, depending on public interest. Only in recent years has thought been given to planning broad park programs to meet specific population needs.

Providing Attractions for All Users. The park planner must be well informed and imaginative to handle the many factors involved in planning a park system for a densely populated region. He must deal with many intangibles for which no measurements are available. A basic job is making a survey of recreation needs. This requires determining the population, its density, growth rate, occupational groups, ethnic make-up, average financial income, and the time available for recreation.

Public recreational facilities are not needed nearly as badly by people dispersed in rural areas or small villages as by those crowded in cities. The free space around dwellings has a great deal to do with public-recreation needs. If ample, the family can cultivate a garden and have a picnic area with reasonable seclusion. Village boys who can play baseball on vacant lots and who have ready access to the surrounding farm and forest land for hiking, hunting, and fishing have less need for elaborate playgrounds than city boys. Office workers have different requirements from factory workers.

Metropolitan areas that have marked racial or national groups have a special problem. Friendly commingling on spacious playgrounds can promote mutual understanding and respect, but people from other lands enjoy also the opportunity to hold folk dances, song fests, and to play games they learned in their homeland. Such artistic expressions often provide entertaining spectacles to native Americans as well as satisfaction to participants. Our own culture is enriched thereby. Imaginative design can provide for the needs of all groups in such a way that the fullest possible satisfaction results.

Financial income of the various groups in the population makes a great difference in their demands for recreation and the types that they can enjoy. People of modest income will enjoy swimming, fishing, hunting, hiking, picnicking, and boating. Those of higher income may wish golf courses, bridle paths, and even public yacht harbors for sailing, motorboating and cabin cruising. Nicety of judgment is required in balancing the needs of the family with lowest income against those with larger incomes. The presence of sailboats and particularly a sailing regatta on a lake adds a great deal to the pleasure of many people, even though they themselves are not sailors nor financially able to indulge in this sport.

The time available has an important influence on the need for recreational facilities and the type that should be provided. The five-day week makes week-end trips feasible. During the summertime when many metropolitan areas are on daylight-saving time, long evenings are available for recreation. Facilities close to the cities need to be provided for constructive use of this leisure time.

Avoiding Conflicting Uses. Age distribution of a population is also important. Older people enjoy the more passive types of recreation. Picnicking, pleasant views or vistas, floral displays, and an effectively landscaped park please them. They may also wish to sit in the sun, to watch the play of younger people, their sailing, swimming, horseback riding, and other more active recreational outlets. Small children enjoy wading pools in summer, larger children want playgrounds, and young adults opportunities for such strenuous sports as baseball, football, soccer, ice hockey, swimming, skiing, and skating. Old and young alike enjoy the solitude of the forest. They thrill with pleasure at the sight of whole hillsides carpeted with spring flowers where only a few weeks before the land was brown. They enjoy hearing and seeing wild birds and observing the animals that live in forests. They seek to identify the many species of trees, shrubs, and herbs that lend variety and interest to our woodlands.

The skillful park planner will try, in so far as he is able, to provide facilities for mass recreation, such as picnic areas, swimming areas, and boating areas, and, at the same time, will try to reserve a portion of his park in as near a natural condition as possible. Here the bird lovers and the seekers for wild flowers can wander off on the little-frequented trails, gaining a certain satisfaction of adventure and exploration as well as an opportunity to become acquainted with the natural forest environment where it has been little disturbed by man. The truest test of successful park planning is to provide for both active and passive recreation, for

FIG. 70. An opportunity to see trout in clear mountain lakes and wildlife in natural settings is a prime attraction of forest recreation. Emerald Lake, Lassen Volcanic National Park. (*National Park Service photo.*)

group activities, family activities, and for the pleasures of solitude (Fig. 70).

The seasonal needs of the population must also be considered. More and more our people are seeking year-round outdoor activities. Warm, light, water-repellent clothing now provides the outdoor enthusiast with comforts in wintertime comparable to those of the warm months. Skiing, snowshoeing, and hiking across the snow-covered landscape brings as much enjoyment as can be had from an equal period spent out of doors during summer.

Promoting Understanding. Park planners have an opportunity to educate as well as to entertain people through the proper development of recreational resources. Historic locations are often made the site of county and state parks and national monuments. Restored battlefields, old forts, and other historic places may enable the imaginative citizen to relive the events of history. Here the utmost scholarship and subtlety are required if such an educational program is to be effective. The public will be educated only in so far as the displays provide them the imaginative thrill that comes from projecting themselves into historic situations.

The national parks offer conducted tours by naturalists, which have become exceedingly popular. Nature talks are scheduled for evening programs, which add greatly to the enjoyment and understanding of park visitors. A properly planned park system for a metropolitan area will normally include parks and historic sites having educational value. It may also include special nature trails with appropriate signs, changed at regular intervals, to teach the people the names of the various wild flowers, trees, and plants, to show them the locations of birds' nests, and other interesting features that might otherwise be passed by unnoticed.

Having sized up the population and the recreational needs of the various subgroups within it, the park planner must next survey existing facilities. In appraising these, he will want to give consideration to their present use, the effectiveness with which they serve the people that now frequent them, and whether or not improvements might be made that would enable them to serve a greater number of people or to serve the present number more effectively. Many parks are overused. Only by decreasing use can these be restored to a standard that will bring to the users the degree of satisfaction desired from outdoor recreation. In surveying existing parks, the recreation planner will oftentimes glean additional information as to the attitudes of the public toward their parks and the types of additional facilities that are most needed.

Projecting a Park System

A park planner must deal not only with the existing population but with the population anticipated 5, 10, or even 15 years in the future. He must therefore take into consideration the potential growth of the region for which he is planning. He must study the surrounding territory with an eye toward possible park locations that will be needed for at least 15 years in the future. It is far better to acquire areas well in advance of their need than to wait to set aside land for park purposes until private development has increased property values and destroyed many of the recreational features of an area. It is far more expensive to restore a lake shore for park use after it has been developed for commercial or residential use than it is to acquire the land well in advance and hold it for park development. The park planner will wish to consider a number of locations and a variety of types of parks which, taken together with existing facilities, will provide a park system tailored to meet the needs of the metropolitan region concerned. He will then establish priorities for acquisition and development, estimate the over-all cost for each unit, and present his findings to the public park boards for their consideration and action.

Detailed Planning of Individual Parks

Every improvement in a public-park area needs to be carefully designed and located. Only in this way can it most effectively serve the purpose intended. First, consideration must be given to over-all planning so that traffic in the park flows with a minimum of congestion and interruption. The major recreational features of the park are first located. Road system, parking area, and trail system must be provided so that the people may move readily into the park from their motorcars to the major points of interest. The trails should radiate from the necessary centers of congestion, such as parking areas, bathing houses, historic museums, and other features. Appropriate signs must be erected to guide park users to the various facilities. These may be simple lettered signs or pictorial signs, but, in any case, they must carry their message clearly so that the visitor is not confused or misdirected. Parking areas, picnicking areas, picnic tables and benches, fireplaces, shelters, bathhouses, toilets, refuse pits, drinking water—all the essentials for the comfort and enjoyment of the people must be properly located, attractively designed, and integrated with other features. With such arrangements the public finds it convenient and natural to use rather than to misuse the park and to enjoy rather than to destroy what has been erected for their pleasure (Fig. 71).

Any park area that is to be used by large numbers of people is naturally subject to wear, and many of the attractive features may even be destroyed. To minimize public misuse of parks, it was customary in the past to post long public ordinances, setting forth in 2,000 or 3,000 words of fine print the various misdemeanors and punishments provided therefor. The posting of such ordinances has been singularly ineffective in attaining the ends desired. It is up to the park planner to try to accomplish this purpose by simple signs, by proper location of trails and facilities, and by informing the public where and how it can satisfy its needs instead of by posting a long list of forbidden activities. A properly planned park will require a minimum of policing. When it is made easiest for the public to do the right thing, they are less inclined to do the wrong thing.

Building the Park

Park construction requires engineering activity in the laying out of roads, trails, facilities, and buildings and in their construction. The park planner and park administrator must understand the basic principles of construction and how to apply these in a park so as to create the proper impression.

FIG. 71. Freedom from overcrowding is a feature of well-planned recreation areas. Bird Lake Camp Ground, Gifford Pinchot National Forest. (*U.S. Forest Service photo.*)

Roadways and trails should be naturalized in so far as possible. Concrete and asphalt are to be avoided and gravel and grass walkways and trails used wherever these will stand up under the traffic to which they will be subjected. Buildings and facilities must be serviceable, simple, yet rugged in design, and easy to service and keep in order. Above all, they must be adequate for their purpose. It is obvious that the park planner and builder must have a wide variety of skills and must know how to choose wisely from the various building materials available in order to construct pleasing and serviceable park improvements.

Maintaining the Park

Park maintenance may involve simple custodianship of the facilities and protection of the park area against fire and misuse. Use of any park area by the public introduces a major disturbing factor that tends to destroy many of the features most desired in a park. Constant trampling wears out grass, compacts soil, leads to erosion, tends to kill trees and other vegetation, oftentimes invites attack of trees by insects and fungi, and consequently requires constant vigilance on the part of the park custodian to direct public use so as to minimize such damage. All parks require a certain minimum maintenance. In rare cases this may be provided by the users themselves if they are well disciplined and truly appreciate the facilities placed at their command. More often a park custodian is employed to keep the park in a neat and orderly condition and to supervise the use of the park by the public.

In addition to simple custodianship, which may suffice for a park that has little use, there is regulation of traffic, assignment of facilities, and the appointment of guards to direct the public in the use of the park area. Where swimming and boating facilities are provided, lifeguards are needed. The park manager will have the guards wear uniforms or carry an appropriate badge so that the public will know that they are in charge of the park and will cooperate with them. Where fees are charged, the collecting of such fees provides an opportunity to point out special features or attractions and to build an understanding between the park authority and the using public.

The most intensive type of park management is that which involves recreational and educational programs, guided tours, museums, nature trails, and other special educational or recreational activities. These are oftentimes found in the national parks and in some of the state, county, and city parks (Fig. 72).

Park Administration

In the foregoing discussion no attempt has been made to differentiate among the various types of parks. Strictly speaking, a forester is perhaps concerned only with large natural park areas, such as national and state parks; yet in these as well as in large city and county parks, such as those of Westchester County, New York, or the city of Denver's mountain parks, the activities of the park planner and park superintendent merge with those of a city-park manager and the national-park superintendent. All have many activities in common and all require the same basic philosophy in planning and maintaining park facilities.

A park organization like the National Park Service or many of the state park services is large in size, spends considerable sums of money, and serves the needs of millions of people. For this reason park administration becomes an important activity, requiring the same type of managerial skills demanded of forest administrators and even business administrators. Park management is, however, primarily public management, and consequently the park administrator draws more on the field of public administration than on that of business administration. The superintendent of the national park has many functions. He must plan the acquisition of land, the development of road and trail systems, the organization of a fire-protection system, the erection of campgrounds and picnic grounds, and the division of the park into ranger districts. He assigns duties and inspects the performance of the rangers and naturalists. He must maintain good relations with the public, both the people that visit the park and those that live in the surrounding territory. He must perform all other duties that are necessary to maintain the park in good condition and provide for economic expenditure of public funds dedicated to this purpose.

The National Park Service

The National Park Service is a bureau of the U.S. Department of the Interior. Administration is headed by the director and associate director who report directly to the Secretary. The field service is administered through four regions each of which has a regional director. The park superintendents who have charge of individual parks report to the re-

FIG. 72. The conducted tour by a ranger-naturalist instructs and satisfies the public while offering opportunities for simple lessons in outdoor manners. Yosemite National Park. (*National Park Service photo.*)

gional director. Service to the individual regions and parks is provided by seven staff divisions in Washington: (1) recreation and land planning, (2) interpretation, (3) historic sites, (4) legal counsel, (5) travel bureau, (6) development, and (7) forestry.

The National Park System dates back to 1832, when Congress put aside four sections of land comprising the Hot Springs Reservation in Arkansas for permanent public use. In 1872 Congress set off the lands of the Yellowstone for a public park for the benefit and enjoyment of our people. This law of 1872 set forth the major policies to be followed in the management of the Yellowstone Park, and these principles have been adopted for subsequent parks as well. It is mandatory upon the Park Service to preserve from injury and exploitation all timber, mineral deposits, and natural curiosities or wonders within the park, to retain the park in its natural condition, and to protect wildlife from wanton killing or commercial exploitation. Fishing is allowed on most areas under National Park Service regulations.

In 1950 the National Park Service administered a system consisting of 190 areas, totaling some 23 million acres. These included 29 national parks, totaling over 13 million acres, 88 national monuments, totaling over 9 million acres, and various parkways and historical, military, and memorial parks and battlefields.[2]

The national parks are more prevalent in the West than in the East, but they are distributed throughout the country. Primarily, they consist of areas of outstanding scenic attraction or regions containing natural phenomena of unusual interest and beauty.

The national monuments are areas containing scientific, historic, or other features that are related to the national interest and welfare. Usually, they are small in size. They are managed in much the same manner as the national parks, due consideration being given, however, to their differences in size, tourist attraction, and purpose.

Despite the wide popularity of national parks the mandate to preserve them from exploitation and maintain them in natural condition has been open to challenge by mining, livestock, power, and irrigation interests. Proposals to build dams that would impair the integrity of national monuments and national parks were live issues in 1951.[1]

Technical Jobs in Park Management and Their Requirements

The jobs of the modern park planner and administrator call for a variety of technical skills. The care for the forests and their protection from fire require the knowledge of foresters, and many foresters find employment in state and national park work; but successful park planning and administration must draw on a variety of additional skills.

These are exemplified in the training requirements for men who enter park work. These have been established by special committees of three nationwide park organizations, the National Conference on State Parks, the American Institute of Park Executives, and the National Park Service. The committees pointed out the need for studying the basic sciences such as botany, chemistry, geology, entomology, and zoology. In addition, they urged the inclusion of specific subjects related to forest areas such as wood technology, wildlife management, pathology, forest entomology, and silviculture. But, most important of all, they stressed the need for a background in landscape design, an appreciation of architecture, a knowledge of engineering and construction methods, and familiarity with plant materials other than forest trees.

Landscape architecture is an important tool of park planning. It has been defined as "the fitting of land for human use and enjoyment." Such a definition is especially applicable to the planning of lands for recreational use. Training that fits men for national, state, county, and municipal park work also fits them to think properly about the development of private land as well. The city planner receives his basic training in landscape architecture as do the men who become camp planners, recreational planners, and park planners.

Drawing is to the landscape designer what writing is to most people, a means of expression. Since landscape planning deals with the presentation of ideas visually, drawing is the best means to the end. Park planners need to use drawings to show others what they plan to do and how they plan to do it. Those called upon to study a drawing are inclined to judge the skill and knowledge of the planner by the ability he has to express himself in the drawing. Inability to acquire reasonable facility in drafting limits the type of work a man can do.

The park manager also must have the ability to understand the wide range of purposes and ideals which will be found among his clients. He will need to express himself well in writing and speaking as well as in the graphic arts. His success will depend largely on how well he can "sell" himself and his ideas to his superiors. Skill in dealing with people is an essential trait for the person interested in park management.

Park management therefore is a special professional field closely related to forestry but with its own specialties in landscape design, land-use planning, and recreational planning. As public needs for recreation increase with growth of population, with concentration of people in cities, and with the specialization of modern industrial life, needs for park specialists likewise will grow. Especially is there need for broad recreational planning that encompasses all facilities, both public and private, that are available to satisfy public needs.

LITERATURE CITED

1. Chapman, Oscar L. 1950. "Annual Report of the Secretary of the Interior" (fiscal year ended June 30, 1950), U.S. Government Printing Office, Washington, D.C. 411 pp.
2. National Park Service. 1949. "Areas Administered by the National Park Service," U.S. Department of the Interior, Washington, D.C. 43 pp.

CHAPTER 14. *Protecting Soil and Watersheds*

NEW YEAR'S EVE in 1933 was a time of destruction, sorrow, and death for people who dwelt in La Canada Valley near Los Angeles. Heavy rains fell for two and one-half days on the steep San Gabriel Mountains, and a flood rushed out of the mountains, carrying silt, sand, and boulders weighing as much as 60 tons. The water with its heavy burden spread over the valley floor, wrecking homes and gardens and blocking highways. As floods go, it was not large nor was the total damage impressive. What was significant about this flood was a contributing cause of it. A few months earlier a fire had swept over the chaparral-covered mountains laying bare 7½ square miles. Rain fell equally on both burned and unburned slopes, but studies made after the flood revealed that, acre for acre, burned land contributed 50 times as much water to the flood flow as unburned land.[4] The area damaged by flood, 3 square miles, was almost half that burned by the fire.

Similar small floods that owed their origin to fires, overgrazing, or other misuse of land have occurred repeatedly, causing concentrated damage (Fig. 73).

Importance of Protecting Soil and Watersheds

Floods during the period 1924 to 1943 are estimated to have destroyed 2.2 billion dollars worth of property. Total direct and indirect flood damage based on 1947 prices has averaged 465 million dollars annually in recent years.[4]

Providing ample pure water for municipalities is becoming an increasingly costly process. Philadelphia alone spends 40 million dollars each year to improve the taste and odor of its water. Water in North Carolina from piedmont farm lands costs three times as much for desilting and chlorination as that from forested Appalachian watersheds. The U.S. Corps of Engineers estimate expenditures needed for water projects for the next 50 years to be 48.1 billion dollars based on 1947 prices.[4] This amount includes expenditures needed for flood control, navigation, hy-

FIG. 73. Floods from poorly protected mountain watersheds devastate villages and fertile valley lands. Mt. Pleasant, Utah. (*U.S. Forest Service photo.*)

droelectric power, irrigation, drainage of wet lands, and abatement of pollution. Water control is expensive.

How Forests Protect Soil and Water

Erosion is a geologic process that goes on continuously wherever air or water moves. Neither man nor vegetation can stop erosion completely. Even well-forested slopes may have streams that become turbid after heavy rains. What vegetation can do is to slow down the process so that soil is formed through mechanical and chemical breakdown of rock more rapidly than it is carried away by erosion. Erosion is then unnoticeable and unimportant from the human viewpoint.

When rain falls on bare soil, individual raindrops strike with considerable force, splashing soil on nearby objects to a height of over 12 inches above the ground and a distance twice as great. The surface soil is quickly puddled; that is, the pores leading to lower depths are sealed. Even light rainfall may cause surface runoff. Suspended silt in the water gives it a murky color as well as abrasive action. Sheet erosion begins where surface runoff starts. The water soon collects into rills that cut small trenches and run together, adding volume to the flow and power to its erosive action. Gully cutting begins. If allowed to proceed unchecked, accelerated erosion will eventually convert unprotected soil into deeply gullied badlands.[2]

Vegetative cover stops the process all along the line. Any form of continuous vegetative cover provides good protection for soil. Natural vegetation is usually better than cultivated crops, because it covers the soil more completely and because the litter and humus help maintain porosity. Forests, because they have heavy crowns and deeply penetrating roots, provide maximum protection against erosion of soil and against rapid surface runoff.

Trees intercept rainfall and break the force of the falling drops. At the same time they retain on their leaves and branches a considerable amount of moisture, especially from showers of short duration, much of which is reevaporated into the air without ever reaching the ground surface. This phenomenon is of benefit in eliminating water that might cause floods but represents a loss of water to feed streams for irrigation and for municipal use. Experiments have shown that tree crowns may intercept as much as 30 to 50 per cent of the total precipitation when this occurs in small showers of short duration separated by periods of clear weather. For long rainfall the percentage interception may still be 10 to 40 per cent, depending on species and density of forest.[5]

After the rain or snow penetrates the tree canopy, it falls, not on bare ground, but on a litter-strewn forest floor. Here leaves, remnants of leaves, twigs, and sometimes a cover of herbaceous plants again intercept the raindrop and prevent soil movement. The resilient, litter-covered forest floor absorbs all the force of the falling drops and allows the clear water to filter slowly through the organic debris and to enter the soil gently without sealing the natural pores, and root, earthworm, and other channels. Surface runoff, therefore, is almost totally lacking from a forest-covered area until after the soil itself becomes completely saturated with water.

Experiments have shown that forest litter and humus have especially effective action in slowing down the rate at which water enters the mineral soil. This action seems to be largely a matter of obstructing the flow of water and holding it back through absorption by the many small particles of litter and humus material. These fine particles tend to swell as they become wetted, and in this manner they absorb two or three times their weight of water. It has been found that a layer of rich forest humus 6 inches in thickness may detain as much as 4 inches of rainfall. Within a period of four or five days more than half of this water is given up to the underlying soil. The litter is then again ready to accept and hold an appreciable amount of rainfall.

The soil mantle itself, if relatively dry and deep, can absorb and detain vast amounts of rainfall. Each foot of soil can absorb from 1 to 5 inches of water. Since this water is retained against the force of gravity, it is available for the growing of plants. This water does not reach underground water-storage reservoirs. The soil acts like a self-emptying reservoir in holding back much floodwater.

The roots serve as a great binding force in holding the soil together and preventing it from moving down the steep slopes. Of particular importance is the binding of the surface 6 or 8 inches of soil. If this is held firmly in place, the underlying layers are less likely to slip. As tree roots are concentrated largely in the upper 6 inches of soil, oftentimes completely permeating it, they add greatly to soil cohesion. For municipal watersheds trees have an especially valuable function in preventing the washing of contaminating materials into the reservoirs. The abundance of small plant and animal life that inhabits the forest floor makes it a great sanitizing agent by breaking down dead material. Toxic and infectious plant and animal remains are broken down completely to harmless chemical compounds by the organisms of the soil. Consequently, contaminating materials seldom are carried by water or other means more than a few inches from where they fall in the forest.

Forest fires destroy the litter and may even convert the humus into bare, reddened mineral soil. Such burned land is highly vulnerable to erosion. Where grazing is heavy, domestic animals compact the soil, cut up the litter so that it quickly decomposes, and make bare paths, down which water runs with erosive force. Studies on pasture lands in Pennsylvania show that runoff from simulated rainfall applied at the rate of 1.4 inches for a one-hour period may vary from none to 80 per cent, depending on the intensity of grazing. Compaction of soil was the primary cause of excessive runoff on heavily grazed pastures. The combined effect of compaction and high runoff greatly reduced pasture productivity.[1]

Timber cutting in itself does not directly change the soil, but absence of an overhead canopy causes soil heating and rapid depletion of humus. Lack of tree cover prevents annual leaf fall to renew the litter. Skidding of logs makes pathways for erosive water. A cycle is thereby set in motion that impairs watershed protection until a new forest grows up and the humus layer is restored.

Use of Water by Forests

Trees require large amounts of water to maintain their life processes. Water is a medium for the transport of minerals needed by the plants. A small amount actually enters into the plant substance in the form of cellulose and lignin through the photosynthetic process. Most of the water absorbed by the trees is transpired into the atmosphere. It has been estimated that a medium-sized elm tree will transpire as much as 15,000 pounds of water on a single hot, dry, clear day. Transpiration is particularly heavy from trees that grow along watercourses, where they have access to an abundant supply. Trees, like men, are prodigal in the use of water when plenty is available and use water sparingly only when it becomes scarce. In the southern Appalachian regions losses of water through evaporation and transpiration from well-drained forested slopes may amount to 15 or more inches per year.[3]

Types of Damage Forests Reduce

Snowslides. In steep, mountainous country snowslides are often caused by the heavy fall of new snow on top of an old crust. Where slopes are very steep, the new snow may slide, and in areas that are subject to frequent snowslides, forests are prevented from growing. Such areas often occur at the heads of very steep valleys and along small side valleys. Protection can be achieved to a limited extent by encouraging tree growth and installing other protective measures on the side walls. The valley heads can be held only by engineering construction which may or may not be justified.[5]

Landslides or Avalanches. These are often frightening affairs. They occur when a great mass of soil lying on a steep slope becomes saturated with moisture and slides over the underlying rock surface. A single severe storm that occurred along the Lehigh River in 1943 gave rise to over 50 small avalanches in a 20-mile reach. These descended onto the railway tracks and, together with bank cutting by the river, caused damage of some $500,000. The valley walls in this case were rocky and covered with tree and shrubby growth, but frequent fires had thinned out the tree cover, thereby reducing their protective influence. It was observed that very few of the avalanches occurred on areas that had a good cover of trees. Avalanches may occur frequently in high mountains where tree cover is lacking or stunted.[5]

Gullies. Where heavy surface runoff rushes across unprotected soil having no permanent stream channels, gullies occur. Those that are actively eroding cannot be stopped by the simple expedient of planting trees. The water which causes the cutting must first be diverted or impeded in its course and the soil stabilized to some extent by other measures. Once this has been achieved, tree growth is possible and permanent stabilization of the gully may be effected.

Silting. The silting and drifting of hardwood leaves into municipal reservoirs can be minimized by planting the borders to conifers. Some cities even control the entire watershed and maintain a forest cover on it (Fig. 74). In the western United States water is precious, more precious than land. Water caught in the high mountains makes possible the irrigation of fertile valley lands that yield rich agricultural harvests. The

FIG. 74. Conifer plantations along municipal reservoirs prevent staining of water by hardwood leaves. Waverly, N.Y. (*U.S. Forest Service photo.*)

yield of water for irrigation is the all-important factor in the economic development of much of the West. Forests play a dual role in western water supply. They are necessary to protect soils and ensure infiltration of water, which is fed into the streams gradually throughout the summer, supplying irrigation water when it is most needed. At the same time, the forests intercept rain and snow and use water for their own needs, thereby restricting the water supply for the valley. The proper balance between these two actions is the objective of the watershed manager.

Flood Damage. Minimizing flood damage is perhaps the most spectacular of all the roles of forests. More than 100 million dollars worth of damage is caused annually along the Mississippi River system alone. The more the land is cleared of forests and the more clean-tilled cropland that exists in the watershed, the more frequent and serious floods become. The more we build up our cities on the great flood plains of the rivers, the greater the property destruction when floods do occur.

Erosion by Wind. Forests protect soils against erosion by wind. The great program of shelter-belt planting in the prairie-plains section of the United States has helped to reduce wind erosion and ameliorated living conditions in this vast area.[7]

Forests stabilize sand dunes along the seacoasts and along the shores of our great lakes. Dunes, once started moving, can destroy large areas of valuable agricultural, recreational, and even residential land unless they are stabilized.

Flood-control Surveys

Congress has authorized preliminary examination of more than 600 watersheds to determine whether or not forestry and other land-improvement measures would be effective in reducing flood damage. By 1950 the Department of Agriculture had completed preliminary examinations on 164 of these 600 watersheds and surveys on 18. Work programs have been authorized for 11 of the watersheds for which survey reports have been completed. It is expected that new authorizations will be made when additional detailed surveys are completed.

Flood-control surveys involve elaborate field work and techniques, that have only recently been worked out. In fact, the whole task of making flood-control surveys is still in the evolutionary stage. Congress has been reluctant to authorize flood-control work programs on an area unless the benefits through reduction of flood damage were sufficient to justify the costs involved. As forestry and other land-use measures may bring in ancillary benefits in the form of timber and recreational use of the land, the problem of evaluating the flood-control program becomes very involved. One of the most useful measures to increase the effectiveness of forests in ameliorating floods is to improve protection against fires.

Foresters who are employed on flood-control surveys must work with large areas. The use of air photos is helpful in mapping forest vegetation. A flood-control survey is an attempt to appraise the effectiveness of various land-use measures applied to a specific watershed.

A flood-control team will have among its members a hydrologist. The forester must be prepared to provide the hydrologist information on rates of absorption and movement of water in forest soils so that he can calculate runoff from storms of various intensities and route flood crests that would result from these.

The final justification of any major public program rests on the benefits it will achieve in proportion to the costs. This means that a very careful evaluation must be made throughout.

Watersheds for Irrigation and Municipal Supplies

Parks, municipal watershed, and strategic flood-source areas are managed with soil protection in mind. The task is by no means simple and varies from one region to another. The job of the protection forest manager in much of the western mountain region is primarily to obtain the maximum water yield from the land. He must, of course, also make certain that this water yield is in usable form and occurs at the proper season of year. This means that he must learn to cultivate those types of forest that intercept and transpire a minimum of water. The forests and other cover must be such that they hold the soil in place and allow the water to flow through the soil rather than over the surface. Special examination of such watersheds must be made frequently to determine the capacity of the vegetation covering the soil to hold it in place and yet not to use too much of the precious water. Grazing and timber use should be permitted to the extent possible without impairing watershed protection.

A forester managing an eastern municipal watershed area will be concerned primarily with avoiding erosion and with keeping a good cover on the land so that the water in the reservoir is clear and of high purity. Contamination and erosion are his two major concerns. He will also wish to obtain from the forested watershed a yield of valuable timber to augment the income of the water department. This can frequently be done to the advantage of the forest and the reservoir system as well. Care must be taken to locate logging roads and skidways where they do no damage, to avoid large clear-cut areas, and to exclude fires. A young forest must promptly take over on areas from which the old trees have been removed. A watershed manager may remove trees along the banks of lakes and streams to prevent their heavy use of water and may use shrubs and herbaceous plants to control road banks, skidways, gullies, and other areas where erosion occurs. He should, therefore, have a wide acquaintance with herbaceous and woody vegetation and their ecologic relationship. He should also understand simple engineering work often needed to stabilize soils, particularly gullies.

River-basin Planning

In 1950 the Federal government in cooperation with the states began to explore comprehensive planning for river-basin development. Such planning is intended to encompass measures for control of floods (Fig. 75). But it is to go further to include water for municipal use and irrigation, power developments, erosion control, pollution abatement, improvement of streams for fishing and recreation, and general development of forest and other renewable resources. The programs envisaged are similar to those projected by the short-lived National Resources Planning Board. The Tennessee Valley Authority program is perhaps the nearest we now have to action of the type proposed. All Federal agencies active in a river-basin area are being drawn into the planning groups, and representatives of the several state governments have been invited to participate.

Such broad-scale planning cannot be quickly consummated, nor can action be undertaken without reconciling many conflicting interests. The

FIG. 75. Flat weirs for measuring debris-laden storm flow. San Dimas Experimental Forest. (*U.S. Forest Service photo.*)

success that awaits such planning will depend in no small degree upon the patience, diplomacy, and leadership displayed by those responsible for the program.

Shelter-belt Planting

The story is told that when President Franklin Roosevelt was caught in a severe dust storm in Oklahoma, he penciled a note asking the Chief of the Forest Service why he could not plant a mile-wide belt of trees from Canada to the Gulf to stop such storms. The Chief was at a loss to answer until one of his advisers pointed out that while one belt of trees would not stop dust blowing, hundreds of them properly placed would go far toward doing so. Thus the Prairie States Shelter-belt Program came into being.

The soils, natural vegetation, and results of past tree planting were surveyed throughout the region. Hardy trees and shrubs suited to the various conditions to be encountered from north to south were selected and propagated. Planting designs were made, consisting of shrubs, low trees, and high trees. Contracts with farmers to fence, plow, plant, and cultivate the trees were drawn up. Between 1935 and 1949, 26,873 miles of tree belts were planted and additional plantings were made to protect farm homes and gardens.[7] Shelter-belt planting has also been active outside the prairie states. Wisconsin farmers have planted 5,942 miles, California fruit-tree growers 2,000 miles, and Indiana muck-land farmers 100 miles.

The general public is concerned about two questions relative to publicly sponsored shelter-belt programs: were the plantings successful, and did the resulting tree belts protect fields and buildings? The answer to the first question is an unqualified yes. The shelter-belt plantings had high survival and rapid growth. The answer to the second question is also positive, though it is difficult to express quantitatively. Shelter belts trap snow, thereby building up soil moisture in a region of deficient precipitation. They prevent soil from blowing on newly seeded fields. They protect crops from hot, dry winds of summer, and they shelter barns, livestock, and the farmers' homes against driving winter storms. Farmers of Nebraska and South Dakota estimate the cash value of shelter belts in crop protection to be $43 to $60 a year per farm. Cotton yields in Oklahoma and Texas are reported to have increased from 8 to 17 per cent, depending on nearness to the protecting belt. The total protected area extends a distance to leeward of 20 times the height of the trees. They therefore have reduced dust storms in numbers and severity.

Forests and Climate

Climate has a great influence on forests. Where annual rainfall is less than 15 to 20 inches in northern lands or 35 to 40 inches in the tropics, forests, as such, disappear, although trees may still persist in sheltered valleys, along stream courses, and as scattered scrubby individuals. Because many regions that supported trees in the past have become arid waste lands following forest removal by man, the question arises: do forests influence climate? The answer to this is that locally they have marked influence on climate but that this influence diminishes so as to be indistinguishable within a short distance outside the forest boundary.[8]

Tree canopies may absorb and reflect 70 to 90 per cent of sunlight, thereby reducing light to support undergrowth and heat to warm the soil. By intercepting substantial quantities of rain and snow, they allow it to evaporate without reaching the ground. They may reduce wind movement to 20 per cent of that occurring outside the forest. Evaporation is decreased and relative humidity increased. Temperature extremes are moderated. Coniferous trees impede snow melting in spring. All these effects are easily measurable.

Forests also evaporate large quantities of moisture, which may equal throughout the year 50 to 70 per cent of all rain that falls. Where forests are extensive, this amount adds appreciably to atmospheric moisture, that must be reprecipitated somewhere. Forests may be as effective as free water surfaces or more so in adding to atmospheric moisture. During summer, because of the cold temperature of the water, Lake Superior dries rather than humidifies air that passes over it. Because air masses travel great distances, it is impossible to assign any quantitative effect of forests on world-wide climate. Their influence is believed to be of minor significance compared with that of oceans.

Induced Rainfall

Recently man is reported to have achieved some success in inducing rainfall by seeding clouds with dry ice or silver iodide smoke.[6] Experiments up to 1951 were by no means conclusive however. Whether cloud seeding resulted in more rainfall than would have occurred naturally is still a controversial question. Also whether cloud seeding robs adjacent areas of rainfall these might otherwise receive is still to be proved. Nonetheless, cloud seeding was tried over forested watersheds to fill New York City reservoirs. It has been attempted in western states as a means of discharging electrically charged cumulus clouds likely to set forest fires. This much is certain: if man succeeds in inducing rainfall, he must learn to use this power with great circumspection. Otherwise he may cause great damage through erosion and floods. The fields for skilled soil scientists, protection foresters, and flood-control experts will be greatly enlarged if rain making becomes practicable by human action.

LITERATURE CITED

1. Alderfer, R. B., and R. R. Robinson. 1947. Runoff from Pastures in Relation to Grazing Intensity and Soil Compaction, *Journal of the American Society of Agronomy,* 39:948–958.
2. Bennett, Hugh Hammond. 1939. "Soil Conservation," McGraw-Hill Book Company, Inc., New York. 993 pp.
3. Craddock, George W., and Charles R. Hursh. 1949. Watersheds and How to Care for Them, "Trees, The Yearbook of Agriculture," pp. 603–609, U.S. Department of Agriculture, Washington, D.C.
4. Frank, Bernard, and Anthony Netboy. 1950. "Water, Land, and People," Alfred A. Knopf, Inc., New York. 331 pp.
5. Kittredge, Joseph. 1948. "Forest Influences," McGraw-Hill Book Company, Inc., New York. 394 pp.
6. Langmuir, Irving. 1950. Control of Precipitation from Cumulus Clouds by Various Seeding Techniques, *Science,* 112:35–41.
7. Stoeckeler, Joseph H., and Ross A. Williams. 1949. Windbreaks and Shelterbelts, "Trees, The Yearbook of Agriculture," pp. 191–199. U.S. Department of Agriculture, Washington, D.C.
8. Zon, Raphael. 1941. Climate and the Nation's Forests, "Climate and Man, The Yearbook of Agriculture," pp. 477–498. U.S. Department of Agriculture, Washington, D.C.

CHAPTER 15. *The Business of Forestry*

A FOREST MANAGER is employed to operate a property in accordance with the wishes of the owner. If the owner is the public, these wishes may encompass many things: watershed protection, public hunting grounds, grazing, recreation, and timber production. If the owner operates a wood-using plant, he will need to meet all or a part of his timber requirements. Private individuals or corporations dependent on the forest as a primary source of income will want the land managed so as to produce a profit, and they will want this profit to be produced as regularly as possible. Return on investment is the major concern of such owners, but profit may be a minor concern to owners primarily interested in recreation or hunting or to wood-using industries whose main source of income is from processing rather than growing timber.

The manager of a retail store can obtain from departments of commerce and other sources information on capital, floor space, inventory, and turnover for a given business volume. How the budget should be distributed among such costs as rents, advertising, labor, store supplies, and miscellaneous expenses may also be learned. The Agricultural Extension Services can advise a prospective dairy farmer on size of herd, land area and quality, and the crops he should grow. He is furnished feeding schedules for dry stock and milkers. Equipment needs and overall capital expenditures can be estimated. Suggested work schedules to budget his time most effectively and appropriate accounting records to determine the financial success of his farming venture are also offered. These serve to minimize business risks.

A man desiring to enter the business of timber growing finds few such comprehensive services to aid him. Public foresters will advise timber owners on tree planting, on silviculture, and even on how to sell timber, but they have little tested business advice to offer. The owner must rely mostly upon his own judgment and upon the opportunities that come his way. The very fact that timber growing is not well understood means that it does offer some good opportunities to those who comprehend the basic business factors involved and are willing to venture in a new field of endeavor.

Basic Factors in Timber Growing

A producing timber property has certain basic features analogous to those to be found in a manufacturing enterprise. Climate, topography, and soil are the timber grower's basic physical factors governing the supply and availability of his raw materials. As such, they are analogous to the location and building that houses an industrial plant. The soil supplies two raw materials, water and mineral elements essential for tree growth. These may be available in ample amounts or in limited amounts that restrict rapidity of growth. The third, carbon dioxide of the air, is so widely distributed that it never need give concern to a forest manager. Sunlight supplies the energy for manufacturing woody substance from minerals, water, and carbon dioxide. Temperature and the efficiency of chloroplasts determine how fast these raw materials can be converted into wood products by the trees. The forest manager, once his location has been set, has no control over climate, soil, or topography. All he can do is to adjust his operations so as to take the best possible advantage of these basic factors.

The Producing Machinery. The production machinery of a forest is the growing stock, that is, the trees of all sizes and ages that occupy the soil. If this growing stock is well distributed over the land, if it is likewise well distributed as to age, and if it is made up of valuable species and well-formed units, it will produce regular crops of high-value products. If the growing stock is sparse or scattered, if it is composed of a high percentage of low-valued species or poor-quality units, it will produce a low volume and a low value of products. Such a forest is analogous to a factory using improperly arranged, worn out, and obsolete machinery. A fire may just as effectively destroy the producing machinery of the forest as might a fire in an industrial plant.

One striking difference exists between the forest and the industrial plant. The trees that grow the wood are both the producing machines and the end product. This fact has confused many people unfamiliar with forest management. An industrial owner would scarcely expect his factory to be productive if he sold off his machinery, allowed fire to gut his factory, and invested no new capital to replace it. But that is exactly what some people seem to expect of forests. They expect land from which the producing machines—the growing trees—have been stripped off and on which the young trees have been killed by fire to produce valuable timber crops with little or no investment of capital.

Forest Valuation. A forest property may be evaluated as an industrial plant on two bases, its liquidation value and its value as a productive plant. Its liquidation value is the value of the standing trees for immediate harvest, or harvest as early as feasible, plus the value of the land and improvements. Young unmerchantable timber is considered as valueless in such an appraisal. This is the basis on which most forest properties have been bought and sold, because the economics of growing timber have been little appreciated by most buyers and owners. The value of a forest as a productive property is based on how much it can yield continuously above expenses. Growth rate is here the all-important factor. This growth when converted to dollar value and after deducting expenses is then capitalized at an appropriate rate of interest to reflect business risks. A property yielding a net revenue of $10,000 annually would have a value of $200,000, assuming a 5 per cent return for capital and risk. The liquidating value of such a property might be considerably less or more than this amount.[3, 4]

Bare forest land that must be planted and tended for 40 or more years before products can be harvested has very low value. In fact, it may have only a negative value for the private investor, because the costs and risks involved may exceed prospective return. Such lands, if extensive, can be restored to forest only by the slow process of nature or by public expense. In the latter case subsidiary values inherent in productive forest land would be expected to offset the short-time deficit in return.

Tending the Forest-producing Machine. The job of operating the forest factory so as to produce efficiently high volume and quality of timber products is the job of the silviculturist. It is through silviculture that we select those producing machines that are most efficient and eliminate those that produce a poor-quality product. Silviculture means making sure that all the land owned is well stocked with growing trees. It involves eliminating the ineffective timber producers, by thinning, improvement cuttings, release cuttings, and other operations, so as to concentrate the growth on those trees that will develop the highest quality and volume of timber.

Organization of the Timber Factory. An industrial plant may have the best of raw materials and the best of modern machinery and may be suited to turning out the finest products in the field; yet unless this factory and its working force are organized so that the various operations of production proceed in an orderly and efficient manner, the enterprise will be doomed to financial failure. The production manager's job in an industrial plant is to organize machinery, labor, and flow of materials so that high-quality products can be turned out with the minimum total cost in manpower, machine power, and factory rent.

The forest manager's job is to organize the forest for efficient use of soil, growing stock, labor, and equipment. He decides what roads, trails,

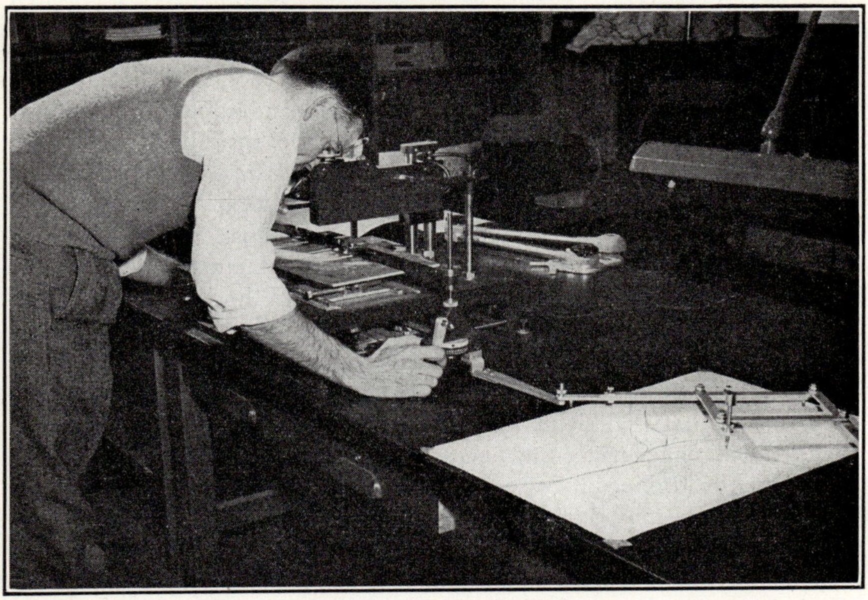

FIG. 76. Making a forest map from aerial photos, with a sterioplotter.

skidways, and other improvements are needed. He locates his protection equipment so that the producing plant can be readily protected against fires and trespass. He decides on an orderly arrangement of forest compartments so that the road and skidway systems will be used to best advantage, thereby minimizing his costs of operation (Fig. 76). He attempts to adjust his growing stock so as to provide a regular yield of products, year by year.[6] He will time his timber-cutting operations so as to take advantage of good markets and will curtail selling when markets are low. He knows that he can store his inventory on the stump and have it progressively increase in value through growth, and he also knows that there is a limitation as to how long he can allow his inventory to accumulate without having it congest his factory so that total production slows down.

The forest manager also decides on what equipment he requires to manage his forest most effectively. This includes personal transportation, protection equipment, logging equipment, road-building and maintenance equipment, and record-keeping equipment. He will find only limited guides available here.

Forest Labor. The industrial manager employs labor to tend his machines that perform various manufacturing operations. The forest manager employs forest labor to protect his forest and to keep it in the best possible growing condition. He also requires labor to prepare his products for shipment, that is, to fell, limb, buck, skid, and haul his logs and other forest products. As his producing plant may cover several thousand acres, he, like many industrial managers, must consider housing his workers. He must decide how many workers he can employ permanently and the extent to which he can depend on temporary labor to speed up his operations during times of good markets. This may mean providing housing for permanent workers and camps for temporary workers.

Markets. One of the major tasks of the forest manager is to find good markets for all the different types of products his factory produces. The forest factory produces a large amount of medium- and low-grade products, a relatively small amount of high-grade products. Some forests, in fact, produce almost no high-grade timber, and even the best of forests, having all age classes represented, will produce a large volume of small-sized timber that is difficult to market. This fact immediately confronts the forest manager with the necessity of deciding how far he should enter into the timber-processing field in order to provide a market for his products. He may decide to operate a small portable sawmill to convert to rough lumber the low-quality logs that will not pay transportation costs to a permanent mill.[2]

It is in the marketing field that forest managers encounter their most serious difficulties. It is here that the greatest financial headaches arise. Nationally, the markets for forest products tend to get progressively better decade by decade, but this may not be true locally. Local markets depend on the presence and economic prosperity of local wood-using industries. Timber is so costly to transport that a good market for fuel wood in New York City means nothing to the Adirondack timberland owner.

The forest manager must always balance carefully the cost of owning wood-processing equipment against that of hiring the use of such equipment. The same applies to logging trucks and tractors. If the forest manager has full-time use for such equipment, he can well afford to own it. If he has only part-time use for it, he might better rent or contract for its use.

Record Keeping. The forest manager obviously cannot know whether his operations are financially successful unless he keeps accurate records, including an inventory of his land, growing stock, improvements, and equipment. He must know whether his land is increasing or decreasing in value, his timber inventory getting smaller or larger in volume, lower or higher in quality. He must keep separately his costs for taxes, protection, general management, logging, sale of products, and timber processing. Costs of unit operations reveal which of his various operations are making money and which ones are losing money and form the basis for intelligent action to produce profits.

Forest Credit

The forest manager must understand how to use money for capital improvements. He may be obliged to borrow. If so, he should know whether the interest he must pay will be offset by the earning value of the purchase he makes. Borrowing money on timber-growing stock is likely to prove financially unwise, since the interest rate on money may equal the percentage growth rate of the forest. An annual growth rate of 3 to 5 per cent is about as high as can be expected in the North.

The astute forest manager will realize, however, that value increases greatly with size of tree and will therefore avoid premature harvesting of young timber. A 60-year-old hard maple tree, 12 inches in diameter, may contain 50 board feet. At 90 years of age it could be 18 inches in diameter and have 200 board feet. Its value will correspondingly increase from $0.50 to $3.00. An understanding of how rapidly trees can increase in value is a part of the forest manager's knowledge and his justification for using credit.

Forest financing also involves forest insurance and insurance required of those who employ labor. Adequate insurance of forest properties against fire is difficult to obtain at reasonable rates. Large forest property owners require no insurance against the small fire but do require insurance against large fires that might destroy the major proportion of their forest.

Insurance against severe damage by windstorm, ice, snow, insects, tree diseases, and animals is even less obtainable. All of these are potential sources of crippling losses, especially to young growth. The owner's best insurance is intensive management that will provide for prompt salvage of whatever losses occur. A hurricane, such as swept over New England in 1938, or a huge fire, such as burned over 220,000 acres in Maine in 1947, can destroy a forest business for the lifetime of any individual owner.

The timberland owner, particularly if he engages in logging or any other timber operations, must carry workmen's compensation insurance and social security for his workmen. He must, in fact, be well informed on all requirements affecting his financial and business operations. Basic knowledge of business law and contracts is essential.

Business Acumen

The over-all success of the forest manager will depend, in the last analysis, upon his business acumen. This means the skill with which he is able to manipulate his forest, his growing stock, his forest labor, his marketing and sales, and his use of credit and insurance so as to maximize financial returns from operating the property. Business acumen in forestry is made up of sound basic knowledge of silviculture, logging, personnel management, mechanical equipment, accounting, budgeting, marketing, and business law. Energy and drive must also be put into managing operations and seeking profitable markets for products. The forest manager must keep an adequate reserve of working capital so that his operations may be flexible. Forced liquidation of valuable growing stock when markets are badly depressed can be disastrous. A man may be skilled in most phases of forest management, but if he neglects any one essential element, the entire operation may fail. The wise manager gives equal attention to all phases of his operation.

The greatest financial hazards to the business of timber growing are market fluctuations caused by business and building cycles (Figs. 77 and 78). Timber markets may virtually vanish for two or more years at a time. This may cause no hardship to the casual forest owner who depends on other income for his livelihood, but for an owner who receives all his income from timber, and who has annual fixed charges for labor, interest, and taxes, a prolonged business depression can be ruinous.

Business cycles cannot be predicted nor their turning point identified, except in retrospect. It seems unlikely that economists in government and in business will have sufficient understanding of the forces that influence business cycles to control them completely in the near future.

How Business Cycles Affect Building Cycles. It is during periods of expanding business activity that manufacturers expand their plants and that workmen have sufficient money to build or purchase homes of their own. Consequently, building activities expand very rapidly during the

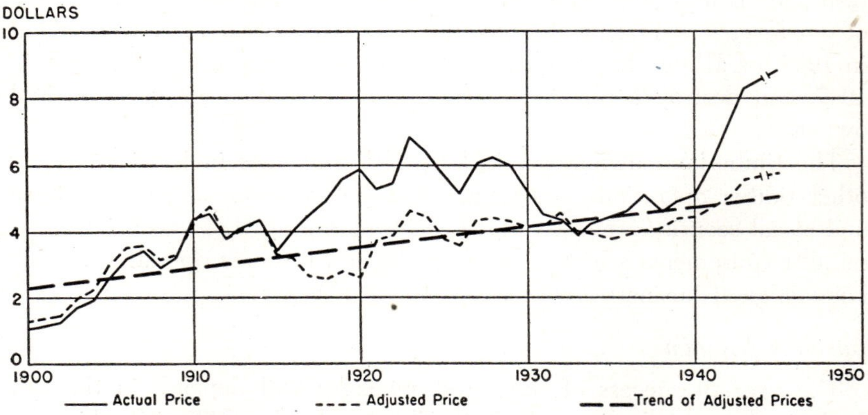

STUMPAGE PRICES IN THE EASTERN U. S., ACTUAL AND ADJUSTED FOR THE PURCHASING POWER OF THE DOLLAR. 1900-1945

(BASIS - 187,350 MILLION FEET BOARD MEASURE)

FIG. 77. (*U.S. Forest Service.*)

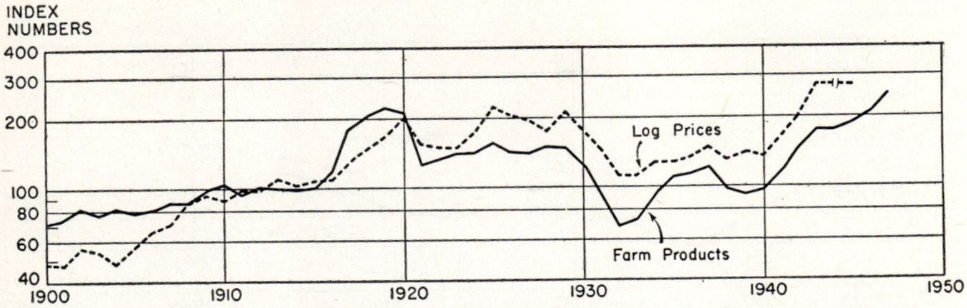

INDEX NUMBERS OF LOG PRICES IN THE EASTERN U. S. AND THE PRICE OF ALL FARM PRODUCTS, 1900 - 1947. 1910 - 1914 = 100 PERCENT

FIG. 78. (*U.S. Forest Service.*)

ascending phase of the business cycle. Building cycles show higher peaks and lower troughs than industrial production. The index of building activity in 1918 at the close of the war was approximately 25; by the year 1925 it had risen more than tenfold to 260. During the same period business activity as measured by industrial production increased only from 60 to 90. By 1932 building had dropped from index 260 back to 20, and industrial production, which reached its peak in 1929, dropped from 115 down to about 55.[10]

Lumber production in 1918 stood at index 125. It rose to 160 in 1920 and fell to 95 by 1922. It reached a peak of 175 in 1925 and in 1932 a trough of 50. Lumber production appears to fluctuate less violently than the building cycle but more violently than the business cycle.

When lumber production fluctuates violently, it obviously influences logging operations, prices of stumpage, and, consequently, the entire forest economy of the country, both public and private. A forester manager should be prepared to make relatively heavy timber harvests during periods of high lumber demand and to restrain operations during periods of low demand. During such periods he might appropriately engage in planting and other activities that build up the value of the forest but do not put products on the market. To do this requires both sound business judgment and ample working capital.

Long-term Trends in Lumber Prices. Long-term trends in American business enterprise have been marked by an increase in abundance and quality of output and a decrease in labor requirement. This has frequently meant a decline in price or at least a decline in man-hours of labor required to purchase the products. This, in turn, has been possible by increasing greatly the productivity of labor through modern production techniques, standardized design, power machinery, and applied research. Living standards have risen thereby. This trend has characterized the

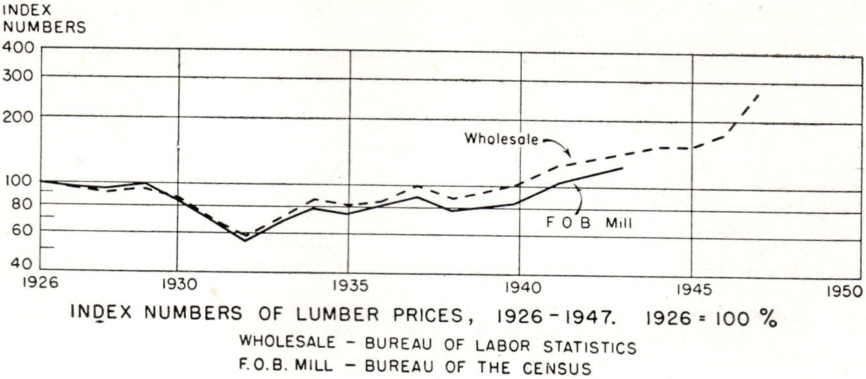

FIG. 79. (*U.S. Forest Service.*)

manufacture of such products as chemicals, steel, motorcars, household appliances, and power equipment.

It has been less true of the lumber industry. During the period 1860 to 1945 lumber prices increased tenfold. During the same period the price of all building materials increased two and one-half times and the price of all commodities one and four-fifths times.[5] As a result the competitive position of lumber in the building and other industries has become weaker. Obviously, one of the major causes of increase in lumber price has been the increase in the cost of labor in logging and sawmill operations without a corresponding increase in output per workman. Where trees are scattered and small in size and the lumber sawed is knotty, it has been impossible to maintain quality of product at the same price as in former years. When good trees are cut and poor ones left, quality declines. As the pressure for cutting extends down to smaller-sized trees, output per worker continues to decline and cost of product to increase. To arrest this trend will require fairly intensive forest management, and certainly much business and financial skill (Fig. 79).

Sources of Information on Forestry Business

The foregoing may have given the impression that anyone who launches a forestry business does so completely in the dark. This is far from the case. Certain phases of forestry are well advanced and a businessman can accept knowledge in these fields with little reservation. Foresters can advise timberland buyers on quality of soil, value of species, and the rate at which they may be expected to produce timber. Books on forest finance present methods that are used in financing forest operations with a discussion of the relative merits of each. The prospective timberland buyer can determine how forests are evaluated and be thus in a position to judge for himself the desirability of a particular forest property for investment purposes. Estimating timber is possible with a high degree of precision. Hence, no one need buy a forest property without knowing what it contains and what growth rate he may expect.

Books on logging give information on how the timber crop can most easily be harvested. Distribution of costs among stumpage, logging, and lumber manufacture are available for average operations.[1, 2, 3, 7, 8, 9] Information is also to be had on how to organize a forest property for efficient management.[6]

It is, after all, in the planning of the timber operations that much of the success or failure of property operation is determined. The plans for operating a forest property may consist of a master plan together with special plans for managing recreation, timber resource, wildlife, grazing, fire control, the road system, tree planting, and other specific operations. In elaborate cases the timber-management working plan will include a discussion of the soils, timber types, growth rates, expected yield, cutting cycles, rotations, markets for timber products, availability of labor, and requirements of local industries. Such elaborate plans are ordinarily prepared only for public forests, but the items contained therein need to be given some consideration by private forest land managers as well. Even more detailed consideration must be given to business and financial planning so that operations are conducted with a minimum use of capital and labor and a maximum long-term financial return.

A great deal of information does exist on various phases of the business of forestry. What is mainly lacking is the integration of all of this information into management manuals that can be effectively used by the individual forest manager. The difficulty is that few properties have been managed on a long-term basis. Hence little information on the financial results of management is available, particularly of the management of properties on which a variety of products are harvested. Operations such as those of the Goodman Lumber Company in Wisconsin and the Crossett Lumber Company in Arkansas, where all types of products the forest grows are processed in the mills of the company, are still few, and detailed financial information on these is not available. It is this over-all task of bringing together all the information from a wide variety of sources that must be successfully performed by the forest manager who wishes to make a financial profit from timber growing.

Unfavorable and Favorable Factors

Unfavorable factors in the business of forestry include the following:

1. The scarcity of concrete examples of successful long-time forestry business.

2. Hazards caused by fire, insects, disease, windstorm, animals, snowbreak, and other factors and the limited availability of insurance against the risk of damage from such factors.

3. A high degree of fluctuation in the national and local markets for forest products in accordance with business and building cycles.

4. The relatively low rate of timber growth, 3 to 5 per cent annually in the North.

5. The lack of appreciation by the public generally of the services the private forest landowner furnishes free to the public; that is, soil protection, watershed protection, some protection against floods, useful wildlife, and public recreation. The public is also careless in respecting the rights of the owner.

6. The present depleted condition of much of the cutover forest lands and the long period that will be required to build these forests up to a high-producing level.

7. Limited government support for owners attempting to improve the productivity of their forest land.

8. Local taxes unrelated to timber yields.

9. The general lack of integrated markets to absorb all the products the forest property produces.

10. The lack of a tradition of long-term forest management and a labor force that appreciates it.

Favorable factors are:

1. An abundance of forest land and good forest-producing soils throughout much of the United States.

2. A wealth of forest species that produce woods of high technical qualities.

3. The general abundance of natural regeneration of valuable trees.

4. Depletion of accessible standing timber with corresponding increase in stumpage values.

5. A rapid increase in forest research that is making possible new uses for wood and more valuable products, justifying a high stumpage price.

6. A high level of knowledge of soils, species, silviculture, and general management practices.

7. A satisfactory level of fire control, largely publicly supported, covering most forest lands in the country.

8. A growing interest in forests on the part of the public, the timber processors, and forest landowners.

9. Favorable consideration by the Federal income-tax law. Growth on forest land is taxed as long-term capital gain, hence the maximum tax is 25 per cent.

10. Widespread public aids in the form of research, statistical information, advice, and service.

Miscellaneous Revenue

The foregoing discussion has omitted consideration of income that may be derived from sale of hunting and trapping rights, water, recreation privileges, grazing rights, use for berry picking, and other subsidiary uses. Where income from such sources can be realized, it may stabilize and augment earnings from timber use, but such resources frequently yield insignificant income. Hunting and even grazing rights which have been allowed the general public without charge may not always be sold for restricted use without creating local resentment. Good will of local residents can mean much in fair taxation, care with fire, and reduction of trespass. The owner looking to timber for his main income will be slow to take steps that might alienate the good will of his neighbors.

All forest land supports game. All serves as a protective cover for watersheds. All protects the soil. Almost all serves to some extent as a source of aesthetic pleasure and perhaps for active recreation for people other than the owners. The management of every property directly and indirectly affects the livelihood of the people who harvest the timber and process it. The forest-land manager, therefore, is much more than a custodian of timberland and purveyor of stumpage. He is, so to speak, the ruler of a small empire. It therefore behooves him to acquire the dignity, restraint, and wisdom of a ruler, because the decisions he makes are bound to affect others than himself. If his decisions are wise, he can gain the admiration and support of his neighbors, his workmen, and his customers. Unless, therefore, a man is willing to assume responsibility for his acts and unless he has or can develop skill in dealing fairly with his fellow men whose interests his actions are certain to affect, he cannot be in the long run a successful forest manager.

The business of forestry is still a pioneering and exciting venture. It calls for a wide variety of skills coupled with energy and determination. Opportunities for amassing wealth are limited, but once the many difficulties are understood and overcome, the business of forestry may prove to be one of the most desirable and stable in our country. Forestry in many European countries is considered both sound and stable. The greatest speculative hazards in forestry are those due to fluctuating timber values caused by uncertain markets, business cycles, and the natural hazards of fire, windstorms, ice damage, insect outbreaks, and disease. Ultimately, insurance may be needed against these major hazards. It is appropriate to consider whether government, which has a major responsibility for protecting forests against fire, insects, and diseases, should not help underwrite such insurance. Government has already considered the desirability of extending long-term credit to timberland owners to enable them to avoid liquidation of their growing stock because of financial stringency.

The general outlook ahead for the business of forestry is bright. Opportunities are perhaps greater now, when hazards are high, than they will be after hazards to private forest management have been minimized.

Forestry, therefore, presents an opportunity and challenge to those who have the capital, interest, and daring to make it a business.

LITERATURE CITED

1. Brown, Nelson C. 1949. "Logging: The Principles and Methods of Harvesting Timber in the United States and Canada," John Wiley & Sons, Inc., New York. 418 pp.
2. Brown, Nelson C. 1947. "Lumber: Manufacture, Conditioning, Grading, Distribution, and Use," John Wiley & Sons, Inc., New York. 344 pp.
3. Buttrick, P. L. 1943. "Forest Economics and Finance," John Wiley & Sons, Inc., New York. 484 pp.
4. Chapman, Herman H., and Walter H. Meyer. 1947. "Forest Valuation," McGraw-Hill Book Company, Inc., New York. 521 pp.
5. James, Lee M. 1947. The Trend of Lumber Prices, *Journal of Forestry*, 45:646–649.
6. Matthews, Donald M. 1935. "Management of American Forests," McGraw-Hill Book Company, Inc., New York. 495 pp.
7. Matthews, Donald M. 1942. "Cost Control in the Logging Industry," McGraw-Hill Book Company, Inc., New York. 374 pp.
8. Simmons, Fred C. 1951. "Northeastern Loggers' Handbook," Agriculture Handbook 6, U.S. Department of Agriculture, Washington, D.C. 160 pp.
9. Wackerman, A. E. 1949. "Harvesting Timber Crops," McGraw-Hill Book Company, Inc., New York. 437 pp.
10. Zivnuska, John A. 1949. Commercial Forestry in an Unstable Economy, *Journal of Forestry*, 47:4–13.

CHAPTER 16. *Social Benefits of Forestry*

THE REAL JUSTIFICATION of a profession is that it produce benefits to society as a whole. The benefits that accrue to the practitioner and his clients are but a part of total social benefits. An example may make this clear. In 1927 a property of 2,200 acres was acquired by a college as a demonstration forest. At the time of purchase most of the timber had been cut off except on a parcel of about 200 acres that had a substantial volume in mature trees. The property was made up of a group of former farms, many of which had been abandoned for agriculture. Only three families were living on the land and these looked to outside sources for most of their income. The college employed a forester to plant the open lands, thin the young stands, and convert the cut trees into lumber, which was sold locally. Cutting was concentrated on the poorer quality trees and the white pines infected with blister rust fungus, and it was kept in line with growth of the old trees. After 20 years of forest management the property was providing full-time work for 11 men and the timber volume, instead of decreasing, was increasing. Land that had been a social liability became a social asset through application of the forester's skill.[3]

The Nature of Social Benefits

Social benefits of forestry include protection of soil and water, amelioration of floods, the fostering of useful wildlife, public recreation, the aesthetic value of forests as opposed to barren wasteland, favorable modification of local climate, supply of raw materials for industry, maintenance of reserve timber for national emergencies, and a source of useful work for human employment. It is the sum of these social benefits that justify public action to support forestry programs. All except the last two have been dealt with in earlier chapters.

Use of Forests and Forest Products in National Defense. The people of the United States have always used natural resources with great prodigality in warfare, and their use of wood has been no exception. Yet when it is recalled how much labor must go into the harvesting,

FIG. 80. Vast quantities of lumber and plywood are used in wartime for cantonments, warehouses, and advanced supply bases. (*U.S. Army Signal Corps, and A.F.P.I. photo.*)

processing, and use of wood for any purpose, we can realize how vital wood is. Lumber for cantonments and for packaging material to ship abroad made heavy inroads on reserve stocks in 1942 (Fig. 80). Requirements of the British and other Allies used much high-grade spruce, Douglas-fir, and veneer-grade birch. Spruce lumber and birch, basswood, and yellow-poplar veneer were used in the construction of aircraft, particularly training planes. Oak and ash went into truck bodies. Two-inch lumber was used in very large quantities in the construction of bases, warehouses, and docks. Lumber to repair bomb damage was also an urgent need. Piling for wharves was in heavy demand. The lumber required to package war material for overseas shipment in 1944 exceeded 16 billion board feet, half of the total lumber produced that year.

In the fighting areas lumber was needed in generous amounts for bridges, road repair, supply dumps, hospitals, and temporary shelters. The abundance of wood available to build up advance bases greatly facilitated the rapid movement of American troops once the offensives got under way.

Paper became scarce during war, and extra effort had to be exercised to keep the pulpwood supply flowing to the mills so that enough paper and paperboard could be manufactured to supply the industries producing munitions and essential civilian supplies.

The U.S. Forest Products Laboratory in Madison, Wis., developed new or improved types of wood products—Compreg, Impreg, Stapack, and molded plywood—all of which found special uses during the war period. Perhaps the most promising material was Papreg, a paper laminate with strength properties comparable to mild steel. It is nonmagnetic and can be molded into various shapes and forms. Compreg proved especially valuable for aircraft propellers, because it made possible the manufacture of hubs denser than the tips and any degree of strength and smoothness desired. Molded plywood shapes and forms were used extensively in aircraft manufacture. Laminated keels and skegs were developed for use in patrol torpedo boats, mine layers, and other wooden craft used during the war period.

New water-resistant fiberboards were designed especially for packaging food and other materials. Special types of boxes were constructed with airfoils so that materials might be dropped from airplanes without the use of parachutes.

Such expanded uses are possible in wartime only if our nation has at all times an adequate reserve of standing timber, including ample high-quality trees of useful species. The timber should be readily accessible to transportation and processing plants to avoid undue use of scarce manpower in times of national emergency, for lumber and most other wood products are too bulky to stock-pile for war use. The most efficient method is to maintain adequate reserves of high-quality standing timber in managed forest properties.

Employment for People. During the great depression of the 1930's any resource that could provide useful work for people was looked upon with high favor. Public leaders found in our undeveloped forests an almost inexhaustible supply of useful work that would add to the future value of the resource. Building needed roads and telephone lines, planting trees, and improving cutover lands figuratively soaked up man-days of labor. Needed work on public forests alone had by no means been exhausted when the programs terminated after a total of nine years of operation.

Of far greater importance than emergency programs has been the year in, year out employment in protecting forests, and in harvesting, and processing products (Figs. 81 and 82).

FIG. 81. Shelton, Wash., home of the Simpson Logging Company, manufacturers of lumber, and Rayonier Plant. The permanence of this community is assured by a 100-year, sustained-yield timber agreement covering company and national forest lands and by integrated use of forest products. (*A.F.P.I. photo.*)

Forestry as a Form of Land Use

The social values that stem from forestry are not easy to grasp in abstract terms. Understanding of the national and world-wide significance of applied forestry can be furthered by examining the situation in a specific county, state, and region. The older sections of our country that have gone through a high stage of industrial development well illustrate the trends. It is here that wealth is abundant, markets for labor and products most highly developed, and evolution in land use well advanced. It is in some of these states also that progressive steps have been taken to change the use of land to fit modern economic conditions. Such a case is Broome County, N.Y.

Forests of Broome County, N.Y. Broome County, N.Y., is one of the more prosperous counties of the Empire State. The valley in which the Triple Cities, Endicott, Johnson City, and Binghamton are located, has been called the Valley of Opportunity. It has been the valley of opportunity for many people; for in these three cities are located some of the most prosperous and rapidly growing industries of the United States.

Outside the cities opportunities are far more circumscribed. Originally, most of the county was farmed. The rural picture is similar to that in

FIG. 82. An attractive frame school building in a western lumber town. (*A.F.P.I. photo.*)

many rural counties of New England, the Middle Atlantic, and the Lake States.

The land area of the county is divided as follows:

	Area, acres
Active farm land	150,000
Idle farm land	103,000
Brush land	9,000
Forest land	161,000
Residential area (lakes and roads)	30,000
Total	453,000

During the past 12 years there has been a substantial increase in forest land and a decline in agricultural land.

The changes in land classes that occurred over this 12-year period are shown at the top of page 278.[5]

The two classifications reveal a limited amount of good land and a substantial decrease during the 12-year period in land suited to farming. The questions naturally arise: why was this land cleared for agriculture in the first place, and what has been happening over the years to drive the farmers off the land?

The early settlers who came to New York found the uplands much the easiest to clear for cultivation; the land was high, away from the

Class and description	Area, acres	
	1938	1950
Unsuited to agriculture, mostly idle or forested	120,636	146,165
Submarginal for permanent agriculture	104,237	136,647
Will support commercial agriculture in normal times	181,674	105,622
Will provide a satisfactory living if well managed	17,723	13,908
Will provide a good living if well managed		976
Residential	30,130	39,526
Publicly owned		11,556
Total	454,400	454,400

"miasma" and mosquitoes that infested the low lands; and natural drainage existed. Consequently, the hill farms were the best farms in the early days. But erosion washed away much of the fertile topsoil and, gradually, the initial fertility disappeared. The advent of machine tillage and other advances in scientific farming added to production of hill lands, but it added the same percentage increase to fertile, easily tilled valley farms. The income spread between the good and the poor farms expanded enormously, and hill farmers were forced out of competition. What once seemed a logical land-use pattern had been completely reversed.

A study of the land that remained in forest revealed that the area supporting sawtimber of medium and high volume was only 13,000 acres. Light sawtimber and high-volume pole timber occupied 60,000 acres; low-value pole stands, seedlings, and saplings occupied 82,000 acres; forest plantations occupied only 3,900 acres.

Timber volume was concentrated in trees less than 14 inches in diameter. The low-value species, beech, elm, and red maple, exceeded in volume, by far, the high-value species of hard maple, white ash, basswood, and white pine. The unsatisfactory stocking and species distribution and the low volume in large tree sizes were the direct result of cutting the most valuable trees in the past and allowing the less desirable to remain standing.

Optimism might be expressed over the fact that the annual growth of sawtimber, 14 million board feet, was found to be almost twice that removed by cutting, 8 million board feet. When it was realized, however, that most of the timber harvested came from the 44,000 acres in sawtimber stands and that relatively little came from the remaining 114,000 acres, the picture was not bright. The growth per acre annually was about 84 board feet. If the land had been properly stocked, it could have easily been twice this amount.

Were Broome County heavily dependent upon its forests as the major economic base, this situation would be tragic. As it is, the county pays out in freight on lumber shipped in from other states an annual bill of over $750,000, an amount exceeding the current income from all the forest lands in the county. It uses 37 million board feet of lumber, of which only 5.5 million feet is grown within the county.

Many lands once in active farms have been bought up by factory workers who commute by motorcar to the cities. Some of the land may be rented but most lies idle. Meanwhile, the demand for road and school services is maintained at a high level and yield from the land is negligible.

Broome County presents an opportunity in forestry. It has a large amount of idle land and land now inadequately stocked with trees. To convert the idle land to timber will require a rather heavy capital investment. The State of New York began to acquire submarginal farm land in Broome County for forestry purposes in the 1930's, but the people of the county prefer to see their lands remain largely in private ownership. Although the state is ready to help private owners build up their forests, this will still require fairly substantial investments that will yield little income for a period of from 20 to 30 years. The question is how many private owners in the county will invest in forestry with such a long wait in mind.

Forest Situation in New York State. The forest situation in Broome County is more or less typical of that in the state as a whole. Idle farm and cutover land is estimated to be 4,000,000 acres, seedlings and saplings, 1,600,000 acres, pole timber, 2,640,000, and light, medium, and good sawtimber, 5,370,000.* Volume appears to be concentrated in small trees and low-quality species and units. Growth is believed to exceed cutting. Timber requirements are high with probably less than 20 per cent produced locally. Freight costs on lumber shipped from other states are estimated to exceed 50 million dollars annually.

NEW YORK'S WOOD-USING INDUSTRIES. New York's wood-using industries employ 137,000 workers and have an annual payroll of over $422,000,000.[1] The workers and payrolls by major industries are: lumber and wood-products establishments, 22,500 workers, $62,000,000; pulp and paper industry, 25,000 workers, $85,000,000; furniture and fixtures manufacturing, 31,000 workers, $101,000,000; ship- and boatbuilding, 17,500 workers, $16,000,000; miscellaneous wood processing, 37,000 workers, $103,000,000. The pulp and paper industry has a capital investment of approximately $19,000 per worker. This was reported to be the fifth industry in New York State in number of wage earners employed and the sixth in amount of wages paid.

New York's wood-using industries have shown a sharp decline in the

* At the time this was written, 1951, data from the U.S. Forest Service Survey were not available.

amount of wood used during the 34-year period 1912 to 1946. In 1912 such industries consumed 1,741,000,000 board feet; in 1919, 1,279,000,000 board feet; in 1926, 836,000,000 board feet; in 1946, 655,000,000 board feet.[4] These are industries that manufacture furniture, toys, agricultural implements, automobiles, musical instruments, household appliances, and a host of other products.

The pulp and paper industry in New York State had 79 paper mills, 35 paperboard mills, and 44 wood-pulp mills in 1951. They represent a total investment of $380,000,000 and did an annual business of $400,000,000. The industry paid almost $30,000,000 annually in taxes and produced annually 1,739,000 tons of paper and paperboard. Twenty-three communities, having a total population of some 20,000 people, were almost wholly dependent on the industry.

From 1917 to 1940 a total of 69 paper and pulp manufacturing industries in the state were dismantled and only 21 new establishments erected. Although the capacity of paper mills showed a slight increase during this period, pulp production declined. The pulp industry has shown signs of unhealthy condition for some time, partly because of the high cost of pulpwood and the dependence of New York State mills on wood imported from Canada and elsewhere. As Canadian provinces are placing restrictions on the export of pulpwood to the United States, the wood supply of New York State mills becomes progressively more precarious. As softwood supplies dwindle further, more pulp mills will be forced to close or convert to using hardwoods still in ample supply. The total woodpulp production of New York State is about ½ million tons per year.

It is obvious, therefore, that the people of New York State have a great stake in their forest lands. These furnish recreation, watershed protection, and timber products for the lumber, pulp and paper, and wood-processing industries of the state, and they also furnish useful employment to many people and pay much of the Empire State's annual tax bill. The forest lands of the state today are producing barely one-half the volume of timber they could provide under good management.

Forest Lands of the Lake States. The situation in Michigan, Wisconsin, and Minnesota is less encouraging than that in New York.[2] Total forest area is 54.7 million acres of which 3.4 million acres is forest scrub unlikely to produce commercial products, 1.3 million is reserved, and 50 million is commercial forest land. The distribution of commercial forest by condition is:

	Area, million acres
Sawtimber	5.9
Pole timber	9.6
Seedlings and saplings	22.0
Poorly stocked and deforested	12.5

Distribution by type is:

	Area, million acres
Pine	4.4
Spruce-fir	9.8
Hardwoods (except aspen)	16.3
Aspen	19.5

Timber volume is relatively low in sawtimber trees, 6.4 billion cubic feet; high in pole timber, 13.0 billion cubic feet; and as much as 3.8 billion cubic feet in cull trees that probably will produce only fuel wood. The desirable white pine, red pine, and spruce make up 4.6 billion board feet, other softwoods 7.4 billion board feet, and hardwoods 28.0 billion board feet.

In contrast to the situation in New York, timber cutting in the Lake States exceeds growth by a substantial margin. Growth is 1.2 billion board feet, cut 1.5 billion. Drain in relation to growth is the heaviest in the sawtimber stands and in the softwood timber.

If current trends continue, the future Lake States forests will have less timber than those of today, depleted as they now are, and the timber in them will be largely aspen and hardwood rather than pine and spruce. Growth, now less than half potential growth, will decline to even lower than the current level.

The social dislocations that follow unrestricted cutting of timber have harassed hundreds of Lake States communities. Lumber operations in their heyday bring a feverish prosperity. Sawmills are erected, followed by homes, schools, churches, stores, and civic buildings. Nearby farms share in the prosperity by providing food for men and horses. When the timber is finally cut out, the mill closes and itinerant workers leave, but the home owners, storekeepers, and service workers find themselves stranded. Community decay sets in unless new industries arise. Many communities disappear completely. Others may maintain a vicarious existence by processing the remnants of the forest or by other means.

The village of Cass Lake, Minn., went through such a boom during the logging of the Chippewa National Forest from 1902 to 1925. Huge timber sales were made by the United States government on what was formerly Indian reservation lands. By 1925 the mills had cut the timber and closed, and the village of Cass Lake began a period of rapid decline. Houses were no longer painted or kept in repair. Stores and hotels closed. Churches were barely able to keep open. The village relied heavily on the limited payrolls of the national forest headquarters and a local Indian office. During the depression years of the thirties the Civilian Conservation Corps and other relief activities were the main source of income.

Meanwhile, the forest administration began seeking markets for the despised aspen and jack pine and for small-sized red and white pines. Production of piling, poles, mine timbers, box lumber, and fence posts was added to the sale of sawtimber and pulpwood. Applied research greatly increased the effectiveness of forestry measures. A cooperative of timber operators provided finances for men with little or no capital. As a result of all these measures, employment spread, stumpage prices increased, and prosperity gradually returned.

Three disasters—windstorms, an ice storm, and severe insect outbreaks—swept the forest from 1938 to 1941. Fifty-six million board feet of timber was killed, but prompt salvage and replanting were carried out so that by 1951 the effects of these huge losses could be detected only by the experienced observer. So effective was the forest management that the annual harvest of timber was increased from 10.2 million board feet in 1940 to 17.6 million board feet in 1950. A further increase in allowable cut is scheduled for the near future.

What has been done on the Chippewa National Forest reversed the decline in timber volume and value that characterize Lake States forests as a whole. Cutting has been extended to include the low-value species and small sizes. The result is a gradual increase in volume and allowable cut with corresponding increases in employment, income, and local prosperity. As local people gained confidence in the stability of their livelihood, civic pride has also grown. The community has changed from a debilitated lumber camp to an orderly, growing village of confident, loyal citizens.

World-wide Examples of Good and Poor Land Use. The thrifty Dutch push back the Zuider Zee, leach out the salt from the soil, and cultivate the land to produce hay, vegetables, flowers, and the many other plants and animal products that give the Dutch the well-earned reputation of being the world's best farmers. The Danes have converted the heath on Jutland into good farms and forests. Through use of forest plantations, shelter belts, proper farming practices, patience, and hard work, they have built up prosperous communities, supporting as many as 25,000 people, in a region that once supported only scattered villages and a few nomadic shepherds. The Swedes have been particularly successful in building up a high standard of living through proper care of their forests, fields, pasture lands, and mineral resources.

Norwegians inhabit a most inhospitable land. The western coast is outstanding for its beauty and ruggedness, but for this very reason it has a minimum of level land on which agriculture and forestry can be practiced. Through fishing and through diligent cultivation of forests and crops on the small patches of moderately sloping land along the fiords, however, the people maintain a good standard of living, educate their children, and live in harmony with their environment. Switzerland

FIG. 83. The hill farm home of a Puerto Rican family. The husband does some farming on their 2-acre holding and works intermittently in the cane fields. The wife, age about 26, has five children. The youngest baby and a boy are not shown. Shifting agriculture is practiced. (*U.S. Forest Service photo.*)

has Europe's most rugged topography, but the Swiss practice most intensive dairying and farming, and they have become renowned the world over as cheese makers. Some of the most intensive forestry to be found anywhere is practiced by the Swiss.

Italy, Greece, Albania, much of North Africa, and the Asia Minor region stand in sharp contrast to the good land-use practices just described. Mountain slopes have been denuded of timber, erosion is severe, and human poverty and misery have become widespread throughout a region that once was the garden spot of the world. Equally striking examples of misuse of land can be found in much of Central and South America, in Africa, and in Asia (Fig. 83). One of the most poignantly beautiful passages of recent literature is the introduction to the book "Cry, The Beloved Country," which describes overuse of land in South Africa.[6]

Forestry's Place in the Land-use Program. It should be clear from the foregoing discussion that forestry has an important part to play in the development of any land area for human occupancy, provided it lies within the region where trees can be grown. The pattern is everywhere the same, whether it be Finland, Sweden, Denmark, Switzerland, Puerto Rico, the Ozarks, or central New York. The rocky, steep, sandy, and otherwise poor soils should be retained in forests that protect the soil and watersheds and provide raw materials for industry. The level, fertile valley lands are suited to cultivated crops and the moderately sloping land to hay and pasture. This is the pattern that has been worked out on a trial-and-error basis throughout much of New York and New England. It is found in other states where settlement has existed for a long time and where the population has made an intelligent effort to adjust itself to its land resources. Once an adjustment is reached, it seems so logical and so natural that it is relatively easy for the people to perpetuate it.

Adjusting Land Use. The task of obtaining radical revision in the land-use pattern of an area is difficult. Once people have undertaken to farm an area, they naturally resist steps taken by outsiders to move them off the land. It is very difficult indeed for any outside agency to take leadership in modifying a land-use pattern that has grown up through a long period of time. The best that they can do is to supply expert information. The forester, of course, is only one technician whose talents must be called upon. He must have the cooperation of the farm manager, soil scientist, soil conservationist, and others in surveying the potentialities and suggesting new patterns of land use for the future.

The agricultural economist plays the key role. It is he whose task it is to size up the capacity of the land to support a population engaged in agriculture. He draws from the agronomist, animal husbandryman, dairy specialist, horticulturist, and soils man the basic knowledge of the productive capacity of the area. To this he adds his own knowledge of farm management, of markets and marketing, of government costs and services, and the information that he can obtain from the forester on forest conditions, forest types, timber production, and wood-using industries. On the basis of the combined information he is in a position to advise farmers and others which lands are unsuited to permanent agriculture, which may be expected to yield only meager income, and which have potentialities for supporting a prosperous agriculture indefinitely.

Land-classification maps are then prepared and made available to the local people. Along with these go standards of practice that will guide each farmer toward a reasonable livelihood from his land. These include advice on the minimum-sized dairy, poultry, orcharding, or other operation, the list of equipment he will need, the capital he will require, the farming methods that he should use, and the probable income that he may expect to attain if he farms his land intelligently and industriously.

The information is made available to the local farmers, the businessmen, county officials, and others. Estimates are made of the savings in county expenses which are possible if families are moved from remote areas, certain roads left unplowed during winter, and schools consolidated.

Adjusting a land-use pattern is as much a problem in political science, rural sociology, and government as it is in forestry and agriculture. It is not, however, always as difficult as it might seem. In the long run most people are willing to change their methods of farming and even to change their homes if they can be made to understand that the change will improve their lot. It is not uncommon for men who actually own land in poor sections of the county to take leadership in land-use planning, zoning, and other activities that will restrict their own freedom of action but direct future farmers to better opportunities.

Public Measures for Improving Land-use Practices. The towns, counties, and state governments have at their command various means by which they can influence land use toward desirable patterns. In some states, for instance, returning veterans from the Second World War were issued pamphlets pointing out the opportunities to engage in farming. The veterans were told what constitutes a good farm, the type of living they might expect from it, and the capital needed to purchase such a farm. They were warned against purchasing poor land and against settling in remote regions where agriculture had failed in the past. These warnings were repeated by bankers and local county agents. The veterans were encouraged to talk over their plans to engage in farming with the county agent so that he could direct them to the better opportunities.

The public can also acquire remote farm lands located on poor soils and retire them to forests. This has been done extensively in central New York, in Wisconsin, in Minnesota, Michigan, and in other states. Land may be acquired by the county, state, or even Federal government. Public land banks, and private banks as well, can do a great deal to encourage proper land use by advising prospective borrowers on suitable farms and refusing to extend credit on farms they believe to be a poor economic venture. The states can authorize counties to zone rural lands for forest and agricultural use. Wisconsin's zoning law enables the counties to zone any areas deemed unsuited to agriculture as forest land. Farmers living in forest zones are free to remain there as long as they wish, but when the land title passes from them to another, farming must be discontinued. The purpose of these ordinances is both to reduce costs of providing roads and school services for remote districts and to provide better opportunities for families.

Once the land has been zoned for forest, it may either be purchased by individuals or corporations engaged in forestry or pass into county ownership. If the county and local governments taxed land in proportion to the cost to the county of providing the services needed to maintain families on such land, many of the remote areas would soon be depopulated.

The state and counties also engage the services of agricultural economists, foresters, and others to make land economic surveys and land classifications. A vast project of land economic surveys was carried out in Minnesota, Wisconsin, and Michigan during the late twenties and early thirties. These served to direct the attention of industries and local people to the resources that existed in their counties. They provided information for intelligent action, and out of them grew the county zoning plan in Wisconsin and much of the state and Federal forestry programs of Minnesota, Wisconsin, and Michigan.

Democratic Approaches to Good Land Use. It is neither feasible nor in keeping with the tenets of American governmental traditions for any Federal, state, or county government arbitrarily to zone an area of land that is already settled, move families from one place to another, and dictate to the landowners what they may or may not do with their land. Such an attempt would be highly resented by any freedom-loving people. The more appropriate approach is for the technical agencies, upon the request of local people, to survey the resources the community offers. With this information at hand and with such suggestions as experts may make for increasing income, the local people themselves can work out a pattern for adjustment to which the majority can subscribe. The nonconformist need not be severely penalized; he need merely be left alone. In the long run he will usually appreciate the wisdom of the changes or, in rare cases, prove the soundness of his own position. In neither case has an injustice been done. In the long run the community will gradually work for its betterment if its leaders are deeply aroused and properly informed.

LITERATURE CITED

1. Anonymous. 1949. New York State Wood-using Industries, a Survey, *Industrial Bulletin,* 28(11):9–14, 34–36; 28(12):9–11, 36. New York State Department of Labor, New York.
2. Cunningham, R. N., and Forest Survey Staff. 1950. "Forest Resources of the Lake States Region," Forest Resource Report 1, U.S. Department of Agriculture, Washington, D.C. 57 pp.
3. Foster, Clifford H., and Burt P. Kirkland. 1949. "Results of Twenty Years of Intensive Forest Management," The Charles Lathrop Pack Forestry Foundation, Washington, D.C. 36 pp.
4. Hoyle, Raymond J., and John R. Stillinger. 1949. "Wood-using Industries of New York," New York State College of Forestry, Syracuse, N.Y. 133 pp.
5. Pasto, Jerome K., and Howard E. Conklin. 1949. "An Economic Classification of Rural Land, Broome County, N.Y.," Cornell Economic Land Classification Leaflet 1, Cornell University Press, Ithaca, N.Y.
6. Paton, Alan. 1948. "Cry, The Beloved Country," Charles Scribner's Sons, New York. 278 pp.

CHAPTER 17. *World and National Forest Resources*

THE CAPACITY of American forests to supply our country's present and future needs is of deep concern to foresters and users of forest products. This capacity will depend on how intensively the forests are managed. It will also depend on our future population, our domestic requirements, our international obligations, our exports and imports of forest products, and the price we are able and willing to pay to produce an abundant timber supply.

Our own situation can be fairly appraised only after a broad review of the forest resources and requirements of other lands. Two highly dynamic factors are involved in such a review, human population and the capacity of the forests to recover from man's use and abuse.

Population and Resources

The population of the world has been growing at a rapid rate during the past 150 years.[14] In 1800 the world population was estimated at 919 million. By 1900 it had risen to 1,527 million and by 1949 to 2,378 million. Not only has world population increased, but the rate of increase has risen during the last 50 years. The increase has been most noticeable in Europe where the population has trebled during the 150-year period, in North and Central America where it has increased 13 times, and in South America where it has increased 17 times. The population of Asia increased from 600 million to 1,199 million. Asiatics still make up more than half of the world's total population.

Demographers are cognizant of the fact that in countries such as India and China, where a large percentage of the population is living at near-subsistence levels, any increase in food supply and shelter tends to stimulate birth rate and reduce the death rate. Industrialization, the growth of transportation systems to permit free movement of people and food, modern sanitation, medical care, improved diets, mechanization which improves agricultural yields and decreases working hours, all tend to prolong human life. A steady increase in population of a country or a continent, such as has occurred in North America and Europe during the past 150 years, is not considered to be the normal trend in human population growth. On the contrary, a highly fluctuating growth rate, such as has occurred in India, is the more normal pattern.[14]

While death rates are declining, so also are birth rates in all industrialized nations. For example, the birth rate in the United States declined from 38 to 23 per thousand during the period 1862 to 1946. Other countries show similar decline—Sweden from 34 to 20; Germany from 38 to 19; Italy from 37 to 20; Rumania from 41 to 22; Japan from 34 to 27. In modern industrial civilization children are an economic drain on their parents. Where birth-control information is freely available, people tend to limit size of family. As a birth rate of at least 15 per thousand is needed to maintain a stable population in the countries having the best food resources and health facilities, it is obvious that a great many countries are rapidly approaching the time when their population will either be stable or decline slightly.

For predicting future population growth, demographers divide the nations of the world into three classes.[14] Class I includes the United States, the United Kingdom, France, Denmark, Norway, Sweden, the Netherlands, Czechoslovakia, Hungary, the former Baltic countries of Estonia, Latvia, and Lithuania, Belgium, Finland, Switzerland, Italy, Canada, Australia, and New Zealand. In these countries a large measure of control is exerted over both death rates and birth rates through sanitation, public health, medical care, and birth control. It is expected that these countries will soon approach a peak in population and thereafter cease to grow.

Class II countries include Spain, Portugal, Greece, Yugoslavia, Bulgaria, Rumania, Poland, South Africa (the white population only), Japan, the U.S.S.R., and possibly also French North Africa, Brazil, Argentina, and Uruguay. In these, both birth rates and death rates have been declining in recent decades, and the death rate more rapidly than the birth rate. These countries are expected to continue growing at a rapid rate for three to five decades.

The remainder of the world is placed in Class III, where neither birth rates nor death rates are under reasonable control. These countries may grow very fast, but, on the other hand, severe disease outbreaks, famine, and other catastrophes are ever staring them in the face. Obviously, class-three nations have great potentiality for growth, but the extent to which growth will occur will depend to a large degree upon the ability to reduce death rate. In Latin America, for instance, population is increasing at a very rapid rate. Kingsley Davis, at the 1948 Inter-American Congress on Conservation of Renewable Natural Resources, predicts that if population growth continues at the 1920-to-1940 pace, Latin America will contain 373 million people by the year 2000.[3] He estimates 280 million as the minimum population, assuming a declining birth rate. Gradually, these and other Class III countries will probably be moved into Class II—the countries having a decreasing death rate while still maintaining a high birth rate. World population seems destined to increase substantially before leveling off at a new plateau.

While population has been expanding at a rapid rate, good agricultural land has been shrinking in area. Erosion by both water and wind has taken a heavy toll of fertile topsoil throughout most of the densely populated regions, thereby reducing crop land. Except in a few countries, forests have been cut with little or no provision for their reestablishment. Sewage disposal in modern nations tends to hasten flow of essential plant mineral elements to the sea. The base for supporting a growing population is diminishing. True scientific agriculture has vastly increased output from the land, yet the potentialities appear by no means to be inexhaustible. Demographers feel that at best science can but provide temporary alleviation of subsistence living while population takes a new surge forward.

Many predict a dire future for mankind, unless world population is rapidly stabilized and a drastic change is made in the way in which farms, forest, and other land resources are used.[1, 11, 12, 15]

In the study of these resources region by region, it should be kept in mind that the press of people on the land is the dominant factor determining use. It may lead either to destructive use or to highly intensive and intelligent use, depending upon the wisdom and self-discipline of the people.

The World Forest Situation

Forest land makes up roughly 9 billion of the world's 36.6 billion acres or about one-fourth the total earth surface. Of this some 1.2 billion acres have already been so badly denuded as to be virtually useless.[5] Some 4.5 billion acres are still relatively untouched and remain to be opened to forest exploitation. Some 2.5 billion acres are at present being cut with little regard for continued productivity, and only 7.5 million acres have been placed under systematic forest management. Looked at from the broad standpoint, therefore, the forests of the world still present a vast, only partially exploited resource of great potential value to mankind.

When we break the problem down further, we see that the opportunities are not as broad as might appear. Industrial demand requires about 85 per cent softwood timber and only 15 per cent hardwood, but only 30 per cent of the productive forest areas of the world produce softwood. North America, Europe, and Russia together have 95 per cent of the world's softwoods. The only other important softwood region is the Paraná pine district of Brazil and Argentina. Both Europe and Russia have high requirements for softwood lumber. The United States scarcely expects to look to them for imports of softwoods to meet our dwindling supplies. Small shipments of Paraná pine from Brazil have been appearing in eastern United States' markets during the last decade, but this lumber is needed within Brazil and Argentina. As South American industry grows and South American standards rise, we can expect Paraná pine to find a ready market close to the source of supply.

Tropical forests furnish us with many rare and useful woods such as mahogany, Spanish cedar, rosewood, satinwood, teak, balsa wood, ebony, lignum vitae, and others. But in the natural tropical forests these valuable varieties are sparsely distributed throughout dense forests of trees having woods that are hard or cross-grained, that check badly when seasoned, or that are nondurable. One tree of mahogany per acre is considered a good logging chance. Logging therefore becomes extremely expensive. Attempts to cut over the tropical forests removing only the highly valuable cabinet woods yield relatively little in return for the effort. Until uses are found for the majority of woods as they occur in the forest, no intensive forest management is possible. Such development, though urgently needed, has been slow to materialize.

Forest Conditions and Practices in Various Sections of the World

Europe.[5] Forestry is practiced more intensively in Europe than elsewhere. Even there overcutting was common before the Second World War, and a considerable amount of overcutting occurred during the war. Since Europe, outside of Russia, has insufficient land to supply total needs, imports are required. Wood is one of the mainstays of the economy of the Scandinavian countries, which are important exporters of wood to central and southern Europe. International trade in forest products was predominantly inter-European trade before the Second World War.

During the war years British forests produced 50 per cent of the nation's consumption. Two-thirds of the coniferous timber and one-third of the hardwood were consumed, and plantations as young as 30 years were clear-cut for mine props. Heroic measures are now being taken to increase to 5 million acres the productive forest area of Great Britain.

War damage to forests of France, Germany, Poland, and other countries was not directly as severe as might have been expected. Heavy damage was confined to a few areas that happened to be in the midst of intense battle zones. The major damage occurred from unauthorized cutting of fuel wood and by deliberate overcutting by Germany of the forests of Poland and other occupied territories. The forests of Germany were not seriously depleted, even though the Nazi government had increased cutting rates throughout their period of power.

Austria and, to some extent, Poland and Czechoslovakia depend heavily on forests for export commodities. In contrast to the forests of Sweden, Norway, and Finland, their forests have been overcut during the war and can furnish additional export lumber only at the expense of long-term production.

Even the forests of Scandinavia and Finland are far from ideal. Large sawtimber, which could be found in northern Finland and Sweden up until the turn of the century, has largely disappeared. These countries have seen many of their large sawmills go out of existence, just as we have in the United States. Today most of the sawtimber going to the mills of these countries is from trees less than 16 inches in diameter. The forests of Finland and Sweden are well located for movement of logs by water transportation, and each country is well supplied with river systems and lakes that flow generally from north to south. Consequently, the mouths of the rivers are ice-free first so that delivery of logs to the mills is facilitated.

Europe is estimated to be cutting 121 per cent of forest growth; yet even with this heavy cutting her lumber requirements are only partially met. To provide housing and lumber for reconstruction and other purposes, vast quantities must be imported (Table 7).

TABLE 7. PRODUCTIVE FOREST AREA AND ITS RELATION TO POPULATION, ACCESSIBILITY, AND KINDS OF FOREST*

Continents	Population, millions	Productive forest land: Total, million acres	Productive forest land: Per capita acres	Classification of productive forests, million acres: Accessible forests: Conifers	Accessible forests: Broadleaved	Inaccessible forests: Conifers	Inaccessible forests: Broadleaved
Europe and U.S.S.R.	578	1,796	3.2	781	227	524	225
North America	201	1,252	6.2	445	381	334	96
South America	103	1,640	15.8	25	734	12	870
Africa	191	756	4.0	5	366	...	385
Asia exclusive of U.S.S.R.	1,224	884	.7	77	353	106	338
Oceania	12	123	10.4	10	49	10	54
Grand total	2,309	6,454	2.8	1,342	2,150	836	1,977
United States of America†	150	472	3.1	173	158	47	5

* From United Nations Food and Agriculture Organization, Division of Forestry and Forest Products. 1948. "Forest Resources of the World," Washington, D.C. 23 pp.

† Included under North America.

For most countries outside of Europe and North America, statistics on forest resources are fragmentary and of unknown accuracy.

Russia. Russia has more productive forest land than any other nation. In normal times she might become a substantial exporter of lumber without overcutting her forests. Important quantities were exported prior to the war in order to obtain foreign exchange, and some has since been exported for that purpose. Excepting Poland, however, Russia suffered greater war damage than any other country and therefore will require for a considerable time in the future all forest products she is able to harvest. Only modest amounts can be expected from Russia to help in the reconstruction of other European countries. Russia is conscious of her forest wealth and is developing plans to open up new forest areas.

As Russia is at present the chief political and military rival of the United States and as she has a large and rapidly growing population, a vast land area, and almost five times the forest area of the United States, it is well that we consider her timber resources in some detail. Russia has 2,370 million acres of forest land or 13.3 acres per capita. Her present cut of timber is less by 2 billion cubic feet than that in the United States. The following statistics from a manuscript by Obet are of interest.[10] The current cut of Russian timber is estimated at 11 billion cubic feet and the current growth at 23 billion cubic feet. The timber cut in Russia in the year 1935 was given as follows: pulpwood, 1.8 million cords; mine timbers, 261 million cubic feet; construction timber, 10,363 million board feet; sawtimber, 13,359 million board feet; crossties, 431 million cubic feet; veneer bolts, 64 million cubic feet; other blocks and bolts, 173 million cubic feet; and fuel wood, 83 million cords.

The chief timber species in Russia and the area they occupy are as follows: larch, 615 million acres; pine, 217 million; spruce, 165 million; cedar, 69 million; fir, 32 million; other conifers, 5 million; total conifers, 1,105 million. Russia has the following hardwoods: birch, 120 million acres; aspen and alder, 32 million; oak, 10 million; beech, 5.5 million; linden, 3.25 million; ash and elm, 2 million; other hardwoods, 79 million; total hardwoods, 252 million acres. In 1939 Russia listed 57 per cent of her forest lands as unexplored.

Despite Russia's high potential in timber production the possibilities of harvesting timber products from the vast forests of Russia are restricted at present. Most of the forests lie in the river drainages that flow into the White Sea and Arctic Ocean. These rivers can be used for timber floating only during the midsummer and fall period. The rest of the year the mouths of the rivers are frozen and consequently unavailable for floating. Rail transportation is still little developed in Russia—57,000 miles in contrast to 250,000 in the United States—and Russian railroads must serve an area three times as large as our own country. Only 12 per cent of the timber is transported by rail and 53 per cent by water. Transportation from Russia's northern forests is so difficult that Russia has been importing large quantities of lumber and prefabricated houses from Finland. It probably never will be cheap to log and transport timber from northern European and Asiatic Russia to the timber markets of Europe. It is expected that increased use will be made of both rail and water transportation in the years ahead, but serious handicaps must be overcome before Russian timber can become a major factor in the world market.

During the war period Russia exhibited far greater industrial and economic strength than the nations of the West thought possible. Given a few years of peace to develop further her vast mineral, agricultural, and forest resources and to industrialize her enterprise, Russia can be expected to bring the vast forests of northern Russia and Siberia into use.

It is difficult today to get accurate information on forest practice in Russia. It has been reported that Russia is turning out annually some 2,500 men trained for forestry and the forest industries. To what extent adequate care is being taken to maintain the Russian forests in productive condition after cutting is not known. It is the government policy, at least on paper, to make certain that adequate protection forests are maintained and that the lands that are cut over for industrial wood reproduce promptly to valuable tree species.

Middle East and North Africa. This region, the seat of powerful empires during Greek and Roman times, presents today an outstanding example of forest destruction and land abuse. The crying need of this area is for reforestation and long-range forest programs, but these measures are handicapped at present by unstable governments and by the general poverty of the countries. Outside help may be needed to rehabilitate this vast land area.

Notable progress has recently been made in Israel.[8] The Zionist movement has brought to Palestine many intelligent and high-minded individuals who recognize that the future of Israel must depend upon proper development of its land resources, a program in which forestry and agriculture are of high importance. The new government of Israel has a forestry division with an active tree-planting program.

Central and South America. Among the countries of Central and South America, only Mexico and Brazil have important areas of softwood timber.[1,7,6,9,13] Mexican softwoods occupy mountain areas in relatively dry regions. The Mexican government has a law regulating cutting practices on forest lands, but she does not have a sufficiently strong forestry organization to ensure effective enforcement. The Paraná pine region of Brazil though small in area is a very important timber-producing forest. It has a total timber stand estimated to exceed 60 billion cubic feet, of which approximately one-tenth is within reach of existing transportation. The Pan-American highway opened up valuable oak forests in Costa Rica, which probably will be exploited in the near future. The more densely populated areas of Central and South America, particularly the Caribbean Islands, the coastal regions, and the populated interior uplands where settlements are concentrated have been subjected to heavy timber cutting. The soil of many of these areas has been depleted because of the destructive form of shifting agriculture commonly found in tropical regions (Fig. 83). Undeveloped forest lands remote from existing transportation occur in most of the South American countries. The development of such areas must await improvement of transportation and studies of wood technology to determine the value of the several woods. The majority of species have as yet not been studied sufficiently to determine their potential uses in northern markets. Much remains to be done to develop forest management and timber use in tropical countries (Fig. 84).

Brazil is estimated to have more standing timber than any other country in the world, but most of the forests are unexplored and unexploited. Shifting agriculture has already made heavy inroads on the forests outside the rain forest zone. Argentina has limited forest areas, of which the 3 million acres of Paraná pine are perhaps the most valuable. Fires have

FIG. 84. Left, a one-year-old teak seedling as lifted from the nursery. Right, the same seedling after root and top pruning for planting. Trinidad, British West Indies. (*U.S. Forest Service photo.*)

caused severe damage, particularly in the areas bordering the Andes. Although the Chaco region contains extensive forests, it is largely unexplored.

Half of Peru is forested; yet before the Second World War Peru imported 50 per cent of the lumber she consumed. Western Peru is practically treeless. The eastern slope of the Andes, particularly the Upper Amazon Basin, is heavily forested and supplied important amounts of *Cinchona* for quinine during the war period. National timber needs are difficult to meet, however, because little timber can be brought over the Andean summit. Most of that harvested must be shipped down the Amazon.

Colombian forests were being explored and mapped in 1949. Potentially, they form a very important resource for the country—particularly for the production of rubber, oils, gums, and nuts. Eventually, they may also produce substantial amounts of timber.

Chile has approximately 40 million acres of forest land, but much of it lies in the southern portion in rugged terrain where logging will be difficult. A preliminary survey of Chile's forest resources was completed during the war period by representatives of the U.S. Forest Service, and a development program was begun in 1951.

Africa. Like South America, Africa has vast areas of tropical hardwoods at present only partially exploited. Logging and other difficulties in this territory are even greater than those in most of South America. Important efforts have been made by colonial administrators to use African forests, but so far the results have been disappointing. We can expect, however, that ultimately these vast areas will be opened to timber use and contribute substantial volumes, at least for local consumption. Prior to the war Africa imported five times as much timber as she exported. As local needs are likely to increase almost as rapidly as exploitation develops, it will be a long while before Africa contributes substantially to the world's timber supply.

South and East Asia. Forestry is highly developed in Japan and substantial development has also occurred in the Philippines (Fig. 85). Some has taken place in Burma, Siam, Malaya, and the Netherlands East Indies. The dense population of this territory and its nearness to substantial wood-importing countries, India, China, and Japan, means that the territory as a whole cannot become an important factor in export of lumber products outside Asia. In both India and China there is need for much forest planting to check erosion and to build up wood supplies. This area as well as the Mediterranean region will need outside help in forestry and in forest products for a number of years to come (Table 7).

Pacific Area. Australia, despite its sparse population, has only a relatively small area of forest land. Before the war Australia imported large

FIG. 85. Plantations of *Cryptomeria japonica* are managed in even-aged stands by the Japanese. (*American Forests.*)

amounts of lumber from Canada and exported very little. New Zealand is much better supplied with forests, having a total of some 750,000 acres of plantations of exotic species and native timber covering approximately 20 per cent of the land area. Timber is exported in substantial volumes to Australia and small amounts to other countries. On the other hand, New Zealand imports Douglas-fir, redwood, and Australian hardwoods (Table 7).

Canada. The forests of Canada are almost one-third larger in area than those of the United States. Much of Canada's forest, however, lies in northern latitudes, where growth is meager compared with that of forests farther south. Unlike the forest land in the United States, much of Canada's is today publicly owned crown lands administered by the provinces. Laws are being considered to grant leases extending over periods up to 100 years. Cutting practices are similar to those in the United States. Private forest ownership totals about 63 million acres. About one-quarter of the drain from Canadian forests is for sawtimber and almost an additional one-quarter for pulpwood. It is estimated that almost one-third the drain of Canadian forests is due to fire, insects, disease, and windstorm. Forest management is on an extensive basis.

Forests of the United States

When Europeans first settled in America, they found a virgin land, abounding in rich resources. The forests were the most valuable and the most extensive that had been discovered. The original forest area was approximately 822 million acres, almost equally distributed between hardwoods and softwoods. In addition, there were some 93 million acres of woodlands interspersed among prairies, plains, and deserts. The virgin country that the Indian knew was made up of the vegetative types shown in Table 8.

TABLE 8. VEGETATIVE TYPES AND AREAS THEY COVERED*

Vegetative types	Area acres	Per cent of total
Hardwood forest	403,000,000	21
Coniferous forest	419,000,000	22
Total forest	822,000,000	43
Woodland	93,000,000	5
Total forest and woodland	915,000,000	48
Desert shrub	267,000,000	14
Tall-grass prairie	304,000,000	16
Short-grass prairie	267,000,000	14
Other grassland	152,000,000	8
Total nonforest	990,000,000	52
Total land area	1,905,000,000	100

* Shantz and Zon. 1924. "Atlas of American Agriculture," U.S. Department of Agriculture, Washington, D.C.

Statistics on forest resources in the United States have always been weak. Attempts to gather them date back to the 1880's,[4] but the first comprehensive study was made by the Bureau of Corporations, U.S. Department of Commerce and Labor in 1909. In 1930 the U.S. Forest Service began a nationwide survey of forest resources, forest growth, drain, and present and potential requirements. This was still incomplete in 1951. Estimates based on survey data and other information were made in 1933, 1938, and 1945. The 1945 estimate by the U.S. Forest Service is believed to be the most reliable. Fortunately, it was paralleled by an independent estimate prepared by the American Forestry Association, with which it is in substantial agreement. The U.S. Forest Service figures are used throughout this section to avoid confusion.

Present Land Use in the United States. Occupation of the land and clearing for agriculture have profoundly modified our country's native vegetation. Forests have been cleared for farm land; prairie lands have been plowed and cultivated. Roads, railways, canals, reservoirs, cities, and villages now occupy substantial areas. Virgin lands have mostly been exploited.[11, 12] The present land pattern is as shown in Table 9.

TABLE 9. LAND-USE AREAS IN THE UNITED STATES AS OF JANUARY 1, 1945*

Present use	Thousand acres	Per cent
Cropland in farms	525	27.5
Pasture and range land in farms	301	15.8
Open range and desert	308	16.2
Farmsteads, roads, waste, etc.	147	7.7
Total nonforest land	1,281	67.2
Commercial forest land	461	24.2
Withdrawn for parks, preserves, etc.	13	.7
Valuable for other than commercial timber	150	7.9
Total forest land	624	32.8
Total land area	1,905	100.0

* U.S. Department of Agriculture, Forest Service. 1946. Gaging the Timber Resource of the United States, Report 1, from "A Reappraisal of the Forest Situation," Washington, D.C.

Forest-land area has declined from 43 to 33 per cent of the total land surface. This decline, though heavy, is small compared with the decline in the volume of standing timber.

Future declines in commercial forest-land area will probably be due chiefly to segregating lands for parks, private recreation-land areas, roads, airports, and other nonagricultural uses. It is an interesting fact that the regions in which abandonment of agricultural land occurred earliest and has proceeded the farthest are those thickly populated eastern states where markets for agricultural products are best and where agriculture has already reached a high stage of development.

Truck gardening, poultry raising, and dairy farming all represent very high uses of land. In the eastern states, forestry must give way to these high-value uses on lands needed for such activities. But all of these three uses require good tillable land or high-quality pasture. Low-quality pasture land has little value unless adjacent to rich tillable land that can supply feed for winter use.

Timber Volume. Although forest area in the United States has declined by only one-third since white settlement, timber volume during the same period has declined by more than two-thirds (Tables 10 and 11). This decline in volume has been accompanied by declines in accessibility, size, and quality of timber, and softwoods in relation to hardwoods.

Regional Distribution of Forests. The forest areas by condition classes are shown for three broad regions in Table 10.

TABLE 10. DISTRIBUTION OF COMMERCIAL FOREST LAND BY REGIONS AND CONDITION CLASSES IN THOUSAND ACRES*

Region	Sawtimber	Pole timber	Seedling and sapling	Poorly stocked and denuded	Total forest
North	48,196	41,445	49,370	31,310	170,321
South	98,377	29,322	22,873	32,694	183,266
West	58,603	24,246	13,309	11,299	107,457
Total	205,176	95,013	85,552	75,303	461,044

* U.S. Department of Agriculture, Forest Service. 1946. Gaging the Timber Resource of the United States, Report 1, from "A Reappraisal of the Forest Situation," Washington, D.C.

It is significant that most of the timber-growing land lies east of the Great Plains—353 million acres as opposed to 107 million west of the Plains. The western states contain 65 per cent of the sawtimber volume and only 23 per cent of the forest-land area. The eastern states contain the lowest proportion of sawtimber area to total forest area. This, of course, is merely an expression of the long-continued heavy demand for forest products in this section of the country.

The estimated decline in standing timber from 5,200 to 1,601 billion board feet is probably conservative (Table 11). No accurate figures are

TABLE 11. ESTIMATES OF TIMBER VOLUME IN THE UNITED STATES*

Volume estimates	*Billion bd. ft.*
Original volume	5,200
Volume in 1909	2,290
Volume in 1945	1,601
Projected volume in 1965 if 1944 drain and forest practices continue	1,166
Volume needed as growing stock to provide annual growth of 72 billion bd. ft. estimated to represent needs by 1965	1,700

* U.S. Department of Agriculture, Forest Service. 1948. "Forests and National Prosperity: A Reappraisal of the Forest Situation in the United States," Miscellaneous Publication 668, Washington, D.C.

available on the original timber volume that stood in the great forests that covered our land. In three southern states that have been twice covered by the nationwide survey, sawtimber has declined an average of 14 per cent over a period of 13 years. It is estimated that timber volume in the United States declined by 40 per cent during a 30-year period. These two estimates show a decline of more than 1 per cent a year in the volume of sawtimber standing in American forests. Or to look at it another way, we have 1,601 billion board feet. We cut and we lose from fire, insects, and other causes annually about 54 billion board feet, and we grow annually 35 billion board feet. The difference, 18 billion board feet, represents a decline in sawtimber capital of 1.1 per cent each year.

By expressing volume only in terms of board feet, the actual situation is obscured somewhat. Only trees and stands of sawtimber size can be cut for lumber manufacture, but smaller-sized trees may be used for poles, posts, pulpwood, and fuel. They also grow into trees of sawtimber size in time. Only 205 million acres bear trees of sawtimber size, but pole timber covers 95 million acres and seedlings and saplings 86 million. The timber volume for sawtimber trees is 316 billion cubic feet and for pole timber trees 154 billion cubic feet. Softwood volume is 320 million cubic feet, hardwood 150 million cubic feet.

Growth, Drain, and Requirements. The Forest Service determined the 1940 growth rate in forests throughout the United States to be 35.3 billion board feet of sawtimber or 13.4 billion cubic feet of total wood. The drain from all sources on the forest was 53.9 billion board feet from trees of sawtimber size or 13.7 billion cubic feet from all trees. Drain on sawtimber as of 1944 was 50 per cent in excess of growth. The drain in cubic feet, on the other hand, was virtually in balance with growth. The reason for this disparity is that throughout the East and the South most of the lands have been so badly cut over that trees of sawtimber size are inadequate to stock the land; hence the growth is primarily on small trees on the cutover lands. In the West the virgin-timber stands are for the most part growing very little. Losses from decay, insects, windthrow, old age, and other causes tend to offset the growth.

The present growing stock of 1,601 billion board feet is short by 100 billion feet of that necessary to produce an annual growth of 72 billion board feet, which the Forest Service considers desirable. This growth goal includes 61 billion board feet of timber needed for lumber, pulpwood, fuel wood, veneer, poles, piling, etc. Losses due to insects, diseases, fire, windstorm, and other causes take another 3.2 billion board feet. To provide a margin for defense, new uses, and export, a total of some 65 to 72 billion board feet of growth would be needed annually (Fig. 86). The potential requirements were built up after a careful study of the amount of wood used for various purposes in our country during the past, the trend of this use up to the present, and the projected future use, assuming a population growth up to about 185 million inhabitants and a full level of employment. These estimates of future requirements make no provisions for increasing very substantially our country's exports of lumber and other timber products to other lands. Should need for such exports expand greatly, as might be the case were a brisk trade developed between the United States and the heavily populated countries of Asia, our export requirements might considerably exceed those provided in the estimate (Table 12).

FIG. 86. Wooden ships have many peacetime and wartime uses. (*Wisconsin Conservation Department, A.F.P.I. photo.*)

TABLE 12. GROWTH, DRAIN, AND REQUIREMENT*

Item	Billion cu. ft.	Billion bd. ft.
Current drain in 1944	13.7	53.9†
Current growth in 1944	13.4	35.3
Deficit	.3	18.6
Future needs plus allowance for loss	15.9	64.2
Margin for export, new use, and security	4.1	7.8
Goal for timber growth	20.0	72.0‡
Growing stock needed for goal	594	1,700
Current growing stock in 1944	470	1,601

* U.S. Department of Agriculture, Forest Service. 1948. "Forests and National Prosperity: A Reappraisal of the Forest Situation in the United States," Miscellaneous Publication 668, Washington, D.C.

† Drain in 1950 probably approached 58 billion bd. ft.

‡ Upper limit.

Any long-range forecast is open to question. Industry representatives have challenged requirement estimates as too high and growth estimates as too low. In 1951, six years after the forecast was made, prices of lumber, business activity, and population were ahead of the forecast. Population increased 19 million between 1940 and 1950 in contrast to 8.9 million the preceding decade and 17 million for the period 1920 to 1930. Forecasts for 1960 have been given at 179 million. This is close to the 185 million estimate of maximum population used in the Forest Service study. Should the high level of business activity and population growth continue, timber requirements are bound to increase.

Future use of timber obviously will be determined by availability and price, which, in turn, will be strongly influenced by the forestry measures applied today.

World Outlook

The forest resources of the United States and of the world seem ample for today, but the long-term outlook is only moderately favorable. It is true that 50 per cent of forests still remain to be tapped. Wide difference exists today in the growth rate of forests in different countries. The highest growth rate on the average is found in Denmark, 104 cubic feet per acre annually. One of the best managed forests of Denmark has stands growing as much as 300 cubic feet per acre annually. This contrasts with an average of 28 cubic feet in Finland, Sweden, and the United States and virtually no net growth on the large areas of devastated and untouched forest land. Moreover, even in countries with fairly intensive forestry, such as Finland and Sweden, foresters estimate that current yield could be almost doubled by adopting the best silvicultural practices known at present. The opportunities of still further augmenting the timber supply through use of better genetic strains is being explored through vigorous research programs.

From the world's 9 billion acres of forest land today, mankind harvests annually 53 billion cubic feet or the equivalent of only 6 cubic feet per acre. Since this cut actually comes from only half the world's timberland, the yield from land that is cut over is about 12 cubic feet per acre per year. If forest productivity throughout the world could be built up to the standard attained in Finland and Sweden on lands now being operated for timber, the average use per capita could be increased from 11 to 26 cubic feet each year. This would be about one-third the present consumption rate in the United States.

One conclusion seems clear: the United States cannot look beyond North America for its future softwood supplies.

The Responsibility of the United States in World Forestry

The United States has an interest in the supply of forest products that can be imported from other countries and the market for forest products that can be exported from our own country. If the interests of improving forest economy throughout the world are pursued, it will also make available to the nations of the world the techniques in forestry and utilization of forest products that have been developed in the United States. We have a particular obligation to those countries which up to now have relatively undeveloped forest lands. If we only seek opportunities to harvest their timber in the same reckless manner in which we have cut over our own forests, we will do these nations a great disservice and incur ill will and suspicion. If, on the other hand, we unselfishly make available to them American technique, American research, and American educational procedures, we may expect these nations gradually to develop their own forestry and to improve their domestic and international economy at the same time. If we as Americans do not give them this aid, they will seek it elsewhere. Our opportunity to perform a world-wide service through forestry was never greater than it is today. Our need for performing such a service has never been more acute.

Through cooperation with Food and Agriculture and other international agencies, we can expect to help spread good forestry throughout all the countries. In this way we can help peoples improve their local economy, incur their good will, and help them build for peace in the years ahead.

World Forestry Organizations

Effort in world forestry is guided by international organizations. Four important world forestry organizations existed prior to the Second World War.

The International Union of Forest Research Organizations, founded in 1893, has been concerned primarily with technical problems in forestry. These include the preparation of an international forest bibliography, standardization of permanent sample plot technique, international studies of tree races and tree seed, and an international timber-testing program.

The International Institute of Agriculture at Rome, organized in 1927, provided for a bureau of forest statistics, but inadequate finances compelled it to reduce this to a silvicultural section. Lack of funds prevented the institute from undertaking more than one or, at most, two forest problems at a time. Consequently, it never served as an over-all world center for dissemination of information on world forestry sufficient to meet international needs.

The Comité International du Bois was organized in 1932 to promote exchange of information on international trade in timber and to promote such trade. It also hoped to coordinate technical research on wood utilization. The Comité made substantial progress up until the outbreak of the Second World War. When Germany occupied Belgium, she captured the files of this organization and transferred them to Berlin, where they were consolidated with those of the Centre International de Silviculture.

The Centre International de Silviculture, sponsored in 1936 by the Nazi government, was used as an agency for international policy by the Germans. By developing wood as a versatile chemical raw material, they hoped to enhance their war potential. Through the prestige of German foresters and forest technicians they sought to win the esteem and support of foresters throughout other lands. This organization functioned up until 1946. An attempt was made to extend its life beyond the termination of the Second World War, but other countries by this time had pledged their support to the Food and Agriculture Organization, and this ultimately took over the files of the Centre International de Silviculture.

The Food and Agriculture Organization grew out of the Hot Springs Conference held in 1943 and became the first international agency of the United Nations group. Forestry was early conceived to be an important function of FAO. An interim-commission report on forestry was published April 25, 1945. This report, entitled "Third Report to the Governments of the United Nations by the Interim Commission on Food and Agriculture—Transmitting the Report of the Technical Committee on Forestry and Primary Forest Products" and reprinted by the Charles Lathrop Pack Forestry Foundation, Washington, D.C., contains a good summary of early attempts at international cooperation in forestry. FAO has branch centers in Geneva, Switzerland, and in Bangkok, Siam. Its headquarters were moved from Washington, D.C., to Rome, Italy, after the 1949 meeting.

The Forestry Division was formed early in 1946 with Marcel Leloup as Director, S. B. Show, Chief of the Division of Forest Management, Egon Glesinger, Chief of the Division of Forest Products. One of its first tasks was to publish a report entitled "Forestry and Forest Products—World Situation, 1937–1946," Stockholm, 1946. This organization, which has received modest support from the United Nations, has performed many valuable functions in helping nations to appraise their forest situation. It has sponsored several international meetings concerned with forestry. It publishes, bimonthly, *Unasylva,* in which forest conditions in various parts of the world are described and world summaries of forest conditions set forth.

Other international organizations in forestry functioning in 1950 on a regional basis included the Imperial Forestry Institute (British Empire); the Pan-American Union, Division of Agriculture and Conservation; Northern Forestry Union (Denmark, Sweden, Norway, and Finland); Pacific Science Association, Forestry Committee; Institute for International Collaboration in Agriculture and Forestry (Russia, Czechoslovakia,

Poland, Yugoslavia, Bulgaria); Caribbean Research Council, Committee on Forestry.

The Third World Forestry Congress. The Third World Forestry Congress was held under the auspices of the Forestry Division of the Food and Agriculture Organization on July 6 to 20, 1949, with Finland serving as host country. Delegates came from 29 countries, including India, Pakistan, Poland, Czechoslovakia, and Soviet Russia. Those coming from farthest away were from Argentina and New Zealand. The total number of members was 406.

The congress was divided into five sections: silvics and silviculture, forest surveys, forest economics including forest policy, forest utilization, and forest industries. Almost equal weight was given to each of the five sections, which is an important departure from earlier congresses in which silvics and silviculture received major emphasis. Many of the papers, particularly in forest economics and forest policy, evoked vigorous debate. For a while it appeared doubtful that the congress could bring forth unanimous recommendations. Differences were compromised, however, in a spirit of utmost amity and cooperation. Much credit goes to the Finnish people themselves and to the members of the Forestry Division of FAO for inspiring all members to rise above the enmities and jealousy that engulf the world and to bring forth progressive recommendations that the delegates of all nations could support. The unanimous recommendations of the congress included the following:

All nations were urged to develop adequate forest programs and policies.

The need for developing tropical forests was recognized, but the congress asked that care be taken in this development to avoid the destructive exploitation that has characterized forest development in other lands.

The congress favored a greatly increased program of reforestation and afforestation, free exchange of seeds and plants, an active program of forest genetics, and the selection of elite trees throughout the world to serve as sources of seed for forest propagation.

Exchange of information on forest-survey techniques, particularly the use of photogrammetry in forest surveying, was recommended.

Nations were urged to give appropriate attention to protecting the headwaters of streams and avoiding soil erosion.

Proper coordination of forestry with agriculture in the tropics and elsewhere was deemed essential. It was recommended that social and economic factors be considered in all programs for exploiting new forests.

Studies for improving the efficiency of forest labor and training programs for forest labor were advocated.

An international congress on tropical forestry was recommended for the near future.

More research was felt to be needed on wood and wood use. Silviculturists were urged to improve wood quality as well as wood quantity.

Integration of industries was suggested as a means of minimizing waste. It was recognized that outlets for waste need to be expanded and that better grading of lumber and emphasis on use of the lowest grade suitable for the purpose should lead toward closer utilization.

The congress urged more extensive pulping of hardwoods and the encouragement of semichemical pulping. Better preservation of wood and better designing of wooden structures were recommended. The latter applied particularly to prefabricated houses that need to be given individuality.

The congress urged the compilation of a multilingual forest dictionary.

America's Forestry Opportunity

Compared with the rest of the world, the United States ranks only as average in available forest area per capita (Table 7). It is using timber almost three and a half times as fast as any other region and is a particularly heavy consumer of both lumber and pulpwood. We used roughly 246 board feet per capita in 1944, which contrasts with some 24 board feet used per capita in Europe in 1937. We are also heavy users of wood pulp and pulp products, consuming three to five times as much as any other nation. In fuel-wood use, on the other hand, the United States' demands are relatively modest, because we have abundant supplies of coal and fuel oil to supplement our wood fuel. For the world as a whole, fuel wood makes up 53 per cent of the wood drain on forests, lumber constitutes 28 per cent of world use of timber, and pulp 7 per cent.

The presentation of nationwide and world-wide statistics on forest areas and timber volume gives no adequate view of local conditions such as were described in Chap. 16. The results of such misuse of forests are worth restating here.

Local timber exhaustion has created ghost towns and caused abandonment of roads, schools, churches, and entire communities. Where fires followed cutting, as in much of the Lake States and South, new forests have frequently been slow to reoccupy the land. Understocking is a common cause of low forest productivity on cutover lands. Local dislocations that timber exhaustion bring reduce local markets for farm produce, increase the cost of lumber because of freight costs from distant sources, and generally have a depressing effect on living standards. Yet despite all the abuse to which our forests have been subjected, they remain today potentially the most valuable forest lands in the world. They are rich in species, are generally accessible, produce high-quality timber, and have trees and stands which grow rapidly in size and volume. They are unsurpassed by any temperate-zone forests and they respond well to forestry measures.

Intensive care is not essential to the restoration of America's forest lands. It lies within our power easily to double our growth rate and

FIG. 87. Heavy logging waste from operations in mature Douglas-fir. Cull and broken logs, small trees, and tops contain much sound material. (*U.S. Forest Service photo.*)

thereby to supply all potential needs we can foresee in the future. If measures as intensive as those applied in northern Europe were adopted, the growth rate would probably be increased three-fold over the rate that now prevails.

We should not think of our forests as a depleted resource that must be husbanded by restricting use of timber. The key to more rapid growth in the future is more intensive use and better forestry applied in the woods today. Productivity is low primarily because cull trees, brush, and short-lived trees of low value occupy land suited to high-quality trees. Use of timber products need not be greatly restricted in the future if due care is exercised in selecting those trees that are cut for today's use. Intensive use wisely directed can increase future forest values faster than nonuse.

Our nation has a wonderful heritage of forest wealth. We can go on as we have during the past 50 years and see this wealth further depleted (Fig. 87). This is the easy course to follow. If we do so, we will see our forest regions become progressively impoverished, industries dislocated, and rural slums spreading a cancerous growth over our countryside that might threaten the nation's economy (Fig. 88). We can take the alternate choice of adopting a vigorous, far-reaching forestry program, looking

Fig. 88. An abandoned farm on Class II land in New York. This farm has neither been incorporated in adjoining farms nor planted to trees. Class II land is unsuited for full-time commercial farming.

toward doubling the yield from our forest lands. This would provide additional jobs to begin with and more jobs as more and more timber reaches marketable size. It would place a firm base of raw material beneath the great wood-using industries of our land. It would add to our national strength and prestige, both in peace and in war, and it would do something more—it would provide incontestable evidence that our American democracy can discipline itself to the extent necessary to ensure an ever-expanding prosperity in the years to come. This is the basic issue at stake, and it should not be lost sight of during the discussion of measures to be taken to this end.

LITERATURE CITED

1. Baker, Richard St. Barbe. 1949. "Green Glory," A. A. Wyn, Inc., New York. 253 pp.
2. Brown, N. C. 1947. "Lumber: Manufacture, Conditioning, Grading, Distribution, and Use," John Wiley & Sons, Inc., New York. 344 pp.
3. Davis, Kingsley. 1948. Population and Resources in the Americas, "Proceedings, Inter-American Conference on Conservation of Renewable Natural Resources, Denver, Colorado, September 7–20, 1948," pp. 88–97, Department of State, Washington, D.C.

4. Dietz, Martha A. 1947. A Review of the Estimates of the Sawtimber Stand in the United States, 1880–1946, *Journal of Forestry,* 45:865–874.
5. United Nations, Food and Agriculture Organization, Division of Forestry and Forest Products. 1948. "Forest Resources of the World," Washington, D.C. 23 pp.
6. United Nations, Food and Agriculture Organization. 1948. Latin-American Conference on Forestry and Forest Products, *Unasylva,* 2(3):93–137.
7. Gill, Tom. 1951. "Land Hunger in Mexico," The Charles Lathrop Pack Forestry Foundation, Washington, D.C. 86 pp.
8. Greenberg, David B. 1949. The Rebirth of Israel, *American Forests,* 55(10):14–16, 44.
9. Huberman, M. A. 1948. Our Good Neighbor's Forestry, *Journal of Forestry,* 46:79–98.
10. Obet, Viktor. 1949. "Logging Operations in the Baltic States and Russia," unpublished manuscript, New York State College of Forestry, Syracuse, N.Y. 118 pp.
11. Osborn, Fairfield. 1948. "Our Plundered Planet," Little, Brown & Company, Boston. 217 pp.
12. Sears, Paul B. 1935. "Deserts on the March," University of Oklahoma Press, Norman, Okla. 231 pp.
13. Shirley, Hardy L. 1948. Obstacles to Sustained-yield Forestry, "Proceedings, Inter-American Conference on Conservation of Renewable Natural Resources, Denver, Colorado, September 7–20, 1948," pp. 564–571, Department of State, Washington, D.C.
14. Thompson, Warren S. 1948. "Plenty of People," The Ronald Press Company, New York. 281 pp.
15. Vogt, William. 1948. "Road to Survival," William Sloane Associates, New York. 335 pp.

CHAPTER 18. *Recent Programs and Progress*

DURING THE PERIOD 1941 to 1945 American citizens concentrated so much thought and energy on bringing the war to a successful conclusion quickly that little thought was given to planning for the future. The planning that was done was predicated upon a sharp decline in business activity and employment when war orders stopped and the 12 million men and women in uniform returned to civilian life.

Planners for depression were met with a period of the highest peacetime prosperity our country has known. Business prospered. Labor won round after round of increases in wages; farmers' incomes soared to a new high; corporation dividends increased. Delayed marriages were consummated, and new families were established, increasing the demand for housing, furniture, motorcars, and a host of consumers' goods.

Price inflation caused serious hardship to teachers and others on fixed incomes, but, in the main, inflation added to business activity by stimulating buying against still higher prices for the future. In the midst of this boom our country in 1950 became involved in war in Korea. A wave of scare buying set off another cycle of price rises. Immediate and projected military expenditures together with predicted shortages gave even further stimulus to business activity. Again the American economy began to demonstrate its high degree of flexibility and immense capacity for production.

The postwar boom brought residential and industrial construction to levels higher than ever reached before. Drain on natural resources—iron, coal, petroleum, nonferrous metals, other minerals, and forests—of necessity was increased. Paper manufacture rose to new records and lumber production approached the 1908 peak.

Timber cutting throughout the war years was in excess of growth. This fact naturally raises certain questions. How will this new demand affect our forests? To what extent have forestry programs in the last decade expanded to minimize depletion of growing stock? Have we an adequate reserve of well-located timber stands to carry us through a prolonged period of bitter economic and perhaps military struggle? A study of 1951 programs will provide some background for answers.

Federal Programs

Federal agencies engage in forest-land management, research, and cooperative effort with states and industries to protect forests and improve the use of forest land and products.

National Forests. The largest single forest program in the United States is that on the national forests. In 1951 these totaled 151 separate forests. The total area was approximately 180 million acres. National forests were found in 38 states and in Alaska and Puerto Rico, and two other states had national-forest purchase units. States having 2 million or more acres in national forests include the following:

	Area, million acres
California	20
Idaho	22.5
Alaska	21
Montana	19
Oregon	17
Colorado	15
Arizona	12
Washington	10
New Mexico	10
Utah	9
Wyoming	9
Michigan	5
Minnesota	5
Virginia	4.1
North Carolina	3.6
Missouri	3.5
Mississippi	2.8
Alabama	2.4
Wisconsin	2.0

The gross area within the boundaries of the national forests is 230 million acres.

National forests are managed on the principle of multiple use under the general directive set forth by Secretary of Agriculture Wilson in 1905 (see Chap. 3). The national forests contained 518 billion board feet in 1945, almost one-third of the total reserve supply. National forests are reported by the U.S. Forest Service to contain 73.5 million acres of commercial forest land. From the national forests are cut annually about 4 billion board feet of forest products, a volume equivalent to more than 10 per cent of the total lumber cut in the United States. The Forest Service estimates that this cut could be increased to 6 billion board feet annually on a sustained basis as more intensive forestry is practiced and as remote areas are opened to use.

Less than one-third of 650 national-forest working circles were being cut up to their sustained-yield budget in 1950. A full harvest of products awaits the development of access roads.

In 1944 Congress passed the Cooperative Sustained Yield Units Act. This authorized the Secretaries of Agriculture and the Interior to sell standing timber without bids to local forest industries. These in turn must agree to pool their own timber with the public timber to provide for sustained yield of forest products to support local communities. They must further agree to follow forest-practice standards as mutually set up. This represented an innovation in cooperation between the Federal government and private industry. The first such unit to be established was worked out between the Forest Service and the Simpson Logging Company, of Shelton, Washington, and is known as the Shelton Sustained Yield Forestry Unit. Provisions are made in the agreement for the Forest Service to reappraise the value of the timber that is being offered to the company and also for the Forest Service to inspect the standards of practice being followed by the Simpson Logging Company.

Opposition to cooperative sustained-yield forestry units has been expressed by independent logging operators, labor spokesmen, and by citizens who fear that local monopoly and community dominance by industry may grow out of such agreements. For this and other reasons the Forest Service and the Department of the Interior have gone slowly in setting up units. Only the Shelton unit was in operation in 1950.

Closely related to cooperative sustained-yield units and authorized in the same law are the Federal sustained-yield units. These commit definite areas of Federal forest land to sustained-yield management to maintain the supply of timber for local industries. Timber sales are competitive but local manufacture is required. In 1950 Federal sustained-yield units in operation included the Grays Harbor on the Olympic, the Big Valley on the Modoc, the Lakeview unit on the Freemont National Forest, and the Vallecitos in New Mexico.

The Service maintains high standards of fire control on the national forests.[11] The area burned annually is less than 0.1 per cent of the total area protected. Effort is also made to control insect pests and disease.

The national forests are important sources of water for irrigating arid regions of the West. Attention is given to managing forests for both water yield and flood protection. The national forests provide range for some 1.25 million cattle and horses and 3.5 million sheep. Big-game herds and impressive numbers of other game animals and birds abound, and public hunting is encouraged. The Forest Service maintains 4,500 camps

FIG. 89. Two years' complete nonuse has caused rapid recovery of forage plants within the enclosure. (*Bureau of Land Management photo.*)

and picnic areas for the public, about 240 winter-sports areas, and has issued summer-home permits to 12,000 people.[6, 4]

Receipts from the national forests have grown rapidly in recent years. For the fiscal year 1949 they totaled 31 million dollars and for 1950, 34 million. Timber sales accounted for 29.4 million dollars, grazing permits 3.4 million dollars, land use and occupancy 0.7 million dollars. All receipts are deposited in the U.S. Treasury. The law provides that 25 per cent of these receipts be turned over to the states in which the national forests are located. The state apportions this to the counties on the basis of their national forest area and the money is used for county purposes. The law also requires the Forest Service to expend an additional 10 per cent of receipts on roads and trails on national forests located within the state from which receipts came.

Forestry in the Department of the Interior. The Department of the Interior has 282 million acres in the United States and 344 million acres in Alaska under its jurisdiction. It thus has more land under its management than any other Federal department. Much of this land bears forests and some has as good forests as any to be found on the national forests. The Division of Land Utilization is responsible to the Secretary for the coordination of the management of lands with the Department of the Interior (Fig. 89).

BUREAU OF LAND MANAGEMENT. The Oregon grants (O & C lands) are lands that reverted to the government when the Oregon and California Railroad and the Coos Bay Wagon Road grants were revested in the Federal government through failure of the respective companies to fulfill their contracts. These lands, administered by the Bureau of Land Management, include approximately 2.5 million acres and contain 50 billion board feet of merchantable timber. Because of the policy of granting alternate sections of land to railroads, many of the O & C lands are intermingled with national forest lands and private timber holdings.

Revestment occurred in 1915. The first law (1916) provided for classification of the land as agricultural, timber, and power site. The timber was to be sold and both agricultural and cutover timberlands opened to homestead entry. Most of the land proved unsuited to agriculture and only a small amount was ever homesteaded. Meanwhile, some timber was sold, but revenue to local counties in lieu of taxes was low. A preliminary adjustment of revenue was made in 1926. Upon original revestment the U.S. Forest Service sought jurisdiction over the lands within national forests. As it became evident that the land was unsuited to agriculture, the Forest Service renewed its claim. Meanwhile, the counties were establishing a strong claim for tax revenue.

Congress in 1937 enacted legislation providing for administration of the O & C lands. The 1937 law provided for cooperation between government and private landowners to operate their combined lands on a sustained-yield basis. This principle was restated and extended in the Sustained Yield Units Act of 1944.

The O & C law requires that 50 per cent of gross receipts be paid to counties in lieu of taxes and an additional 25 per cent be set aside to reimburse the counties for taxes accumulated before the O & C law was enacted. Controversy continues, since two different Federal departments have jurisdiction over lands in checkerboard pattern on which local counties have heavy tax claims. Private lands similarly distributed through both national forest and O & C lands greatly augment administrative difficulties. Despite these complications public officers on the ground have sought to divide responsibilities so as to minimize overlapping jurisdiction. The O & C lands are managed for permanent sustained-yield forestry very much as national forests are. The timber is sold to private timber companies operating in Oregon. Cutting is restricted to 650 million board feet per year.[1] For the fiscal year 1950 receipts from revested lands were $4,005,156.

In 1950 the Bureau of Land Management extended forest management to 3 million acres of commercial forest land and 25 million acres of woodland on public-domain lands.[1]

INDIAN RESERVATION FORESTS. Indian Reservations have been set aside in many states of the Union. Some contain valuable timber properties. Timber on the Menominee Indian Reservation, Wis., exceeds 1 billion board feet. Other reservations having large timber reserves are the Klamath with 4 billion feet, the Warm Springs with over 3.5 billion feet (both in Oregon), the Colville with almost 4 billion feet, the Yakima in Washington with 3.5 billion feet, and the Tahola with 2.9 billion feet. In 1950 the annual timber cut on reservations approximated 10 per cent of the cut from national forest lands. Lands are held in trust for the Indians by the Federal government. The forests are managed in the interests of long-term production on a sustained-yield basis.

NATIONAL PARKS. The national parks and national monuments in the United States totaled, in 1951, 184 separate units with an area of 23,926,444 acres. Many of these include heavy timber stands, which, however, are permanently protected as examples of unmodified wilderness areas. Timber cutting and grazing of domestic livestock are excluded except from lands acquired with this right reserved for the owner's life.

ALASKAN TIMBERLANDS. The interior forests in Alaska total some 144 million acres, supporting white spruce, Alaska white birch, cottonwood, willow, and tamarack. These forests are drawn upon heavily by the local population for fuel, lumber, poles, and other products. In the past, fires set to drive mosquitoes from camp, or for other reasons, burned unchecked. Even in 1947 the area burned was estimated to exceed 1 million acres.[7] Organized protection was begun in 1939, and expenditures now are in the neighborhood of $200,000 annually. No organized management exists, because, except for local use, the forests are still untouched.

WILDLIFE REFUGES. The Fish and Wildlife Service has jurisdiction over 282 wildlife refuges, aggregating some 18 million acres. These include migratory-waterfowl refuges, colonial nongame bird refuges, big-game refuges, general refuges, and refuges for special research. Game is protected on the refuges. Controlled public hunting and trapping is being introduced on some refuges as a part of the management program. Many refuges include considerable areas of forest land, which is managed in the interest of the wildlife. Returns to counties in lieu of taxes are 25 per cent of gross receipts from sale of surplus products.

The Tennessee Valley Authority. The Tennessee Valley Authority has a Division of Forestry Relations. Since more than 54 per cent of the Tennessee watershed is wooded, forests play an important role in the over-all program of the Authority. Consequently, the utilization of timber, control of forest fires, reforestation, and forest-land management are encouraged by the TVA. The agency itself has a considerable area of forest land around the reservoirs, for which the Forestry Division is directly responsible. It operates two forest nurseries and carries on research in forestry and forest utilization. Special effort is made, through cooperation with existing extension agencies and farm foresters, to promote forestry among all landowners of the Tennessee Valley.

Federal Cooperative Programs. Federal cooperative and other research programs will be dealt with in Chap. 24. Federal cooperative programs with the states will be dealt with under state programs.

The U.S. Forest Service has a Division of State and Private Forestry authorized to cooperate with private timberland owners in promoting better forestry. The major portion of the work of this division is channeled through the state forester. Upon request the Forest Service does render expert service to large forest landowners. This usually takes the form of advice on location of plants, special suggestions on management programs and policies, help on special forest-survey tasks, distribution of new information on mechanized logging, sawmill operation, lumber seasoning, waste use, and other problems in which specialists in the Forest Service can render unique help to industry. The Service prepares various informational bulletins. It administers the naval-stores conservation program for the Production and Marketing Administration, a program which provides benefit payments to naval-stores operators for following prescribed conservation practices. Through the several states the Federal Service assists in farm forestry programs.

SOIL CONSERVATION SERVICE. The Soil Conservation Service is responsible for stimulating good forest practices on farms in soil-conservation districts. Farmers cooperating with the Soil Conservation Service agree to follow a unified farm plan that includes measures to improve the productivity of the woodland as well as measures to improve farm practices and to conserve soil. Steep pasture lands and eroding fields are often retired from agriculture and planted to trees as a means of checking erosion. The Service has supervision over lands acquired by the Resettlement Administration in its efforts to move farmers off poor lands. These are mostly managed on long-term leases by state forestry organizations.

State Forestry Programs

State forestry organizations manage state forest lands, provide fire control on forest lands throughout the state, except national forests and parks, administer timber management, marketing, and processing programs under the Cooperative Forest Management Act, conduct forest research, operate a forest-land purchasing program, operate nurseries, provide planting stock for state lands and for private forest landowners, conduct important control operations against destructive forest insects and tree diseases, carry on educational programs to stimulate interest by the public in better forestry, particularly in the prevention of forest fires,

FIG. 90. The annual harvest from a 53-acre compartment selected as typical of farm wood lots of the deep South. (*U.S. Forest Service photo.*)

and in a number of states regulate timber-cutting practices on private lands.

Under the Clarke-McNary Act the states receive Federal aid in fire control, distribution of planting stock, and extension forestry. They receive help in farm forestry and other programs under the Cooperative Forest Management Act (Fig. 90). Although the Federal government is authorized to aid states in purchasing lands for state forests under the Fulmer Act, no appropriations have yet been made for this purpose. The cooperative programs are under the general guidance of the U.S. Forest Service, but states have wide latitude in administration.

The four major state forestry activities are fire control, management of state forests, service to timberland owners and wood processors, and regulation of forest practice on private lands.

Control of fires is by far the largest and most costly task. The Federal government appropriated 10 million dollars for cooperative control in the fiscal year 1951, and this sum was matched more than twice over by states and private companies. The funds were expended to erect and man lookout towers, to employ forest guards, fire wardens, state fire-control officers, and fire fighters, and to purchase fire-fighting equipment. In 1951 protection against fire was provided on all but 70 million acres of forest land, most of which lay in the southern states.

State-owned forests and parks totaled 14,848,000 acres in 1949. Four states had about three-fourths of this area. Michigan had 3,860,000 acres; New York 3,026,000; Minnesota 2,094,000; and Pennsylvania 1,767,000 acres. Every state has at least some forest land. These lands are managed for timber production, recreation, public hunting grounds, and watershed protection.

Activities under the Cooperative Forest Management Act may be appreciated by examination of the program of New Jersey. Owners desiring help to sell timber sign a contract with a state forest officer in which they agree to carry out a forestry program. The state estimates the timber, marks the trees to be cut, and refers the owner to timber agents. The timber agent, for a fee equal to 10 per cent of the stumpage value, makes a sale contract with a timber operator, supervises the sale, collects the money, and reports to the owner, accounting for each tree cut. If the owner violates the timber agreement by cutting unmarked trees within five years of the date of agreement, he is obliged to pay the state $2 per thousand board feet for all trees marked. The cooperative program began in 1938. By 1949 half of the timber cut in the state was handled under the program. This is believed to be the best record for any state in obtaining voluntarily a good standard of forestry on private lands.[8]

By 1951 some 15 states had undertaken to regulate cutting practices on private lands. Maryland has enacted the most comprehensive forest-practice act. It provides for forestry boards to regulate cutting practices and to license timber operators. These boards also set up minimum forest-practice standards for their counties. Washington and Oregon have minimum cutting practices specified by law. Owners must obtain the state forester's approval of cutting practices that depart from these standard minima. Oregon may require a performance bond of owners not following approved practice to ensure that the owner will reforest his land either naturally or by planting. If the owner fails to do this in reasonable time, the state may plant the land at his expense to the limit of the bond required. Maryland and Washington laws have been tested in the courts and upheld. California has a law similar to Maryland's with two important exceptions: forest practices are determined by committees and approved by vote of owners holding two-thirds of the land. The state forester is empowered to require compliance with these rules. New Hampshire in 1949 enacted a 10 per cent yield tax on timber cut, which substitutes for the forest real-estate tax. Owners cutting their timber according to forest-practice standards set up by the state are eligible for a 30 per cent rebate on their yield tax. It is hoped that this will prove to be a strong inducement to meet minimum standards of state forest practice. Diameter-limit cuttings have been established in Idaho, Minnesota, and Nevada. Seed trees must be left in Louisiana, Minnesota, Mississippi, and

Virginia. Massachusetts requires that owners notify the state of their intent to cut timber, and the state forester then makes a plan for cutting, which the owner is encouraged, but not obliged, to follow. New York provides technical advice and assistance in marketing timber to forest owners who agree to follow the standards set up by forest-practice boards in the state. Cooperators under the Forest Practice Act are also provided with trees for planting open lands. No penalties other than withdrawal of future state aid are attached to the New York State law. Missouri, like New Hampshire, lowers taxes for those who follow approved cutting practices. Colorado has a board authorized to regulate cutting practices, but definite steps to carry out regulations were still to be taken in 1949.[4]

Appropriations for forestry are an important item in a number of state budgets. States appropriating more than a million dollars annually for state forestry activities include California, 10.6 million dollars; New York, 3.1 million dollars; Pennsylvania, 1.3 million dollars, plus capital improvement money for postwar forestry work; and Wisconsin, 1.3 million dollars.[4] Several other states have expenditures in the neighborhood of $500,000 annually and higher. Income from state-managed forest lands in recent years has been as high as $736,000 in Washington, $455,000 in Montana, and $97,000 in Idaho. Eastern states recording receipts in excess of $50,000 annually include Pennsylvania, Ohio, Michigan, and Florida.

Colleges of Forestry

Colleges of forestry and universities with forestry departments manage college properties for research, demonstration, and education in forestry. They carry out important research in all phases of forestry and forest products. Many either directly or through the state agricultural extension service engage in spreading forestry knowledge among landowners, school children, and the general public. Extension forestry is the function of the state extension services and will be covered in Chap. 23.

County and Community Forests in the United States

County and community forests are found in 43 of the 48 states of our Union. Those not having community forests have extensive national forests, with the single exception of Kansas. Community forests for the entire United States numbered 3,358 in 1951 with a total area of 4,503,323 acres. Wisconsin led all other states with 2,387,000 acres in county and municipal forest. Minnesota had 672,000 acres. Other states having over 100,000 acres in county and municipal forest were Oregon with 239,000, Michigan with 135,000, and New York with 150,000 acres.

Almost a third of the 1,121 municipal forests are held for municipal watersheds. These generally are restricted in the amount of use the public may make of them, since recreational use is usually inconsistent with maintaining a sanitary water supply. Watershed forests may be used for timber production, but special care must be taken with logging operations to avoid creating conditions conducive to erosion and pollution.

School forests total 1,279. These have served a great many useful purposes. They provide recreation for school children. More important, they serve as a center for instruction in biology and in manual arts. Imaginative teachers have even used school forests as an aid in instructing the students in music and in mathematics. In some cases school forests have brought the schools revenue that has been used for special projects, providing supplementary educational opportunities for the students and giving them a keen interest in forestry and conservation.

Organizations, with Boy Scouts perhaps leading, have a total of some 96 forests. Some of these yield substantial revenue from timber operations and provide campgrounds and places for other scouting activities.

Most communities permit free public recreational use of their forests, but charges may be made for special services. On some forests special recreational buildings may be rented at nominal fees by organizations desiring exclusive use for picnics. Fees for fuel wood, parking, and other use of the county property are sometimes levied, providing money that helps to offset operating expenses.

County forests in Wisconsin and Minnesota are much more than recreational areas. These were established by the counties as a means of rehabilitating abandoned cutover land that was tax-delinquent. Under the Wisconsin laws the state was willing to pay to the counties in lieu of taxes 10 cents per acre annually to help support local government. The state provided an additional 10 cents per acre annually to cover expenses of improving and developing the forest. When the forests are harvested, the state recaptures at least a part of its investment through a 50 per cent severance tax. Ultimately, it is expected that both the county and the state will find these investments financially profitable. The annual return from Wisconsin's county forests was $150,000 in 1948. Because of the heavy demand for pulpwood in this region, it is to be expected that the revenue will increase very rapidly within the next few years.

Foresters are employed by many counties and communities. They are especially needed by communities with municipal forested watersheds of large areas and by cities, such as Chicago, that operate extensive forest preserves.[3, 4]

Industry and Other Private Forest Programs

Group Programs. The forest products industries through associations have engaged in a wide variety of forest programs.

The American Forest Products Industries (A.F.P.I.)[10, 4, 5] sponsors the tree-farm program. Under this program, which is entirely voluntary, the A.F.P.I. cooperates with the several states to publicize and certify tree farms. A tree farm is an area of forest land which the owner pledges to operate for permanent timber crops. High standards of fire control are required. Owners also agree to meet other minimum standards of forest practice, including protecting land from destructive grazing. Standards for certification vary somewhat from state to state as does also inspection for compliance. Independent appraisal finds results good. The movement began in 1941 and by 1951 embraced 23 million acres of forest land (Fig. 91).

The A.F.P.I. has a widespread educational program, that includes supplying forestry literature to public schools and other agencies. It gets

FIG. 91. Selective cutting in a southern tree farm. Union Saw Mill Company. (*A.F.P.I. photo.*)

out a great deal of publicity on fire control and on the need for care of forests. It sponsors the Keep America Green movement, which is channeled through state Keep Green committees, with special programs for each state. This movement has become highly effective in some of the western states. The A.F.P.I. also sponsors the More Trees program. The A.F.P.I. is supported by lumber plywood, pulp and paper companies, and by timberland owners.

The American Turpentine Farmers Association has had an active program of forest conservation since 1936. It has been particularly effective in stimulating research and in spreading good turpentining practices among its membership and others engaged in naval-stores operations.

Various other private associations are engaged in stimulating good forest practice on forest lands throughout the country. The Southern Pulpwood Conservation Association carries on a campaign to promote good forest practices by member mills and by owners who sell pulpwood to these mills. The Western Pine Association has long aided in fire-control programs and conducted research in forest products. It also has taken leadership in insect control and forest management.

A group of pulp and paper companies in Wisconsin sponsors the Trees for Tomorrow program. They hold conservation camps for school children, distribute tree planting stock, encourage good forest practice on farm woodlands, and in other ways try to improve the care of woodlands in the state.

Other important groups and associations sponsoring forest programs include the American Forestry Association, American Bankers Association, U.S. Chamber of Commerce, U.S. Junior Chamber of Commerce, Friends of the Land, Izaak Walton League of America, American Paper and Pulp Association, West Coast Lumbermen's Association, and a number of other regional and national associations. As of 1949 it was estimated that well over 1,000 organizations were interested in our forests. These include 200 trade associations of the woods-products industries, a large number of sportsmen's organizations, park associations, outdoor-recreation groups, fire-protective associations, bird clubs, Boy Scouts of America, 4-H Clubs and Future Farmers of America, American Legion posts, labor unions, garden clubs, and many others. All have done effective work in forwarding forest conservation.

Individual Industry Programs. The activities of individual companies are equally impressive. Some 20 or more railway companies employ foresters to encourage good forestry in the regions they serve, realizing that what adds wealth to their territory will bring traffic to their road. The programs of the Illinois Central and Seaboard Air Line railroads have been particularly impressive. The latter features a conservation train and includes movies, talks, demonstrations, field days, and other promotional activities.[2]

Power companies and many of the pulp and paper companies have similar programs. Of special note is the policy followed by several pulp and paper companies of supplying technical forestry service to farmers and other owners of small forest properties.

Forestry programs of individual lumber and pulp and paper companies have been expanded very rapidly since 1945. As a result, the number of foresters in private employ was estimated to be 4,400 in 1951 and action on the ground was impressive. Excellent cooperation exists between timber operators and state forest marketing services such as prevail under the cooperative forest-management and other programs. Industry as well as government can justly view its forestry programs with pride in tangible accomplishment.

Forest Practices by Ownerships

The discussion on forest practice that follows is based primarily upon the standards revealed by the Forest Service reappraisal study in 1944 and 1945.[11] The number of active foresters doubled between 1945 and 1950. A large part of this increase occurred in foresters employed by industry and for consulting or service work. It is believed, therefore, that substantial improvement in forest practice occurred between 1945 and 1951. The broad trends and relationships that were revealed, however, are believed to be characteristic of 1951 conditions.

Public Forest Holdings. Public agencies, because of their long-term interest in the land and also because many of them control lands specifically for forest-management purposes, do a good job of managing their forest properties.[11] They control 116 million acres of commercial forest land. The best practice on the whole is found on the national forests, Table 13. Other Federal agencies, among them the Bureau of Land Management, the Indian Service, the Fish and Wildlife Service in the Department of the Interior, land-managing agencies in the Department of Agriculture other than the Forest Service, and the Military Reservations, control some 15.4 million acres. Most of these do a good job of managing their forest land. The Army and Navy, being concerned with military use, are not always able to manage their forest lands in the interests of timber production. State and local governments control some 27 million acres of forest land. Practice varies widely from state to state and within states among local governmental units. State forest lands that are under the jurisdiction of a State Forest Service are usually well managed. Those held by other state agencies are less likely to be well managed. County forest lands are oftentimes well managed provided the counties have a definite interest in forestry. Tax-delinquent lands that have reverted to

TABLE 13. OWNERSHIP OF COMMERCIAL FOREST LAND AND CHARACTER OF FOREST PRACTICE*

Ownership	Area owned, million acres	Per cent of total	Character of forest practice, per cent: High order and good	Fair	Poor and destructive
National forests	73.5	16	80	19	1
Other Federal	15.4	3	43	32	25
State and local government	27.1	6	47	10	43
Total public	116.0	25	67	19	14
Pulp companies	14.8	3	33	49	18
Lumber companies	36.6	8	25	27	48
Farmers	139.1	30	4	23	73
Other private	154.5	34	5	30	65
Total private	345.0	75	8	28	64
Grand total	461.0	100	23	25	52
Private:					
Large private	50.7	15	29	39	32
Medium private	32.9	10	8	31	61
Small private	261.4	75	4	25	71
Total private	345.0	100	8	28	64

* U.S. Department of Agriculture, Forest Service. 1946. The Management Status of Forest Lands in the United States, Report 3 from "A Reappraisal of the Forest Situation," Washington, D.C.

counties are seldom well managed unless the county has a regular program of forest land management.

Large- and Medium-sized Private Holdings. Private forest landowners control three-fourths of the commercial timberland in our country or 345 million acres. The large- and medium-sized landholders, those controlling 50,000 acres and more and those controlling 5,000 to 50,000 acres, are far better managers of lands than the 3.2 million farmers who control about 139 million acres and the 1 million other owners of small forest properties who control about 122 million acres.

Among large owners the pulp and paper companies have shown most interest in good forest management and have the highest standards of forest practice. Mill owners have felt that they must protect their source of raw material through control of forest land. Pulp and paper companies not only practice good standards of timber cutting on their own lands, but they encourage farmers and others who deliver material to their mills to maintain equally satisfactory cutting standards.

In the western states and in part of the South, individual lumber companies have commendable programs. They are managing their lands on a sustained-yield basis, and they have endeavored to stimulate good forestry on the lands tributary to their mills. Some of the most outstanding examples of intensive forestry in the country are to be found on the lands of lumber companies. A notable example is the Goodman Lumber Company property at Marinette, Wis. Here the company maintains a sawmill, a veneer mill, a wood-chemical plant, and a pulp mill so as to use all types of material that their land produces. The company has been on a sustained-yield basis for 20 years. The owner is convinced that his income from sustained-yield forestry has been equal to what he would have obtained had he operated on a liquidation basis, and his mill remains a going concern rather than junk in the midst of gutted timberlands. The cutting standards on the property are based on the best silvicultural information to be obtained from the Lake States Forest Experiment Station and other forestry agencies operating in the region.

Portable-mill owners of the South and North have displayed little interest in high standards of forest practice. The small permanent mills are usually located in territory where they can obtain logs from a fairly wide radius, or else they operate intermittently and with whatever material they can obtain from woodlands nearby. Few of them attempt to adopt good cutting practices unless the landowner insists on it.

Coal and other mining companies own considerable areas of timberlands in the East, South, and West. These are primarily interested in subsurface rights and have often been negligent of the timber that grows on the lands. Even though considerable quantities of timber are used in the mines, the companies have usually preferred to buy this from outside sources instead of attempting to grow it on company lands. A number started forestry programs when times were good and mine timbers scarce but abandoned them when depression struck the industries.[9]

Programs by mining companies started under the stimulus of the Appalachian Hardwood Manufacturers Association offer promise of permanence. The Oliver Iron Mining Company in Minnesota and the Cleveland Cliffs Mining Company in Michigan have had forestry programs for a long period of time.

Most wood-chemical companies have had unstable profits and unstable forestry programs.

Railway companies have been large holders of timberland in the West and moderate-sized holders in the East. Most of them have chosen to lease their timber rights to operating companies rather than to operate their own lands, but a few have excellent programs on company lands.

Private investors, among them trained foresters, own some well-managed forest properties. These make up as of 1951 only a small part of total private forest land, but their number is growing as investors and landowners learn more of the potentialities for sound investment in forest lands and as foresters acquire capital and leisure to manage their own properties. As of January 1, 1950, 691 members of the Society of American Foresters owned forest land. These included employees of Federal and state forest services, of forestry schools, associations, and industries. The total area owned outright was 206,000 acres. In addition, members had part ownership in 2,748,000 acres. Ownership by foresters not members of the national society is unknown.

Opportunities for expanding such ownership are still good, provided the foresters can procure sufficient capital and are willing to be patient long enough to build up the growing stock on the land they acquire. The difficulty that now confronts such individuals is in finding forest properties that can be operated immediately. Most are so understocked with timber that 15 to 30 years must elapse before any appreciable income can be had from the land.

Farm and Other Small Holdings. The major forestry task confronting our country is to improve forest practice on the 345 million acres of land in private ownership, particularly on the 261 million acres owned by farmers and others who control less than 5,000 acres. It is here that the poorest forest practices prevail and that destruction of growing stock is proceeding most rapidly.

Many obstacles confront the owners of small forest properties. Financial stringency is a widespread difficulty. This impels men to cut the forest whenever merchantable products can be removed, even though that may be the very time when trees are growing most rapidly. Where no regular market exists, owners are obliged to sell when operators are in their territory or to defer sales indefinitely. Small properties frequently have so little timber that the operator feels he must cut all an owner has to be justified in negotiating a contract. Individual owners are very much at a disadvantage in bargaining with a timber operator and often must accept his cutting methods or forgo a sale. The scattered nature of small ownerships, absentee ownership, inability of adjacent owners to pool their holdings into units of manageable size, depletion as a result of past cutting, and high preponderance of cull trees and low-grade species are further difficulties that impede forestry on small holdings. These are also reasons why large companies find such properties uneconomic to acquire.

Such lands for the most part seem destined to remain in small ownership, particularly those areas that are a part of going farms or are so isolated as to be uneconomic for management by large holders. Furthermore, the American people should consider carefully whether, in the long run, it is wise to allow most of their forest acreage to drift into the hands of government on the one hand or large private landholders on

FIG. 92. The farm forester learns of a new homemade winch for log skidding. (*U.S. Forest Service photo.*)

the other, thereby gradually eliminating the small owner from the forestry scene. Americans have cherished deeply the thought that our country is a land of opportunity, where the energetic individual can by his own efforts build up an independent enterprise. The opportunities to build up such an enterprise in forestry appear to be relatively good for men with patience and fortitude, proper technical background, practical experience in logging and sale of timber products, and a reasonable amount of backing for capital expenditures (Fig. 92).

LITERATURE CITED

1. Chapman, Oscar L. 1950. "Annual Report of the Secretary of the Interior" (fiscal year ended June 30, 1950), U.S. Government Printing Office, Washington, D.C. 411 pp.

2. Craig, James B. 1950. A Railroad Crusades for Forestry, *American Forests,* 56(2):6–9, 38–39.
3. Duthie, George A. 1949. Community Forests, "Trees, The Yearbook of Agriculture," pp. 394–398, U.S. Department of Agriculture, Washington, D.C.
4. Gill, Tom, and Ellen C. Dowling, compilers. 1949. "The Forestry Directory," The American Tree Association, Washington, D.C. 420 pp.
5. Gillett, Charles A. 1950. Citizen and Trade Associations Dealing with Forestry, "Fifty Years of Forestry in the U.S.A.," pp. 285–298, Society of American Foresters, Washington, D.C.
6. Granger, C. M. 1949. The People's Property, "Trees, The Yearbook of Agriculture," pp. 299–304, U.S. Department of Agriculture, Washington, D.C.
7. Heintzleman, B. Frank. 1949. Forests of Alaska, "Trees, The Yearbook of Agriculture," pp. 361–372, U.S. Department of Agriculture, Washington, D.C.
8. Moore, E. B., and A. N. Lentz. 1951. The Cooperative Forest Management Program in New Jersey, *Journal of Forestry,* 49:31–34.
9. Shirley, Hardy L. 1949. Large Private Holdings in the North, "Trees, The Yearbook of Agriculture," pp. 255–274, U.S. Department of Agriculture, Washington, D.C.
10. Timber Engineering Company. 1944. "The Forest Industries Blaze New Trails," Washington, D.C. 32 pp.
11. U.S. Department of Agriculture, Forest Service. 1946. The Management Status of Forest Lands in the United States, Report 3, from "A Reappraisal of the Forest Situation," Washington, D.C. 29 pp.

CHAPTER 19. *Proposed Programs of Forestry*

THE POLICIES governing use of any major raw material, such as forest products, affect everyone dependent upon the national economy. The nation as a whole needs timber for national defense and to house our growing population. The forest landowner needs a forest program that will stabilize markets for the products from his woodland. Labor is concerned because some 3.3 million workers are directly or indirectly dependent upon the forest industries for their source of employment and forest industries need a dependable wood supply.

Forest Policy Is Controversial

Public forest policy is controversial because it deals with economic relationships among people and between people and government. Opinions on government responsibility for natural resources vary widely. Some maintain that natural resources originally belonged to all the people and that they should be used for the benefit of all, not the profit of a few. At the opposite extreme are those who hold that natural resources should be managed by private individuals and companies with a minimum of government interference. They declare that self-interest will lead the individual owners to develop the resources for the benefit of all citizens and that private management is cheaper and more desirable than public ownership and development. The first theory carried to extremes would lead to nationalization of natural resources; the second to a complete laissez-faire policy. All shades of opinion lie between these two extremes. Controversy is often sharp and sometimes even bitter, for few people voluntarily relinquish economic advantages. Attempts to extend government control over private timberlands are vigorously launched by government and just as vigorously opposed by private interests.

Few foresters can escape such controversy. However, the area of agreement among those informed on the forest situation exceeds that of disagreement. All recognize that a far greater forestry effort is needed than is now being exerted. A few hold that existing programs, given ample time for development, will meet our country's forestry needs. Others believe firmly that new measures must also be used.

Unattained Forestry Goals

The need for additional effort on existing programs and for new public forestry measures can be appreciated by first reviewing goals still to be achieved. These include the following:

To place fire control on a satisfactory basis. Large areas are still inadequately protected.

To check losses from insects and diseases that now destroy more timber annually than does fire.

To halt further decline in sawtimber growing stock which is now occurring at about the rate of 1⅓ per cent per year.

To prevent clear cutting of immature timber where this is wasteful of wood and of growing space.

To make full use of the timber-growing possibilities of the national forests. Only 10 per cent of the timber harvested comes from national forests, yet they make up 16 per cent of the total commercial forest land and contain 32 per cent of the nation's sawtimber volume.

To provide rehabilitation for some 60 million acres of poorly stocked and denuded forest land.

To solve satisfactorily the economic problems confronting owners of small forest properties.

To reduce the accident rate in logging and lumber manufacture to levels comparable with those in other leading industries.

To increase the output per worker in lumber manufacturing as rapidly as other industries have improved their labor output.

To provide stable year-long employment to forest workers.

To prevent the growth of rural slums in many forest regions.

To keep lumber prices in line with prices of other manufactured commodities.

To provide optimum control of erosion and stream flow on forest and range watersheds (Fig. 93).

Programs Generally Supported

Representatives of the majority of agencies interested in forestry agree in supporting certain basic programs. These include the following:

Adequate Federal, state, and private expenditures should be made to provide effective protection of all forests and watershed lands from fire. This will require aggressive development of cooperative protection under the Clarke-McNary Act with financing in keeping with current costs and requirements and extension of organized protection to all forest lands not now receiving it.

Protection against insects and disease under the Forest Pest Control Act of 1948 and other Federal and state laws should be developed under a plan

FIG. 93. Overgrazing of headwaters has caused this check dam to be filled to overflowing with mud. (*U.S. Department of the Interior, Bureau of Land Management photo.*)

providing for joint Federal, state, and private effort. Early detection and control before serious outbreaks occur is the objective.

Educating the public to the importance of forests, the need for care in use of forests, and the reasons for expending public moneys to support forestry measures is a never-ending job. It is a task that public and private agencies must shoulder together. Need exists for a cooperative program to avoid public confusion. Educating the landowners to follow practices that will sustain production of timber is also a continuing task requiring public and private cooperative effort. The new cooperative forest-management act provides for such an educational program and for extending it to include the operators of timberland and the processors of forest products (Fig. 94).

Private owners should be encouraged to hold forest lands that can be profitably managed for continuous production of timber crops. The public should take over lands that are incapable of profitable management in private ownership as well as lands on which costly measures must be taken to protect watersheds or other public interests.

Taxes on forest land should be adjusted so that they are neither confiscatory nor so light that they exempt the forest owners from contributing their fair share to the support of government.

State forestry organizations should be supported in all states on a basis that will ensure competence in the management of state forests, in the protec-

FIG. 94. Demonstrating results of good and poor cutting practices to leaders among foresters and laymen.

tion of all forests from fire, and in carrying out state programs that deal with private forests and forest products.

Research should be encouraged as the basis for increasing the productivity of forest lands in both products and services and for expanding the utility of forest products.[6]

Steps should be taken to provide at reasonable cost insurance for forest owners against the natural hazards of timber growing. These include damages due to fire, insects, disease, heavy windstorm, ice, and snow.

A concerted effort should be made by public and private agencies to induce landowners to plant barren and poorly stocked forest land.

The U.S. Forest Service should rapidly extend its management program so that the full productive capacity of the national forests can be developed to add to the nation's timber supply and ease the drain on private lands. Other public forest lands should likewise be developed for sustained yield of forest products and services.

Efforts need to be expanded to realize more fully the social benefits of forestry—recreation; wildlife, water and soil protection; stable employment at good wages, and prosperous, permanent forest communities. Private owners now provide these services, at least to some extent, at little or no cost to the public. Little objection is made to extending such social benefits provided that, in doing so, owner-management objectives are not upset.

Not all forest owners are in a position to undertake sustained-yield operations. Not all forest communities can be made permanent. The remaining resource is too meager to permit this. Despite the best intentions not every timber-operating company is financially in a position to diversify its output so as to maintain current employment in the face of diminishing raw material. How far government should go in forcing sustained-yield management is questionable.

Stable year-long employment at good wages with stable and moderate prices for forest products is also a general goal of industry, labor, and government. Labor has emphasized safety in their programs.[2] The high accident rate in logging and lumber manufacture is ample cause for their doing so. All of these goals are interrelated. Labor is the major item of cost in logging, lumber manufacture, and wood processing. Measures that can increase output per worker and extend working days throughout the year will tend to stabilize communities, employment, and wages. By attracting high-class workmen, these measures will also tend to reduce accidents. Better worker training then becomes possible as well as rapid mechanization.

Size of operating unit is an important factor. Small operators often work intermittently, can afford only modest or second-hand equipment, have limited resources for training workers, and often must cut corners to survive. These should profit from industry-wide training programs and services such as the American Pulpwood Association provides.

Issues on Which Disagreement Exists

Aids to Private Owners and Timber Operators. Representatives of the forest-products industries would go slowly in extending public credit and services to private forest owners. They feel that public funds and credit should not be used to support private timber holdings that are uneconomic, either because of size or management practices of the owner. They favor discontinuance of benefit payments to farmers for forest practices under the Production and Marketing Administration.[4]

The U.S. Forest Service occupies an intermediate position, favoring public credit for timberland holdings where owners meet acceptable management standards. The Service has not advocated vigorously the extension of direct financial assistance to forest owners of the type extended to farmers under the Production and Marketing Administration and Soil Conservation Service.

Organized labor has advocated a very liberal use of public subsidies for all timber owners and operators. These would cover road building, forest-improvement measures, safety measures, and timber cutting and operating. Organized labor also advocates liberal loans to owners, in-

FIG. 95. Spruce forest at timber line on Mt. Washington, N.H.

cluding loans based on work done by the owner himself to improve his own woodlands.[2]

Public Ownership of Forests. Lands too low in productivity to pay costs of protection and management should be publicly owned and managed (Fig. 95). This is what the U.S. Forest Service advocates. The National Lumber Manufacturers Association's position on public ownership appears to be similar to that of the U.S. Forest Service with one important exception—the association favors exchange of public and private holdings to block up lands and transfer of public forest lands to private ownership and Federal land to state ownership where the public interest will be served. The Forest Service does exchange public land for private land but more often exchanges stumpage from national forests for private land. The Service has opposed vigorously every attempt to transfer national forest lands to either states or private companies. The past transfer of forest lands on the public domain to both states and private ownership almost invariably led to forest depletion. This the Forest Service seeks to avoid in the future.

Actually, a wide gap exists between the forest-products industries [4] and the U.S. Forest Service [9] on public forest ownership. Because of the growing scarcity of high-quality private timber holdings, the industries would welcome control over some of the better quality national forest lands. Where national forest lands are checkerboarded with private lands that owners are managing on a sustained-yield basis, a plausible case can be made for such a transfer.

The Forest Service, on the other hand, has been obliged to hold national forests for half a century with only limited opportunities to demonstrate high-standard, sustained-yield management. Now that private operators are ready to undertake national-forest logging on a large scale, the Service wants to demonstrate good forestry over these vast areas. Furthermore, the Forest Service would like to block up its holdings within national forests and to establish new national forests on large areas of low-quality forest land unlikely soon to be reforested in private lands. The industry prefers that additional public forests be mainly state-owned. State foresters generally support this position. Middle ground is represented by the American Forestry Association, which advocates for each state a permanent planning committee to determine an appropriate distribution among private, state, and Federal holdings.[1]

Regulation of Private Forest Practice. The industries favor independent state action on regulation of private forest practice without any Federal interference or intervention.[4] The forest industries supported enaction of the regulation laws in Washington and Oregon. They have also supported the voluntary laws in California, Massachusetts, and New York. They have vigorously opposed bills in the Congress that would give the Federal government control over private forestry.

The American Forestry Association [1] favors state laws to avoid unnecessary destruction of growing stock and young timber and to establish practices that will maintain continuous forest production. The Forestry Association and the Society of American Foresters and state foresters feel that such regulation should be state rather than Federal.

The U.S. Forest Service favors a Federal-state cooperative regulation plan that would prohibit premature and wasteful cutting and extensive clear-cutting, protect the forest, and reserve sufficient desirable growing stock to keep the lands reasonably productive.[9, 8] The Forest Service favors such action by states but believes that the Federal government should have authority to regulate private forest practice within a state which fails to act effectively within a reasonable time.

Organized labor and others [2, 5] have advocated nationwide Federal control over private forest practice. Labor would require that no trees be harvested unless marked in advance by foresters employed by or licensed by the U.S. Forest Service.[2]

Various bills to empower the Federal government to regulate private forest practice have been introduced from time to time since 1920. None has passed either house of Congress.

Proposals have been advanced, from time to time, that would lead to Federal control over forest practice through various leasing and lending activities.[7, 8] These have been vigorously opposed by the forest industries.

New Legislation Required

Most of the proposals for future forest programs actually involve the strengthening of existing public and private programs rather than new legislation. The many proposals included in the comprehensive study of forest policy by a joint committee of the Society of American Foresters and the National Research Council that reported in 1947 also involve primarily more liberal financing of existing programs.[6]

New legislation will be needed if the public is to sponsor insurance of forests against natural hazards and to extend credit to forest owners. Although the property tax may be administered fairly on forest land, it is not always so administered. Clarifying legislation would help in many states. The yield-tax laws of some states appear satisfactory to owners but result in irregular income for local governments unless cushioned by state payments to counties on an annual basis.

Aids to small private forest owners that will enable them to rehabilitate barren forest land and to develop understocked forest properties for high yield are not adequately provided by existing legislation. At present only owners who have an income from other sources can engage in such activities. Public financial aid for planting, drainage, and other improvements of the type provided in Norway, Sweden, and Finland are still to be worked out in America. These may be the best answer to getting forest production reestablished on scattered small holdings.

Regulation of private-forest cutting practices is provided in only some states at present. Changes in some existing laws and enaction of legislation by additional states are needed. The role that the Federal government is to play in this field is still to be determined.

Legislation or at least substantial appropriations will be needed to improve working conditions and output of forest, sawmill, and wood-processing workers. The type of program needed is what the Scandinavians call forest work studies. These cover encouraging young people to engage in forest work, improving living and community life of woods workers, training in handling tools and equipment, insurance, and social security. Such programs are sponsored jointly by industry and government in Scandinavia.

The President has authority, subject to congressional veto, to reorganize the executive branch of the government. As of 1951, administration of Federal forest lands valuable chiefly for commercial timber production was split between the U.S. Forest Service and the Bureau of Land Management. The same is true of range lands. The President has asked the Secretaries of Agriculture and the Interior to work out a plan for consolidating administration, especially on lands having a checkerboard pattern of jurisdiction.

Legislators might well give consideration to providing landowners who permit public recreation and hunting on their lands some aids or tax concessions. The same applies to owners of strategic watersheds. Authority for extending special aids to owners of watersheds that need improvement for public benefits may be needed. Regulations governing forest practice on such watersheds might well precede nationwide regulation of private forests.

Forestry Programs in European Democracies

Perspective can be had on the proposed programs for the United States by reviewing practices that have developed in European democracies up to and after the Second World War.

Switzerland passed a basic law in 1902 prohibiting clear felling, except by permission, restricting grazing and other harmful practices in mountain forests, grouping private protection forests for management, and requiring reforestation within three years of forests destroyed by fire, avalanches, or windstorm.[3] *Belgium* and *Holland* have less strict laws but endeavor to encourage private forestry by tax concessions.[3]

France had mild regulation of private forestry until the country was occupied by Germany in the Second World War. The Vichy Government established strict regulation adopted from the German pattern. Some relaxation occurred with reestablishment of the republic in 1944. In 1946 France passed legislation directed toward making the nation self-sufficient in forest products. This provides public help to landowners but requires definite standards of practice.[3]

Since 1945 *Great Britain* has made strenuous efforts to build up her forests. Her objective is to increase the total forest area of the country to 5 million acres. The program includes government acquisition of land for forests, general planning of land use throughout the nation, and an intensive program of private forestry. Private forest owners are urged to dedicate their forest lands to permanent forest management. The owner who dedicates his land agrees to make timber production the permanent objective of land management, to manage his land under a plan approved by the Forestry Commission, to employ skilled supervision, and to keep adequate accounts. The commission's foresters will help provide the skilled supervision and are available to advise the owner in all his operations.

The government provides financial assistance in two forms. The government will reimburse the owner for 25 per cent of his excess of expenditures over receipts for any one year or will provide a direct payment for forest planting. As forest planting is very expensive in Great Britain because of the necessity of fencing against rabbits and the clearing of heavy brush, the planting subsidy has been placed at £12 per acre. Additional grants are available for maintaining the plantations after they are set out. The owners have been given a five-year period in which to dedicate their lands. Those failing to do so in this period of time may, at the option of the commissioner, have their land taken over and managed by the forest commission, but the commission had up to 1950 failed to exercise this option. The dedication program has moved at a relatively slow pace. Only some 25,000 acres were dedicated or in process of dedication after four years of operation.

In 1950 Britain's program was modified in four directions to make it more attractive to landowners. Disputes between the owner and the forestry commission over forest practice may be referred to a special investigating committee for recommendation, so that owner representatives gain a voice in administration; provision is made whereby the commission may release an owner from the obligations of dedication; the government has increased the grants for planting and plantation maintenance; a scheme of reimbursement of owners has been worked out that does not require the owner to keep accounts. Upon the granting of these concessions, landowner associations have urged their members to dedicate woodlands.[7]

Denmark, a country fervently devoted to democracy and with a high sense of individual freedom and initiative, requires owners to follow good forestry practices. All Danish forests are reserved. Owners are obliged to manage them as high forest, using approved silvicultural principles. Hardwoods may not be diminished in area. State forest officers inspect private forests and advise owners on management. Small owners who do not employ a forester look to the state inspectors to mark trees for cutting. Owners may join a cooperative, subsidized by the state, and hire a forester to manage their lands. Owners of less than 10 years tenure may not cut timber for sale without state approval but may cut for domestic use. Approval for sale may be given for one to three years. This precludes speculation in woodlands. Despite the authority lodged in the Danish forest officers, they act with discretion, and little friction arises between the forest owners and the public foresters.

Finland and *Sweden* have good forest practice on private lands and have a wholly democratic system of regulating cutting practices. This is exercised through local forestry boards that employ county foresters and have the right to dismiss them. These boards get financial support from the central government for carrying out their forestry program and have considerable latitude in deciding what this program is to be. They determine cutting-practice standards, and they supervise the payment of state subsidies for forest draining, tree planting, forest improvements, and other purposes.

The Swedish law of 1948 differentiates between vigorous and nonvigorous forest stands. Vigorous stands are those that are growing at such a rate as to produce wood more profitably than the land would produce if harvested and reforested. Such stands may not be cut other than by thinning without the consent of the Forestry Conservation Bureau. Forest stands that fail to reproduce after cutting must be brought into a satisfactory condition within reasonable time.

In Sweden the county forestry board consists of three members, the chairman appointed by the Ministry of Agriculture, one representative elected by farmers' organizations, and the other elected by the county officers. The landowners are free to cut their timber provided they stay within the minimum practice standards established by this board. Farmers who do not comply are investigated by competent county forest officers and, if delinquent, are warned. If the delinquency persists, they are brought before the county forestry board and placed on probation, and if this measure fails, they are then forbidden to cut any timber from their land unless this is first marked for cutting by the county forest officer. Such restriction usually remains in force for five to ten years and even longer. Failure to comply with this ruling of the board subjects them to court action.

These procedures are more gentle than those proposed in several bills that have been introduced into our Congress. They have, however, been effective in getting good practice on the forest lands of Sweden and Finland. Local forest officers are the first to admit that practice is far from perfect. Finnish foresters believe that if the best type of silviculture known in the country were universally adopted, timber yields could be doubled. Perhaps the same could be said of Sweden. This, however, would require the marking of all trees by thoroughly trained forest officers. The Finnish and Swedish programs have the merit of being nationwide in application with a maximum of flexibility on a county-by-county basis and with completely democratic formulation and administration of regulations. Familiarity with the functioning of Finnish, Swedish, and Norwegian public regulation practices should prove helpful to those in other countries interested in activating public forest programs.

Most of the proposed programs for forestry in the United States have their counterparts in laws now in force in other democracies. In general, the educational and assistance approach has met with wider acceptance and produced better results than strict public controls. An outstanding development has been the sharing of administration of government controls and aids among landowners, the general public, and government.

The Role of Foresters and Citizens

Men with forestry training are expected to inform themselves on the forestry situation in the United States and the measures that need to be taken to place it on a sound basis. They are not expected to be unanimous in supporting any specific methods for achieving this objective. Obviously, the opinions of foresters may be expected to differ, just as do the opinions of other men. It must be emphasized, however, that public programs of forestry on a nationwide scale do not come into being without a great deal of effort on the part of many people. As foresters justify their existence on the basis of their ability to provide adequately for the nation's timber supplies and other forest resources and services, it is up to foresters to take the lead in advocating public measures that will attain these ends. Honest differences of opinion will be respected and are honorable. It must be kept in mind that the nation's economic resources of which the forests are one must be protected, and that in protecting these resources we should strengthen rather than weaken democratic processes. The experience of such nations as Finland, Sweden, and Denmark present ample evidence that such an objective can be achieved if we have the patience and determination to strive for it.

LITERATURE CITED

1. American Forestry Association. 1946. A Program for American Forestry, *American Forests*, 52(8):352–356, 387.
2. American Forestry Association. 1947. "Proceedings, American Forest Congress, October 9, 10, 11, 1946," pp. 86–102, Washington, D.C.
3. Anderson, M. L. 1950. "State Control of Private Forestry under European Democracies," Clarendon Press, Oxford. 112 pp.
4. Black, S. R. 1947. Program of Forest Industries Council, "Proceedings, American Forest Congress, October 9, 10, 11, 1946," pp. 37–45, American Forestry Association, Washington, D.C.
5. Clapp, Earle H. 1949. Public Forest Regulation, *Journal of Forestry*, 47:527–530.
6. Graves, Henry S. 1947. "Problems and Progress of Forestry in the United States," Society of American Foresters, Washington, D.C. 112 pp.
7. Moore, Barrington. 1950. Trends in British Forestry, *Yale Forest School News*, 38:69–71.
8. Raushenbush, Stephen. 1949. "Our Conservation Job," Report 4, The Public Affairs Institute, Washington, D.C. 64 pp.
9. Watts, Lyle F. 1949. A National Program for Forestry, "Trees, The Yearbook of Agriculture," pp. 757–760, U.S. Department of Agriculture, Washington, D.C.

CHAPTER 20. *The Profession of Forestry*

LAW, THEOLOGY, and medicine are known as the "learned professions" or "three professions." These are characterized by devotion to public service, a high standard of ethical conduct, and use of special knowledge and skills requiring years of training for mastery. Their members support professional societies devoted to exchange of information and promotion of scholarly research. Engineers, architects, educators, foresters, and others have followed the pattern of the three learned professions in their standards of education, service, and devotion to professional betterment.

The Society of American Foresters

The stature of a profession is best gauged by the activities and ethical standards of its professional society. On November 30, 1900, seven men met at the home of Gifford Pinchot to organize the Society of American Foresters.[2] The original constitution stated its objective: "To further the cause of forestry in America by fostering a spirit of comradeship among foresters; by creating opportunities for a free interchange of views upon forestry and allied subjects; and by disseminating a knowledge of the purpose and achievements of forestry." This objective has been faithfully adhered to by the society as its membership increased from 7 to 7,000.

Just as forestry is nationwide in scope, so also is the society with its 21 regional sections, its 9 subject-matter divisions, its 23 working committees and its growing numbers of local chapters. These provide opportunities for interchange of views and for young and seasoned members to meet on common grounds in the service of the profession. The society's publications, the *Journal of Forestry*, the proceedings of meetings, and the several books that have resulted from society activities furnish the chief means for recording professional education, achievement in research, and administration.

By invitation the society affiliated with the American Association for the Advancement of Science in 1913 and thus gained a voice in the council of this association. Since the formation of the National Research Council in 1917, the society has been represented in the council's Division of Biology and Agriculture. It is represented on joint committees with the American Society of Mechanical Engineers and the American Geophysical Union. It is affiliated with the American Institute of Biological Sciences and the National Resources Council of America.

The society has sought to promote forestry by encouraging needed legislation, by protesting political sponsorship for forestry positions, and by upholding forest officers in the discharge of their duties against political pressure. It has set up requirements for society membership and from these, established standards upon which schools of forestry and their curriculums are accredited. It makes studies of state administration of forestry upon requests by the states.

Although its members are loyal, the society is no supine, self-laudatory body. Healthy differences of opinion have characterized its meetings from the first. Debates on such issues as public regulation of private forestry have rocked the society since 1910, and this matter is still a vital issue. These honest and sharp differences of opinion prevent complacency and smugness. In recent years such debates have been tempered by an urbanity, mutual respect, and sense of humor that have prevented the society from splitting apart.

Code of Ethics for Foresters. In 1948 the Society of American Foresters adopted a code of ethics to guide professional conduct. This merits the serious study of every student of forestry. The canons as drawn up and adopted by referendum vote [4] are reproduced below.

Preamble

The purpose of these canons is to formulate guiding principles of professional conduct for foresters in their relations with each other, with their employers, and with the public. The observance of these canons secures decent and honorable professional and human relationships, establishes enduring mutual confidence and respect, and enables the profession to give its maximum service.

Professional Life

1. The professional forester will utilize his knowledge and skill for the benefit of society. He will cooperate in extending the effectiveness of the forestry profession by interchanging information and experience with other foresters, and by contributing to the work of forestry societies, associations, schools and publications.

2. He will advertise only in a dignified manner, setting forth in truthful and factual statements the services he is prepared to render for his prospective clients and for the public.

Relations with the Public

3. He will strive for correct and increasing knowledge of forestry and the dissemination of this knowledge, and will discourage and condemn the spreading of untrue, unfair, and exaggerated statements concerning forestry.

4. He will not issue statements, criticism, or arguments on matters connected with public forestry policies without indicating at the same time on whose behalf he is acting.

5. When serving as an expert witness on forestry matters in a public or private fact finding proceeding, he will base his testimony on adequate knowledge of the subject matter, and render his opinion on his own honest convictions.

6. He will refrain from expressing publicly an opinion on a technical subject unless he is informed as to the facts relating thereto, and will not distort or withhold data of a substantial or other nature for the purpose of substantiating a point of view.

Relations with Clients, Principals, and Employers

7. He will be loyal to his client or to the organization in which he is employed and will faithfully perform his work and assignments.

8. He will present clearly the consequences to be expected from deviations proposed if his professional forestry judgment is overruled by nontechnical authority in cases where he is responsible for the technical adequacy of forestry or related work.

9. He will not voluntarily disclose information concerning the business affairs of his employers, principals or clients, which they desire to keep confidential, unless express permission is first obtained.

10. He will not, without the full knowledge and consent of his client or employer, have an interest in any business which may influence his judgment in regard to the work for which he is engaged.

11. He will not, for the same service, accept compensation of any kind, other than from his client, principal, or employer, without full disclosure, knowledge, and consent of all parties concerned.

12. He will engage, or advise his client or employer to engage, other experts and specialists in forestry and related fields whenever the client's or employer's interests would be best served by such actions, and will cooperate freely with them in their work.

Relations with Professional Foresters

13. He will at all times strive to protect the forestry profession collectively and individually from misrepresentation and misunderstanding.

14. He will aid in safeguarding the profession against the admission to its ranks of persons unqualified because of lack of good moral character or of adequate training.

15. In writing or in speech he will be scrupulous to give full credit to others, in so far as his knowledge goes, for procedures and methods devised or discovered and ideas advanced or aid given.

16. He will not intentionally and without just cause, directly or indirectly, injure the reputation or business of another forester.

17. If he has substantial and convincing evidence of unprofessional conduct of a forester, he will present the information to the proper authority for action.

18. He will not compete with another forester on the basis of charges for work by underbidding through reduction of his quoted fee after being informed of the fee quoted by a competitor.

19. He will not use the advantages of a salaried position to compete unfairly with another forester.

20. He will not attempt to supplant another forester in a particular employment after becoming aware that the latter has been definitely engaged.

21. He will not review the work of another forester for the latter's employer without the other's knowledge unless the latter's connection with the work has been terminated.

22. He will base all letters of reference or oral recommendation on a fair and unbiased evaluation of the party concerned.

23. To the best of his ability he will support, work for, and adhere to the principles of the merit system of employment.

24. He will not participate in soliciting or collecting financial contributions from subordinates or employees for political purposes.

25. He will uphold the principle of appropriate and adequate compensation for those engaged in forestry work, including those in subordinate positions, as being in the public interest and maintaining the standards of the profession.

Rules of Construction

These canons of ethics shall be broadly construed in the interest of the public welfare and professional advancement.

They shall be deemed duties, obligations, and responsibilities of the professional forester to all those with whom he comes into contact in the course of his professional work and life.

These canons shall be construed so as to disregard any differentiation between professional foresters in public or private employ. The standard of professional conduct exacted by these canons of ethics from professional foresters makes no distinction as to place or kind of employment.

Because the foregoing canons consider the usual and foreseeable circumstances governing professional conduct only, experience may require modifications or additions.

Men entering professional work, therefore, enter into a covenant between themselves and the general public that they will devote themselves diligently to their chosen profession, adhere to its canons of ethics, place public service above private convenience and gain, and strive to promote the good of the profession and to enlarge its services to mankind.

Obligations of a Forester

A man entering forestry accepts certain benefits that carry with them obligations.

The technical knowledge he acquires has been painstakingly built up by countless foresters since the beginnings of the profession. He is privileged to use this knowledge; it is his duty to pass it on more complete and better tested than he received it.

His education is provided largely at public expense in publicly supported or endowed universities. Training in forestry is costly compared with that of a teacher or graduate in liberal arts.

Foresters, as a group, receive higher than average income. Mason in 1937 [6] reported the average wage in the United States to be $1,500, the average salary $2,500, the average forester's pay $3,000. In 1949 a survey of alumni of a forestry school revealed a median annual salary of those employed in forestry of $4,878. This was $74 higher than the income of graduates of the same college who had left forestry. Although salary scales were much higher in 1951, the same relationship between pay of foresters and that of other workers appeared to prevail.

Forestry flourishes only in those countries where it receives public support in whole or in part.

Men entering forestry colleges, therefore, accept an obligation to apply themselves diligently and to follow the profession as a life career, circumstances permitting. This means that they will devote their best efforts to the profession, follow its code of ethics, keep abreast of new developments in their field, give generously of their time and effort to advance the profession by serving its society, and helping new members to become established. It means also that they will place duty above personal convenience, especially when their duty calls for protection of life and property, and it means that they will be true to their professional ideals in the interests of ensuring abundant forests for posterity.

Age and Size of the Profession of Forestry

Forestry is still a young profession in America but one that is already firmly established. It has now proved its value by half a century of service.

The growth of the profession in the United States is well illustrated by the number of men granted degrees from colleges of forestry (Table 14). In studying this table it should be kept in mind that many of the men receiving the master's degree also took a bachelor's degree in forestry. The total men trained is less than 20,821. A number have been removed from the profession by death, incapacity, and other causes. Total employment in 1950 was estimated to be 13,000.

TABLE 14. TOTAL NUMBER OF DEGREES GRANTED IN FORESTRY BY SPECIFIED YEARS CUMULATIVE SINCE 1900 WHEN FIRST DEGREE WAS GRANTED*

Year	Bachelor's degrees	Master's degrees	Total
1900	1		1
1910	211	264	475
1920	1,433	597	2,030
1930	3,835	1,080	4,915
1940	10,183	1,879	12,062
1950	17,701	3,120	20,821

* Adapted from Guise, Cedric H. Statistics from Schools of Forestry for 1950: Degrees Granted and Enrollments, *Journal of Forestry*, 49:7–12.

Forestry may be compared in size with these other professions:

Foresters	13,000
Chemical engineers	12,000
Architects	22,000
College instructors and professors	76,000
Mechanical engineers	86,000
Physicians and surgeons	200,000

Enrollment in forestry schools in 1950 totaled 6,536 and the peak graduating class numbered 2,321. During the same year 275 men received the master's degree and 29 the doctor's degree.

The profession as of 1951 was still growing vigorously. The fields of employment were expanding in numbers and the intensity of work within each field was increasing. No ceiling on the size of the profession was in evidence.

Employment in Forestry

Beginning Jobs. The forestry graduate may find his first job in any one of the fields discussed in the preceding chapters. Irrespective of field, beginning jobs are likely to involve hard physical work and to afford only modest pay. Apprenticeship is essential before a man can assume full responsibilities. He must learn how to perform not only technical work but also much of the common routine work of laborers, mechanics, and artisans. His objective is to improve efficiency, safety, and over-all performance. By actual participation in such jobs he can best acquire the skill to train labor, prescribe safety regulations, inspect the work for compliance to standards, and figure the costs of the several operations.

When called upon, every forester must be prepared to perform emergency services in fire fighting, rescue work, the leading of expeditions into remote areas, and similar duties for which his training and experience especially qualify him. Descriptions of various entering jobs in the land-management phases of forestry are given by Allen.[1] Additional information will be given in Chaps. 21 and 22.

The basic science of forestry may be well mastered in college but skill in its use requires experience. The skilled forester can identify trees at a glance, without opportunity to examine leaves, buds, and twigs. He can use simple surveying instruments with surprising accuracy. He can select the proper tree to harvest without involved theoretical consideration. He can identify lumber species and grade at sight. He can find his way in the woods with minimum reference to compass or landmarks. The acquisition of these skills and judgment is a part of the apprenticeship training. Not until they have been acquired is he ready for the responsibilities of higher jobs. Only then is he truly in a position to render to his employer service at the professional level.

Intermediate Jobs. Intermediate jobs are those of forest ranger, park ranger, farm forester, timber buyer, timber appraiser, superintendent of a small-sized logging operation or sawmill, lumberyard foreman, supervisor of a unit operation in a pulp and paper mill, a large sawmill, or a woodworking factory, instructor in a college of forestry, and assistant to project leaders in research. It is in such jobs that men are called upon to exercise to the maximum their technical knowledge. They are also required to plan, organize, and supervise the work of others, and they begin to acquire a certain knowledge of the over-all planning and operations of the agency or enterprise (Fig. 96).

Intermediate jobs in forestry require a substantial amount of individual initiative, individual judgment, and the ability to plan and carry on work with only general supervision. The tasks of such foresters require that they spend time dealing with men as well as with trees and physical improvements. Accomplishments here are measured by what the forester's unit accomplishes. Teamwork and leadership become important. Judgment and power of decision are developed. Delegation of responsibility and authority and checking on performance begin to occupy the forester's mind.

Many men will find their greatest usefulness and satisfactions in such jobs. Save for a few staff specialists, research men and teachers, the national forest rangers and their equivalents in state and private forestry perform the real technical work in forestry. It is they who establish plantations, tend the forest crop, harvest its products, and provide for regrowth. Advancement beyond this level depends mainly on administrative skill in organizing and directing the work of others.

Advanced Jobs. Men in advanced jobs are called upon to deal with more than one specialty. Examples of such positions are those of supervisors of national forests and national parks, state district foresters, for-

FIG. 96. Foresters may live and work in isolated headquarters. Bonita Ranger Station, Lolo National Forest. (*U.S. Forest Service photo.*)

esters for pulp and paper companies, lumber companies, or large estates, superintendents of pulp mills or paper mills, logging superintendents, managers of small sawmills, managers of veneer-manufacturing plants, city foresters or city park superintendents, consulting foresters, project leaders in research, and teachers in colleges of forestry. These positions require up to 10 or more years' experience, a broad appreciation of overall objectives, ability to coordinate the work of two or more divisions, and the capacity to deal with the public and to interpret policies and programs to division chiefs. Or they require expert knowledge of difficult processes, which forms the basis for making important decisions governing the expenditure of large amounts of public or private money and the commitment of an agency.

Top Jobs. Top jobs in forestry include such positions as those of state foresters, U.S. regional foresters, directors of forest experiment stations, division chiefs and Chief of the U.S. Forest Service, top foresters in other Federal agencies, secretaries of important associations dealing with forestry and forest products, professors and deans of colleges of forestry, world authorities on special subjects in forestry, top forestry positions in the Food and Agriculture Organization, and managers and presidents of large corporations engaged in forestry, timber processing, landscape, and related work. These jobs require a maximum of ability to coordinate the work of others. Top administrators must deal effectively with the general public, high government officials, legislators, executives of other corporations, boards of directors, stockholders, and others. Executives are required to outline objectives and set the broad policies that guide the development of the organization, to delegate authority and responsibilities for major functions, to coordinate these major functions, to cooperate effectively with other agencies having related responsibilities, to plan long-range programs, to maintain high working morale, to draw out from the organization the best thoughts of all the workers, and to synthesize these into working procedures to improve effectiveness. High capacity for performing detailed work is useful, but not a necessity, for such men. It is their job to give general direction and to set the tone of the organization.

Jobs beyond the beginning positions in forestry usually present so many opportunities and tasks so diversified that they challenge to the utmost the ability of the most brilliant and hard-working forester. Always there are more tasks than can be performed and more opportunities for helpful service than there are funds or time to meet them.

Monetary Rewards. Earnings of foresters compare favorably with those for average jobs in other professions. Opportunities for high salaries, $10,000 per year and up, are fewer for foresters than they are for engineers, architects, businessmen, and lawyers. Forestry is not a profession to make a man wealthy. Federal salary scales for foresters are on the same basis as those for other professional men. U.S. Forest Service jobs fell into the following salary ranges in 1951:

Junior forester	$3,410 to $4,160
Assistant district ranger	4,205 to 4,955
District forest ranger	5,060 to 5,810
Assistant forest supervisor	5,940 to 6,940
Forest supervisor	7,040 to 9,360
Assistant regional forester	8,360 to 9,360
Regional forester	10,800 to 11,800
Assistant chief of forest service	12,000 to 12,800
Chief of forest service	14,800

Teachers at colleges of forestry could expect about the same entering salary. State salaries have been slightly lower. Foresters entering industrial employment may be obliged to accept a lower entering salary than those paid by the Federal government, but promotions in industry are usually more rapid than in government and high-paying jobs more abundant.

The salaries reported by foresters in January, 1950,[3] showed that nearly 40 per cent of the members of the Society of American Foresters received salaries in excess of $5,000. The actual groupings were as follows: under \$2,500, 2 per cent; \$2,500 to \$3,499, 15.5 per cent; \$3,500 to \$4,999, 37.8 per cent; \$5,000 to \$9,999, 35.3 per cent; over \$10,000, 3.6 per cent; and salaries not indicated, 5.8 per cent. Of the Federally employed foresters nearly half received salaries between \$5,000 and \$9,999, but less than 1 per cent received in excess of \$10,000. Salaries of \$10,000 or more per year were received by 8 per cent of industry-employed, 12.4 per cent of self-employed, and 18.6 per cent of association-employed foresters. The lowest paying agencies were states and schools, though the schools report 43 per cent receiving \$5,000 or more. As this analysis included recent graduates as well as mature foresters, it is to be expected that a large percentage were still in the beginning salary ranges.

Men entering jobs in the wood-using industries and particularly in pulp and paper manufacture have in recent years commanded higher salaries than those entering field forestry. This is due in part to the fact that opportunities are somewhat greater in these industries than in land management. If the work of a forester has sufficient appeal to prove attractive, the pay will usually prove adequate for a satisfying life.

Working Conditions in the Field. Men taking employment in forest-land management should expect to fill positions that require a large amount of time in the woods. They will thus gain practical experience and develop an understanding of field problems that can be obtained in no other way and that are essential in the more advanced positions. This is particularly true for those positions carrying administrative responsibility.

A variety of tasks may greet the forester on his first job. It is possible that he will be concerned with such work as cruising or marking timber, making range surveys, or reforesting denuded lands, or he may be detailed to conservation work in soil erosion or flood-control areas. If he takes a position with a private company, he may plan and build logging roads and camps, supervise cutting crews, carry on cost studies, act as a timber buyer, or scale logs at a sawmill. On most of these jobs he will live in the forest or in a small community (Fig. 96).

If he shows considerable promise, the period of apprenticeship may be relatively short, but for most men it will occupy a number of years. As the number of foresters and competition increase, the training period will tend to lengthen. During the first years of professional work the young forester is apt to be moved from one place to another at rather short intervals. After a time he may expect to become more permanently settled as maturity brings advancement to a position of greater responsibility. Most foresters, however, find that a considerable period of each year must be spent away from home, generally in the field. Trips may be for a day or for several weeks. Even with the extension of good roads much of the forester's travel for many years to come will be arduous. The time is not likely to come when it will be easy.

To urban dwellers a forester's life may appear idyllic. Much of it is spent out of doors working with nature. Surroundings are usually pleasant, frequently inspiring. In his daily work the forester may travel to remote places that city dwellers see only on brief vacation trips. Because the forester is trained to be a close observer of trees, forest animals, and plants, he finds much to interest him. His calling leads him to think in terms of resources, human welfare, employment, and industries. He sees the land, not only as it is, but as it was in the past and as it may become in the future. Foresters meet interesting people. Being members of a relatively small profession, foresters readily win each other's confidence and friendship. For most, their profession is both a means of livelihood and a hobby.

But there is another side to a forester's life. Fire fighting is grueling work. Smoke blinds the eyes and sears the lungs. Heat hastens exhaustion. Labor is likely to be unremitting from an hour or more before dawn until well into the night. Such is the forester's lot until the fire is brought under control, even though it may take weeks. Timber cruising, inspection of timber sales, planning logging layouts, and a host of other field tasks by foresters must be performed in fair weather or foul, in comfort or in discomfort.

The vigorous youth has a tendency to make light of such hardship, but he should learn that if he becomes a forester, he accepts these hardships and discomforts for a long time ahead. Stand-by fire duty on fair week ends, long workdays, and long work weeks with no extra pay for overtime must be expected. As long as he remains a forester, no one is entirely exempt from such duty. The fact that foresters in high positions may be called on to perform such tasks at only widely spaced intervals only makes the hardship the more keen when it must be endured.

Life-insurance companies recognize the forester's profession as a hazardous one and require extra premiums for forest rangers and those who perform similar tasks. These premiums are based in part on the normal hazards of working alone in remote localities where frequent chances of injury or overexposure exist. Logging and sawmilling remain among the most hazardous occupations in America. Foresters share some of these hazards. In addition, they must undergo the health hazard that follows the exhausting work of fire fighting, rescue work, and other emergency tasks. Venomous snake bite and Rocky Mountain spotted fever from tick bites are other hazards a forester must bear. The greatest cause of injury has been motor-vehicle accidents.

Wife and Family. Beginning jobs in field forestry frequently involve long periods away from headquarters or isolated living on the part of the forester and his family. Men who enter the profession after they have already acquired a wife and family often do so with considerable handicap. Beginning jobs that will provide them sufficient income to support a family and sufficient time at home to meet the obligations family life entails are difficult to find. Still, field training is essential for future success.

The hardships which the forester himself must endure are often less disturbing than those his wife may encounter. It may be her lot to live in remote places, to wait at home while her husband is away for long periods, to bear and train her children with limited household and community conveniences. She must look forward to spending many years in small communities where schools, hospitals, stores, and amusements may be meager. Keeping intellectually alert and finding congenial companionship are often more difficult for her than for her husband. She may, however, find intellectual stimulation in the classic writings that have expressed man's highest aspirations through the ages, and she can seek outlet for her sympathy and creative energies in service to her family and community. What it means to be a forester's wife is ably expressed in "The Pine Tree Shield." This book was written by Mrs. Flint, the wife of a man who devoted his entire career to forestry.[5]

The forester's wife who can reconcile herself to inconvenience, isolation, hard work, community service, and intellectual discipline can become an inspiration to her husband and friends. She can teach her children the essential values of land, water, trees, and grass and the importance of living a life devoted to more than narrow self-interest; and she can help her husband to rise in his calling to those higher jobs that involve the utmost in breadth of feeling, consideration for people, and devotion to the public interest. If she cannot face these responsibilities and these handicaps, neither she nor her husband is likely to be happy, and the careers of both of them are likely to fall short of their ambitions.

Although the foregoing applies particularly to the field-going forester, it applies more or less equally well to the forester engaged in logging, in sawmill operations, in small wood-processing plants, and in pulp and paper mills. The forest industries are mostly located near their source of raw materials. Frequently they are the major industry in the community, often the only industry. This places on technically trained men special burdens for community leadership and service in addition to high standard of professional work.

Satisfactions Offered by the Profession of Forestry

A career in forestry offers many recompenses other than monetary rewards. First of all is the privilege of working with living things, with trees, with wildlife, and with people. The forest is ordinarily free from the nervous tension of a noisy factory or a busy office. Life in the open is far less nerve-racking, and often it brings the forester in contact with scenery of outstanding beauty, with shy wild creatures, and with the open-forest environment in which our ancestors developed their vigorous faith in our country and love of freedom and democracy. Most satisfying of all, forestry is important work. The fact that foresters realize this and seek to increase the services of forests to mankind lends a dignity and a sense of inner satisfaction that may escape the industrial worker.

LITERATURE CITED

1. Allen, Shirley W. 1947. "Handbook of Information on Entering Positions in Forestry," 2d ed., Society of American Foresters, Washington, D.C. 56 pp.
2. Allen, Shirley W. 1950. The Society of American Foresters, "Fifty Years of Forestry in the U.S.A.," pp. 272–284, Society of American Foresters, Washington, D.C.
3. Anonymous. 1950. Employment Analysis Shows Industry Increase, *Journal of Forestry,* 48(6):444.
4. Anonymous. 1948. Canons of Ethics for the Profession of Forestry, *Journal of Forestry,* 46:626–627.
5. Flint, Elizabeth Canfield. 1943. "The Pine Tree Shield," Doubleday, Doran & Company, Inc., New York. 251 pp.
6. Mason, Earl George. 1944. "Functional Curriculum in Professional Forestry," Studies in Education and Guidance, Bulletin 1, pp. 1–100, Oregon State College, Corvallis, Ore.

CHAPTER 21. *Employers of Foresters*

EVERY YOUNG MAN entering a profession has certain potentialities and latent powers. Whether these powers are gradually developed as his years of service increase or atrophy because of frustration or complacency depends to a large extent upon the opportunities presented to him by the agency for which he works. Occasionally, ambitious young foresters have found their professional development impeded because their immediate superior or agency head had a narrowly circumscribed concept of his field of service. Such men are not likely to hold back an able young man in a large agency that has many channels for advancement. Contacts between foresters are frequent and intimate, so that good men need not remain neglected. Nor need they be satisfied with their current achievement. Alert foresters are constantly seeking to enlarge the scope of their work and their field of service.

Duties and employment conditions vary in each agency that employs foresters. Each has advantages and limitations. The graduate should seek employment with agencies whose objectives he can freely accept and that provide opportunities for professional development, and preferably in an organization that is growing. A small and static organization or a large organization that is no longer expanding offers ordinarily less opportunity for advancement and for self-expression than a young and expanding organization.

Large agencies usually offer better long-run opportunities for promotion and for wide experience than do small agencies. This, however, is not always true. Some small agencies may cover a wide variety of fields and may bring the worker into contact with many other opportunities. Such a case is employment with a small but highly respected firm of consulting foresters. The broader the field of service that a given agency encompasses, the wider, in general, the opportunities for advancement and for self-development. However, agencies having a narrow scope of work are sometimes confronted with the necessity of rapidly expanding their activities. When they do so, they offer exceptional opportunities for their personnel. Finally, a forester would do well to consider the past performance of a prospective employer in terms of the number of professional men he has employed and the opportunities he has placed before them. Some agencies are naturally progressive, employ many technical men, and place them in positions of responsibility. Others may see in the technician merely a man to perform specialized chores and keep him in a position of minor responsibility throughout his life.

The Society of American Foresters made a survey of employment of their members in 1950 and from this estimated the employment connections of all foresters.[2] Similar studies had been made in 1934 and 1937. The results of these three studies, as shown in Table 15, reveal some interesting information. Federal employment provided 55 per cent of all member jobs in 1934. By 1950 only 38 per cent worked for the Federal government. Schools also showed a decrease in percentage employed. State agencies maintained their relative position. Industrial and self-employment made large gains. By 1951 industry employment was estimated to be 4,400 men, thus exceeding any other group listed. Industries furnished 36 per cent of all forestry jobs.

TABLE 15. AGENCIES EMPLOYING FORESTERS IN THE UNITED STATES*
(Foresters Employed by Years)

	Number of Society of American Foresters members			Estimated number of all foresters	
Agency group	1934	1937	1950	1950	1951
Federal	1,147	2,516	2,420	3,500	3,550
Industry	220	368	1,560	3,100	4,400
State	333	486	1,080	2,200	2,500
School	246	307	550	600	600
Self-employed	.....	.....	350	400	450
Association	130	277	140	175	200
Local government	.....	.....	80	100	100
Unclassifiable	.....	.....	228	450	450
Totals	2,076	3,954	6,408	10,525	12,250

* Data furnished by the Society of American Foresters from a canvass of membership and projection to cover all foresters

The table reveals that employment by all agency groups is still increasing, with the possible exceptions of the Federal group and the association group.

The Federal decrease may have been due largely to the curtailment in employment that resulted from discontinuing the Civilian Conservation Corps and other emergency agencies that used many foresters. The most rapid increases are found in the industrial and self-employed groups.

Conditions of employment in the several agency groups will be discussed in the following sections.

Foresters in Federal Agencies

The Society of American Foresters' 1950 employment study referred to above also provided information on employment of society members by the individual Federal departments and bureaus. The departments with largest employment were: Agriculture, 84 per cent; Interior, 10 per cent; Executive Office of the President, 3 per cent; and military establishments, 2 per cent.

The bureaus employing substantial numbers of foresters were the Forest Service, Soil Conservation Service, Bureau of Entomology and Plant Quarantine, Bureau of Indian Affairs, National Park Service, Bureau of Land Management, Bureau of Plant Industry, Soils and Agricultural Engineering, and the Fish and Wildlife Service.

Federal Personnel Practices

Recruitment. Entrance into Federal agencies other than government corporations such as the Tennessee Valley Authority is through the civil service registers. Examinations are of two types, assembled and unassembled. The assembled examination usually requires an aptitude test and proof of professional training in the various fields of employment. Unassembled examinations are held for the higher positions in government service. These require the applicant to submit a complete history of his professional training, experience, publications, and other technical writings, and a record of past employment and salaries. Under regulations in effect in 1951, honorably discharged veterans had their examination grades increased by five points, and disabled veterans by ten points. With these additions they competed on a total-grade basis with nonveterans for civil service jobs paying in excess of $3,000 per year.

Probation Period and On-the-job Training. All Federal civil service appointments at the professional level are for a probationary period of one year. During this period the appointee's performance is carefully checked. At the end of the year he is recommended for permanent employment or for dismissal. Men employed for seasonal work are only employed as needed and discharged when the work terminates.

During the probationary year and following it, special on-the-job training may be gained. The purpose of early training is to acquaint the new employee with his superiors, his associates, the functions of the agency, the part he is to play in accomplishing these functions, and the work standards and methods used. Special schools for new workers may be held during working hours.

Training schools of short or long duration are also used to prepare men for advancement to new assignments and for undertaking any new and difficult job. The Soil Conservation Service, Forest Service, Fish and Wildlife, and other agencies have commendable on-the-job training programs.

Formal training schools may not be operated by all divisions of these agencies. Since training of new employees is still a matter in which the immediate employment officer has wide latitude, the training offered varies widely from one field office to another and from one Federal agency to another.

On-the-job training continues throughout a man's career. Transfers and progressively more difficult assignments are major methods used to train men for higher posts. Conferences and field inspection serve the same end.

Some agencies prepare tentative career plans in consultation with the employee. These are adjusted to each man's aptitude, interest, personality, and responsibilities. Transfers and promotions are then worked out with the career plan in mind. Career plans are not a means of regimenting men but, rather, of setting goals toward which they may strive with the helpful consideration of their official superiors. Career plans are usually reviewed annually in conjunction with performance ratings.

Performance Ratings. Civil service procedures formerly required that annual efficiency ratings be prepared for each employee. In 1950 the law was modified to allow considerable latitude among departments in establishing their own performance-rating plans, subject to approval by the Civil Service Commission.

The plan adopted by the U.S. Forest Service involves developing performance requirements and discussing them with employees, systematic appraisal of performance in relation to requirements, and improved supervision through better understanding of standards and objectives by employees. Three ratings are given—satisfactory, unsatisfactory, and outstanding. Ratings of unsatisfactory or outstanding must be supported by justification statements. Employees rated as outstanding are considered for promotion or merit awards. Those rated unsatisfactory may be given special training, reassigned, or, as a last resort, separated from the Service.

High level of performance by employees is recognized to be a mutual responsibility of the employer and employee. Performance ratings serve as an objective measure of success in fulfilling this responsibility. The rating may be discussed by the employee with his superior. Such discussions, which are informal and strictly confidential, have proved to be helpful in showing a man where he stands and what he may do to improve his opportunities for advancement. Similar performance ratings are used by other Federal agencies. Procedures for appeals of ratings are also set up.

Salaries and Promotions. Graduate foresters may accept subprofessional positions but usually enter government service in grade GS-5. Only men with considerable experience enter in higher grades. The salary ranges for each grade are shown in Chap. 20. Periodic increases up to the top of salary brackets are automatic for employees whose performance ratings are sufficiently high to permit advancement.

Sick and Annual Leave. Federal employees are entitled to 26 days' annual leave with full pay and to 15 days of sick leave with full pay. Sick leave in excess of 3 days must be certified by the attending physician. Sick leave is cumulative up to 90 days and annual leave up to 60 days.

Stability of Employment. Employees accepting temporary positions in government service have no stability of tenure. They are usually employed for a specific length of time and at the termination of the work are discharged. Employees in permanent jobs usually have a high degree of stability in their employment, though this also varies from agency to agency. Appropriations for specific lines of work may vary widely from one year to another. Work may be curtailed or even discontinued. When this happens, the employees are discharged if they cannot be transferred to some other agency or line of work. No guarantee exists that an employee will receive a transfer in his current grade and salary.

The key jobs in such Federal organizations as the U.S. Forest Service, particularly ranger and supervisory jobs, have a high degree of stability and are unlikely to be affected by ordinary cuts in appropriations. Research jobs, on the other hand, have varied widely. In these, relatively little security in tenure has existed except in the key positions of the more important agencies. When new jobs are undertaken, the staff is considerably expanded, and when the work is curtailed, employees are reassigned. In such fluctuations the less effective members of the organization are eliminated along with the newest recruits. Misfits theoretically are weeded out before the end of the probationary period. The organization is constantly obliged to train new men. It is difficult to build up a competent research organization, able to carry on the work with the greatest efficiency, with fluctuating budgets.

Opportunities for Self-expression and Self-development. Even in beginning jobs encouragement is given the individual to study his job critically and to make suggestions for improvements. Advanced jobs in almost all fields of Federal service offer foresters wide latitude for self-development and self-expression. Much responsibility and authority are delegated to the field men. They may have almost complete responsibility for technical standards, for programs of work, and for maintaining good relations with the public. They must, to be sure, conform to over-all laws and regulations governing the handling of Federal property and funds and the management of civil service personnel, but with these restrictions, wide latitude exists for developing new methods and new relations with the public they serve.

Retirement. Permanent Federal employees are obliged to deposit 6 per cent of their salaries in a retirement fund. Retirement is optional at ages 55 and 60 if the employee has 30 years of service and at age 62 with 15 years of service. Retirement is compulsory at age 70, provided the employee has 15 years of service. Annuities cannot exceed 80 per cent of an employee's highest 5-year average salary. The present retirement law provides survivor benefits for widows and children of employees who die in service. The law also provides disability retirement benefits for employees who have five or more years of service and are not able to perform the duties of their positions. Retirement rates for men with 35 years' service in 1950 provided for 52.5 per cent of average highest salary over 5 consecutive years.

Some Advantages and Shortcomings of Federal Service. Contrary to frequently voiced opinion, Federal employment in forestry and other fields is not ideal. The civil service organization is cumbersome in operation and poorly equipped to serve as an ideal recruiting agency. The more or less rigid classification system it imposes minimizes glaring inequities but by no means eliminates them. Most technical jobs have some unique requirements, and administrators have little discretionary power in adjusting salaries. Federal professional workers receive meager aid from unions in redressing grievances, compared with that available for private employees. Since the Federal government and many states forbid their workers to strike, public workers may be obliged to accept working conditions not tolerated by progressive private employers.[3, 5] During periods of inflation salaries of public employees traditionally lag behind those for private employment. Despite the efforts of civil service, political sponsorship still exists in some Federal jobs. Where widespread, it tends to impede efficiency, to frustrate conscientious workers, and to add unnecessary costs to public work.

Progress has been made slowly in improving public employment. Civil service is gradually being extended. Reforms advocated by the Hoover Committee on Reorganization of Federal Activities would greatly improve Federal employment and would give work agencies far greater authority in handling personnel.[1, 3] Patronage employment is gradually diminishing. Overtime work is no longer allowed as a day-to-day requirement without appropriate compensation. Organization of employees is permitted, even though activities of such organizations are circumscribed. Perhaps most satisfying of all to Federal employees is their realization that they are working for the common welfare of our country. This enables them to bear with some degree of equanimity the barbs of private workers, the castigation of officious Congressmen, and the slow advances in pay and expense money granted by economy-minded appropriation committees.

Prospects of Future Expansion. Expansion may be expected in Federal forestry work but will probably be less rapid than it has been in the past. Federal management of forest land seems unlikely to extend to substantially greater areas than it now covers. National forests, national parks, national monuments, and game refuges are rapidly approaching a mature stage. State management, on the other hand, has a long way yet to go and private management still farther. Expansion in Federal activity, therefore, is likely to occur primarily as cooperative work with states and with private owners.

The following description of work by agencies applies to conditions as they existed in 1951. Reorganization of Federal agencies has been more or less constantly under way since 1933. Greatest changes have occurred since 1949 when the Hoover Committee recommendations became available,[3] and further substantial changes were being considered in 1951.

The U.S. Forest Service

In 1951 the U.S. Forest Service employed a total of some 2,600 professionally trained men, about 2,400 of them foresters. This, the largest agency employing foresters in the United States, is a highly decentralized organization. Ten separate field regional offices are maintained for administration of Forest Service activities and 13 offices for forest research. The 10 regional forest offices have under their jurisdiction the national forests, each with its independent headquarters. They also supervise Federal grants-in-aid to states for fire control and cooperative forest management. The forest and range experiment stations have branch work centers and individual lines of work. Consequently, the organization, although large and far-flung, offers wide opportunities for display of individual talents and initiative.

Living and Working Conditions. On national forests and at research centers foresters usually live in small communities. Much of their work is in the field. Often it is strenuous and frequently it requires long hours. Overtime for technical workers is rarely paid. Throughout the fire season men are expected to remain on call 24 hours a day and seven days a week, but extended periods away from home are seldom required.

Men working in the forest and range experiment stations may have their headquarters in a university or other large community but carry on field work in remote stations while traveling extensively over a fairly large territory. Research officers are frequently obliged to be away from home for extended periods, varying from a few days up to periods of three to four months. If located on a university campus, they usually enjoy some or all academic privileges and the stimulating atmosphere a university community provides.

Men working in state and private forestry usually have their headquarters in regional offices, where they share working conditions with other regional-office field inspectors. The regional-office headquarters are located in cities. Travel away from headquarters may be of three or four weeks' duration. Men may be sent to Washington on details lasting for extended periods. Many spend as much as half or more of their time away from their permanent headquarters.

The Forest Service expects each employee to devote himself wholeheartedly to the assignments given. He is expected to place the good of the Forest Service above his personal convenience, even though this may entail some sacrifice. Often men have been obliged to move from one community to another where the only living quarters they could obtain were homes that had to be purchased at inflated prices. Major personal requirements are seldom overlooked. The Service tries to work out transfers in the interest of health or family needs and career development.

During the period 1900 to 1950 the Forest Service has generally been able to recruit from the best of forestry-school graduates. Competition within the Service has been keen and advancement has often been slow. Men with outstanding ability, however, still have a chance to reach relatively high positions at early ages. Forest experiment station directors have been appointed from men in their thirties and regional foresters from men in their mid-forties.

Many leaders in Federal, state, and private forestry have been selected from among men who spent their early years in the U.S. Forest Service. The organization's long tradition of devotion to the public good, inspiring leaders, and high standards of achievement have developed excellent *esprit de corps*. Friendliness is universal. Loyalty and devotion are expected; initiative, imagination, and skill in administration are needed. To the men who can unstintingly provide such qualities of service and leadership, the Forest Service offers high satisfaction and substantial rewards beyond those of salary alone. There is every reason to expect the U.S. Forest Service to remain a friendly, imaginative, devoted, and hard-working organization, in which young men can receive splendid training for future forestry careers.

Other Agencies in the U.S. Department of Agriculture

The Bureau of Entomology and Plant Quarantine. This bureau employs forest entomologists and forest pathologists for research and control activities. In 1947 Congress authorized the bureau to make surveys to determine incipient insect and disease outbreaks and to undertake control measures preceding specific appropriations. This authorization will lead to steadier employment in control and research work. Employment in many ways parallels that in the U.S. Forest Service. The bureau uses fewer foresters, however, and may have its top positions filled by men trained primarily in entomology or plant pathology. For men desiring to specialize in these fields, it offers excellent opportunities for satisfying service.

The Bureau of Plant Industry, Soils, and Agricultural Engineering. This bureau has a *Division of Forest Pathology* engaged in conducting research on fungi and other agents causing diseases of forest trees. It is somewhat smaller than the *Division of Forest Insect Investigations* of the Bureau of Entomology and Plant Quarantine. Its research opportunities parallel those in the forest experiment stations and Forest Products Laboratory, where many of its personnel are stationed. They observe the same working hours and have the same working conditions as the Forest Service men with whom they associate. The division has attracted able, devoted men. Since it is small, competition for advanced jobs is keen. Occasionally men have been transferred to high administrative posts in the U.S. Forest Service.

The Soil Conservation Service. Established in 1933, this Service has expanded rapidly. A large task still lies ahead before good soil practices are the rule on farms. Woody vegetation has an important function to perform in helping to stabilize farm lands and farm income. The Soil Conservation Service employs foresters both as technical experts to plan and to supervise farm forestry work and as over-all conservationists to help in farm planning and other conservation activities of this large and important organization. Several foresters have risen to important positions in the Soil Conservation Service. They have found in it conditions of employment more or less parallel with those to be found in other agencies of the Department of Agriculture. Reorganizations in 1951 have transferred some of the forestry activities of the Soil Conservation Service to the U.S. Forest Service. Future expansion of forestry in this agency may be limited.

Production and Marketing Administration. Foresters have been employed by the Production and Marketing Administration to supervise conservation programs of turpentine farmers. Others have been employed by the Farm Credit Administration in appraising farm woodlands as a basis for loans and to check on forest practices on farms having loans. A forestry program was a feature of the work of the discontinued Farm Security Administration in its efforts to aid farmers unable to obtain credit from regular sources.

Agriculture Extension Service. This Service employs one or more foresters in most states that have a substantial area of forest land or a shelter-belt program. These men are paid in part from Federal funds and in part from state funds. Their work will be discussed in Chap. 23.

Foresters have been employed from time to time in other agencies of the U.S. Department of Agriculture, including the Bureau of Agricultural Economics and the Secretary's office. Any large job requiring the service of foresters is usually referred to the Forest Service.

The U.S. Department of the Interior

The U.S. Department of the Interior employs foresters in a number of its bureaus for carrying on forestry, wildlife, grazing, and other land-management activities.

The National Park Service. Foresters may be employed as park rangers and may rise to be park superintendent and to higher positions. The park ranger is responsible for managing his district in much the same way that a national forest ranger is responsible for his ranger district.

A national park superintendent has duties paralleling those of the supervisor of the national forest. He must protect the public interest in concessions that are made to private companies operating transportation or hotels within the boundaries of the national parks. He is responsible for collecting fees, if those are required, and for making certain that no violations of park regulations occur. As national parks are Federal territory, any offense committed within a national park is a Federal offense, and, consequently, the park superintendent, park rangers, and other park officers are, in fact, law-enforcement officers within the boundaries of the national park.

The Planning and Construction Division of the Service supervises architectural and landscape design and construction. It is responsible for approving all major improvement on national park areas. The Recreational Planning Division supervises nationwide planning for recreation in cooperation with other Federal and state agencies. It investigates proposals for new developments. Through publications, conferences, and other devices it promotes recreational developments by others.

Men with training in landscape engineering are especially needed in the planning divisions. In fact, these offer the best outlets for men trained in landscape engineering and park management. Extended travel is often required of men in these divisions. They must have the capacity to appraise areas for historic and scenic merit and must understand how to develop such areas for public use without jeopardizing the major feature which gives them outstanding merit.

The Park Service also has a Forestry Division responsible for supervising forest-fire, insect, and disease control activities on national park lands and for other forestry activities. Men with training in general forestry may handle such work. General foresters are also useful in the

Lands Division, which supervises land acquisition, especially where forest lands are involved.

Conditions of employment in the National Park Service are similar to those in the U.S. Forest Service. They require considerable amounts of field work and a high degree of responsibility on the part of individual officers. The Service offers a fine field of employment for foresters who enjoy working with people and who are interested primarily in aesthetic rather than commercial values in timber.

The Fish and Wildlife Service. The Fish and Wildlife Service carries on the work formerly handled by the Bureau of Biological Survey in the Department of Agriculture and the Bureau of Fisheries in the Department of the Interior. The functions cover research, control, and management activities involving game birds, other birds, mammals valuable for game and fur or for other reasons, commercial and sport fishing, fur seals, whales, and other marine mammals, shrimp, lobster, shellfish, reptiles, and amphibians. The Service employs both foresters and forest biologists. Foresters are employed as refuge managers and for directing forestry operations on 800,000 acres of timberlands under intensive management for wildlife. Because of their special training in land-management techniques, foresters have made excellent refuge managers.

Through the Pittman-Robertson and Dingle-Johnson laws the Fish and Wildlife Service administers public funds collected from special excise taxes on sporting goods and ammunition amounting to several million dollars annually. This money is distributed to the states for cooperative projects that involve primarily research on game and fish management and the establishment and maintenance of conditions conducive to an abundant wildlife population on public and private lands and refuges. One feature of the program is the encouragement of farmers and others to plant trees and shrubs especially valuable for cover and food.

The conditions of employment in the Fish and Wildlife Service, particularly for men engaged in research, parallel those of men in the U.S. Forest Service. Extensive field work is often required. This may involve travel for considerable distances from headquarters and long periods of field observation uninterrupted by return to headquarters. This is especially necessary for men assigned to studying bird migrations, nesting habits, or conditions for winter survival. Field work is often very arduous and may involve moving camps progressively from day to day. For men who enjoy wildlife and outdoor work, who are skilled in observing and interpreting animal signs, and who have good fundamental knowledge of animal anatomy, physiology, ecology, and habits, the Service offers splendid opportunities (Fig. 97). Because of its wide contact with state organizations, the Fish and Wildlife Service can prepare men for high posts in state work.

The Bureau of Land Management. The Bureau of Land Management is responsible for managing the Federal range, the interior forests of Alaska, the Oregon and California Railroad and Coos Bay Wagon Road revested grants, and for miscellaneous other land-management functions not handled by other agencies of the Department of the Interior. Under the Taylor Grazing Act the Department of the Interior is responsible for managing 142 million acres of public lands to be administered for the use of range resources and the advancement of the livestock industry. An additional area of between 40 and 50 million acres is available for leasing and exchange purposes and for entry under the Homestead Act. This also is administered by the Bureau of Land Management. Within

FIG. 97. Decadent trees become hosts to insects and dens for birds and mammals. Holes in a 20-inch hard maple made by the pileated woodpecker in search of insect larvae. Such work is an indication of serious insect damage. This woodpecker does not damage sound timber.

the grazing lands are approximately 28 million acres of forests and woodlands. Timber on this land is subject to management by the Bureau of Land Management. The bureau employed 80 foresters, 30 range managers, and others with forestry training to carry out its functions in 1951. Primarily, it is interested in grazing on the sagebrush and other lands of the West. Foresters with a background of range experience or with special training in range management may find useful employment in this bureau. They should be especially interested in the problems of the livestock industry. They need to know the ecology of range plants and the amount of grazing these plants can undergo without depletion of their forage value or the exposing of the soil to serious erosion.

The interior forests of Alaska total some 144 million acres, of which approximately 112 million acres are in forest. These lands require protection from fire and ultimately should be developed for timber and other use. The bureau has limited funds for protecting and managing this valuable resource. The recent interest of our nation in establishing military bases in Alaska and in the general development of the territory justifies increasing attention to the valuable forest resource in the interior lands. The bureau will need men trained in forestry to plan the development and use of this vast resource.

The O & C lands are administered by the forestry division (see Chap. 18). Since these lands are interspersed with private holdings and with national forest holdings, management has been difficult, and only 25 per cent of the gross receipts is available to pay management costs. The bureau limits its management activities to those essential for proper protection and utilization. The entire O & C administration is in the hands of technically trained foresters, and the task they are performing involves extensive forest management on a sustained-yield basis. Employment and working conditions in the O & C administration are similar to those in other bureaus of the department. The administration pays primary attention to the timber resource.

The Bureau of Indian Affairs. Indian reservations in the United States have a total land area of some 56 million acres. A special division of forestry and grazing within the bureau is responsible for administering the grazing and forestry activities on some 46 million acres, of which 16 million acres contain forests and woodlands and about 6.5 million acres are classed as commercial forest land. This land has a timber stand of some 30 billion board feet. Some Indian forest lands have been under management since 1910. Some 16 billion board feet of sawtimber have been removed, and more than 68 million dollars have been received from sale of timber from Indian lands. Indian reservations cutting 1 million board feet or more have a total allowable cut of 424 million board feet.

Practice on the different reservations varies considerably. The Menomonie Indian Reservation at Neopit, Wis., has one of the largest sawmills operated by the Federal government. A sawmill on the Red Lake Indian Reservation in Minnesota manufactures about 5 million board feet of lumber annually. The Navajo tribal sawmill produces about 15 million feet annually. Foresters are responsible for sustained-yield forestry and timber operations and for range management on reservations. It is estimated that the Indian land will sustain an annual cut of more than 550 million board feet.

Problems of range management are acute in some of the western Indian reservations, particularly in the Southwest, where Indians own considerable numbers of horses and cattle. Since the Service makes it a practice to employ Indians whenever qualified men can be found, men of Indian parentage can look forward to filling many of the forestry jobs in the Indian Service. A total of 115 foresters and range conservationists were employed by this bureau in 1950.

A few forestry jobs exist in the territories and insular possessions of the United States. Most of these come under the jurisdiction of the Department of the Interior. In general, the jobs are somewhat limited in scope and offer relatively minor opportunities for advancement. The forestry activities in Puerto Rico are organized under the U.S. Forest Service.

Other Federal Agencies

The Tennessee Valley Authority. The Tennessee Valley Authority employs foresters and experts in forest products in its Division of Forestry Relations. These work with people and industries located in the valley in efforts to improve woodland management and to increase the markets for forest products. TVA foresters are responsible for the woodlands and forest plantations on Federal lands around reservoirs. The TVA channels its programs through local, state, and Federal agencies operating in the valley. These include the forestry departments, extension services, fish and game services of the seven valley states, the U.S. Forest Service, and the Fish and Wildlife Service. The Division of Forestry Relations has three major branches: (1) The Forest Development Branch is responsible for watershed protection, woodland management, fire control, reforestation, and timber utilization. (2) The Forest Investigation Branch makes resource and industry surveys in forest management, economics, influences, protection, and silviculture. It also studies forest utilization and may from time to time operate pilot plants in efforts to develop new uses for forest and wildlife resources. Much of its work with woodland owners and operators is educational. (3) The Fish and Game Branch strives largely through landowner education to build up and use wildlife resources. It is particularly concerned with effects of other TVA programs on wildlife, and its activity is closely integrated with woodland management.

The Tennessee Valley Authority has its own recruitment service, which screens prospective employees. It seeks men with general sympathy for the aims of its program. The number of jobs is limited, but opportunities for a high level of service are afforded.

The United States Treasury. The Bureau of Internal Revenue employs a few foresters to review income-tax returns of large timberland holders and forest industries. Foresters also have the responsibility of preparing policies to govern the taxing of incomes from forests. They must devise methods to cover depreciation of growing stock and other financial considerations involved in determining the actual revenue such properties produce. These jobs are few in number. They require men of mature ability, well versed in accounting and principles of economics.

The U.S. Department of Commerce. The U.S. Department of Commerce maintains certain statistics on forestry and forest-products industries. These services are performed by the Bureau of Foreign and Domestic Commerce and the Bureau of Census. The information gathered covers employment, wages paid, and production and value of products for the wood-processing industries. A limited number of foresters are employed in the department and other work is performed by the U.S. Forest Service. The Defense Production Administration in the Department of Commerce has a Forest Products Division, in which several foresters are employed as products specialists.

The Department of Labor. The Department of Labor has from time to time employed foresters to help in its study of labor conditions in the lumber, pulp, and other wood-processing industries. Only men of mature judgment are needed and relatively few jobs are available.

The National Security Resources Board. The National Security Board has a forest-products division to collect information on our nation's forest resources and the wood-using industries as a means of judging war potential and as a means of determining how this potential can best be realized in the event of war. It employed no foresters in 1951.

National Production Authority. The National Production Authority had in 1951 pulpwood, lumber, and other forest-products divisions in which foresters were employed. The Economic Stabilization Agency had similar divisions in which men with forestry and industrial background worked. These agencies are considered temporary. Some of the fact-gathering activities and other defense jobs in forestry and wood products have been assigned to the U.S. Forest Service.

Other Temporary Agencies. Various emergency agencies are set up in times of crisis to deal with a wide variety of public problems. Among those that have used foresters in recent years are the War Production

Board, the Office of Price Administration, the War Assets Administration, the Civilian Conservation Corps, the New England and Texas Timber Salvage Administrations, the Works Progress Administration, the Economic Cooperative Administration, and Military Government. All these are obliged to deal with forest properties or forest products. It appears likely that our country's need for foresters to help implement its foreign policy will require the services of a few men of mature ability and wide experience for an indefinite period.

State Employment of Foresters

State conservation departments use foresters in fire control, in management of state forests, in providing aid to private timberland owners, either directly through state programs or through the Federal cooperative forest-management program, and to provide control against tree-destroying insects and diseases. They also provide planting stock for farmers and other landowners. Some of them give limited or substantial services to the wood-using industries. The activities covered and the conditions of employment vary widely from state to state. Some states maintain a civil service system for all forestry employees with standards comparable to those of the U.S. Civil Service Commission and with retirement and other benefits in line with Federal standards. At the opposite extreme are states in which foresters must obtain political endorsement before appointment. Foresters in such states may be asked to participate either directly or indirectly through financial contributions in the support of the political party in control of state government. Fortunately, during recent years many states have removed their forestry departments from the patronage system.

Only a few states provide formal training courses for their employees, but a number of them do have one or more annual conferences. The forest personnel of the state is brought together for a three-day session or longer and given special instruction in specific programs of work.

The salaries and opportunities for promotion vary widely from state to state. In some the salaries, salary increments, sick and vacation leave, and retirement are comparable with those in the Federal service. Promotions to higher jobs are possible at fairly regular intervals. The individual state forester may be given wide latitude to plan and develop programs in the way he deems best. Tenure has been exceptionally stable in a number of the northeastern states. Service periods in excess of 30 years are not unusual. Such states provide some of the most attractive forestry positions to be found in the land.

The prospects of expansion of state forestry are also excellent. It seems likely that state forestry will grow faster in the immediate years ahead than Federal forestry if only for the reason that it has farther to go to reach its goal. The *esprit de corps* in state forestry organizations depends on freedom from political patronage and upon the decision, energy, and devotion of the administrative head. On the whole, men in state employ enjoy prestige equal to that of foresters in Federal employ. The opportunities for service are equally great and the satisfactions the service brings are just as rich.

The state park services employ foresters as do the Federal park services. They also employ men specially trained in the field of landscape design and recreational management. This work is almost certain to expand as public demands for parks increase.

The state highway departments sometimes use foresters to plant roadside trees and hedges for snowbreaks, to help maintain tree growth along parkways, to stabilize embankments and road cuts, and to beautify the highways. A few highway departments have developed picnic areas and other public recreation grounds at strategic locations along the highways. These are under the jurisdiction of the highway department and offer a field for special service by foresters skilled in landscape design.

The state fish and game departments may employ foresters to manage state lands acquired for game production and public shooting grounds. They may also employ forest biologists as wildlife managers. Frequently, foresters are used to handle the multitude of duties involved in protecting and furthering the abundance of wildlife. Employment in the state parks, state fish and game departments, and state highway departments is likely to be under conditions similar to those in the state forestry department.

State agricultural experiment stations, extension services, and colleges as employing agencies will be discussed in Chaps. 23 and 24.

Cities, Counties, and Communities

City foresters care for shade trees and parks. They handle tree planting, pruning, line clearance, cleanup after wind damage, spraying, and other activities. Municipalities having their own watershed frequently employ foresters to manage these. A few cities have park and other properties outside the city limits that require services a forester can perform. As noted in Chap. 18, counties and communities in the aggregate maintain control over considerable areas of forest land. These furnish opportunities for foresters either by direct employment or by providing service through state or other employment.

Strong city governments, such as those of New York City and Buffalo, have large park departments with employment under civil service providing benefits similar to those in state service.

Small cities and counties offer variable employment conditions. Political sponsorship is sometimes required. To avoid misunderstanding and disappointment, foresters should investigate such openings carefully before planning to accept such employment as a life career.

Private Industries

Private work in 1951 employed 4,400 foresters. Since the industries are far from saturated, additional opportunities may be expected to develop.

Recruitment methods for the various private industries differ widely. Some of them send personnel officers to colleges of forestry to interview promising seniors and to make definite offers of employment. Others have little understanding of their need for professional foresters and, consequently, employ them only when the professional man appears at their door and offers his services, usually in some minor capacity. Beginning jobs with private industry, therefore, are likely to be mainly labor jobs. Some may involve clerical work, such as timekeeping and scaling, simple field jobs, such as running compass lines and estimating timber, or simple surveying tasks. Few industries employ training officers. The majority expect the man to obtain his training through performing various menial jobs and through shifting from one job to another in the organization. The alert individual who sees in these jobs an opportunity not only to learn how the jobs are done but to inquire into new methods and new possibilities can be expected to rise in his organization.

The apprentice period for industrial foresters may be long or brief. As foresters rise to positions of executive leadership, they tend to shorten the apprentice period for others. One private company employed as many as 135 foresters in 1950. These handled jobs varying from that of compass man to that of wood-procurement officer for huge pulp mills. Many types of work may be performed by foresters: silvicultural work on company and other forests, including planting, thinning, and marking trees for harvest; protection against fire, insects, and diseases; timber harvesting, including delivery to company mills; complete land-management services; various jobs in mills; and selling forest products. Line operating jobs in logging, milling, remanufacture, and selling are most abundant and offer the best opportunities for advancement. The need for forestry information by industries adds to employment opportunities.

Beginning salaries are usually low, generally at the level of the unskilled or skilled laborer. Promotions are likely to be slow to begin with but rapid thereafter. Ultimate salaries are generally considerably above those in public employment.

Individual industries differ widely in their vacation and sick-leave policies. Some have none; others have well-planned programs.

The stability of employment likewise varies from industry to industry and from company to company. In general, both the lumber industry and the pulp and paper industry show wide fluctuation in production and employment. The lumber industry is particularly prone to fluctuations in output. Lumber production varies on a nationwide basis and also in separate regions and individual companies, depending on local business conditions and the availability of labor, standing timber, and capital. The forester contemplating employment would do well to investigate the resources back of a company and its record for stability in employment.

Being too hard to please in these matters, however, is likely to result in the forester's not obtaining private employment at all. The lumber industry during the past decade has shown considerable advance in its efforts to place operations on a sustained-yield basis. Consequently, conditions in the future may be expected to be more stable than in the past.

Opportunities for self-development and self-expression are very wide for foresters employed by industry. The very fact that employment of foresters has grown so rapidly in the past is indication that wide opportunity exists, rapid promotion is possible, and attainment of jobs of high importance may be anticipated.

A major difficulty to be encountered in small industries is nepotism. An industry owner's natural desire to give employment to members of his family may place an outsider at extreme disadvantage. The only cure may be to seek a new employer.

Private industries vary widely in their retirement programs. Some have generous ones; others have none other than those provided by public social security programs.

Jobs with Pulp Companies. Men with general forestry training are employed by pulp and paper companies to manage company forest lands, to run woods' operations, and to promote forestry among farmers and others. In the West large company holdings are usual. In the North and South, company holdings often are inadequate to supply the needs of the mill. Considerable pulpwood is bought. Foresters are often employed as pulpwood buyers, and in a number of mills foresters have complete responsibility for all pulpwood procurement. They contract with timber operators and purchase timber from farmers and others.

The forest is the source of raw material for the pulp mill. In the past, foresters have felt that many pulp company managers considered their timber holdings simply as a source of cheap raw materials. Little concern was felt for the condition of the forest land, the logging camps, or the timber workers. This attitude had changed decidedly for the better by mid-century. As a consequence pulp and paper companies are proving to be among the most stable private forest owners and, on the average, sponsor the best forestry programs.

The pulp and paper industry, in contrast with the lumber industry, is still expanding. Consequently its periods of depression generally are of short duration. Individual industries may be forced to close in periods of intense competition, but usually these are the ones with limited or expensive pulp supply, with obsolete equipment, or with inefficient management. Because the individual companies are relatively large, pulp and paper concerns tend to offer more stability than the lumber manufacturers.

The pulp and paper mills offer the best opportunities for employment of men trained in pulp and paper manufacture. Many have good personnel programs, plan for the proper training of the individual men, and have career ladders by which men may advance to high positions within the organization. Beginning jobs are generally in the laboratory. Intermediate jobs involve management of a section of the mill. Advanced jobs are those of mill superintendent, general manager, and even vice-president in charge of production or sales. Foresters must compete with chemical engineers for jobs in the paper industry. Special skill in combining engineering, industrial chemistry, and business management to make a profitable enterprise is essential in top jobs.

Lumber Companies. Lumber companies, on the whole, have lagged behind pulp companies in forest management. Many lumber companies are small, not too stable, and often inadequately financed. This is an indication of opportunities that exist for better integration and better practice throughout the lumber industry. Men of energy, business acumen, ambition, and willingness to take a few chances can find in the lumber industry excellent opportunity for exercising their skill. Beginning jobs are likely to be those such as scaling and cruising or various simple jobs on logging or milling operations. As men become skilled and show that they have an understanding of the output to be expected of loggers and millmen, they may hope to rise to positions of foremen and managers. The top jobs include woodland managers and woodland superintendents. Stability of employment depends in large degree upon the individual company and its policies with respect to a continuous timber supply.

Wood-processing Industries. Furniture factories, wood-turning mills, planing mills, sash and door mills, casket factories, and the many other plants that manufacture consumer goods from wood have been slow to employ men with technical training, particularly men with an understanding of the technical properties of wood. Some are becoming aware of the fact that wood technology plays an important role in their business. A few employ technicians, notably the larger furniture companies, plywood companies, piano manufacturers, and others who process a considerable amount of wood and who must employ good machining and gluing techniques to get satisfactory results. The opportunities in the wood-processing industry are wide, and these are expanding almost daily as research opens new fields of endeavor. Seasoning wood, machining

FIG. 98. An integrated sawmill and wood-turning establishment employing about 30 men, Cherry Valley, N.Y.

it to exacting tolerances, gluing it, and finishing it are all tasks in which a man well versed in wood technology can be of real help to an industry.

Foresters will obtain few jobs, however, by writing letters to such companies. Most wood-processing industries are small and have only a vague idea of what use could be made of a man trained in wood technology and forest utilization. They see their job as mainly one requiring engineering skill or the services of a chemist rather than those of a wood technologist. Wooden furniture that gives unsatisfactory service is evidence of lack of technical skill applied in the manufacturing processes.

Woodworking plants are generally small and highly competitive (Fig. 98). The forester who enters their employment must know how to fit into the organization as it exists and then how to prove his worth to the company by gradually suggesting technical advances that will pay off in the long run. This field, like that of logging, may be expected to absorb more and more foresters in the future as young men with imagination and energy work their way up toward top management. Skill and versatility in applying technical knowledge are required. A young man must be alert to keep abreast of new developments that occur in these fields and to see opportunities for their industrial application. Many advances in wood technology involve expensive techniques, which can be applied only if the added cost can be carried by the products. The forester who accepts a job in such a company should develop a good knowledge of costs and find out the extent to which the company's product can be improved and, through such improvement, gain wider acceptance by the public. To obtain higher posts, business and managerial skill are essential.

Suppliers to Wood-using Industries. Dry-kiln manufacturers, wood-machinery manufacturers, glue companies, and others that serve the wood-using industries frequently find men with forestry training useful in their research and sales organizations. To be successful with such a company, a man should understand wood and the problems of the wood-using industries. The individual jobs are somewhat circumscribed, but for men with an over-all grasp of industry needs the opportunities for high service are good. The competition foresters are likely to encounter will come from men who know machinery or glue rather than from men who know what machines or glues are expected to do for the wood-fabricating industry. For example, metal-to-wood bonding came into use only after a wood technologist was drawn into the field. It was he who knew what an adhesive must do to bond wood to metal. After he wrote the performance specifications for the adhesive, the chemist produced it.

Manufacturers of logging equipment and sawmill machinery likewise may employ foresters, as demonstrators, salesmen, and research men. Men who understand the forest and forest work are in a position to specify the type of equipment needed to do a good job in the woods. Forest workshops have developed such important items as the Osborne fire finder and the Byram haze cutter, both of which are used by fire lookouts. They also developed the first bulldozer. The idea was conceived by a forest officer who was having difficulty in building narrow trails in mountainous country (Fig. 99). The wide use of this earth mover today is testimony to the soundness of his original concepts. Another example is a logging tractor which combines the features of tractor and arch. This is called the tomcat logging tractor and was developed by foresters and loggers rather than by tractor manufacturers.

Private Estates. Relatively few large privately owned estates are at present managed under forestry plans with a technical forester in charge. The record of such estates in the past has not been very encouraging. The difficulty has been due in part to the lack of skill of the early foresters in making a financial success of forest-land management during a time when abundant timber resources were available throughout the land and were cheaply harvested by clear-cutting methods from other private forests. Two properties in the Adirondacks have provided more or less continuous employment for forest-land managers since 1898. Private forest estates in Maine have been under the continuous management of consulting foresters for long periods of time. Most forest estates, however, have not been intensively managed, and, consequently, the actual number of foresters engaged by them has been small.

The time for change may be near. The growing scarcity of timber and the increase in stumpage and lumber prices presages a growing market for timber products at fair prices. Since income from timber may be treated as a long-term capital gain for income-tax purposes, wealthy investors should be attracted to forest land. Private forest estates should, moreover, prove progressively more profitable in the years ahead.

The main weakness in the past has been a change in interest with change in ownership. This has occurred, in part, because many of these forested estates were acquired as recreational and hunting areas rather than as income properties. Some owners wished the forester to provide sufficient income to cover taxes and operating expenses and the forester has sometimes been able to meet them only by overcutting the timber stand. On the other hand, owners not interested in income have consid-

FIG. 99. Bulldozing a forest road through a western pine forest. Peshastin Lumber and Shingle Company. (*Caterpillar Tractor Company, and A.F.P.I. photo.*)

ered the forester a luxury. When ownership changed, the forester might either be required to overcut the property or be dismissed. Inheritance taxes also handicap transfer of a private estate to heirs.

Alert foresters willing to take a chance and bold enough to present the situation in a factual manner to an estate owner, promising neither too much nor too little, can expect to find desirable employment in such jobs. A few foresters have held such positions throughout most of their active career. Their jobs have been among the most satisfying for foresters.

Cooperatives. The small woodland holdings of the eastern states comprise a very large acreage where profitable forestry could be practiced. The farmers and other small owners, however, are not foresters, and few are likely to make a success of their forestry problem if left to themselves. They are in need of help in marketing timber and in building up their forest property. One hope of improvement is cooperative timber production and marketing associations. Trained foresters, employed by the cooperatives, direct the operation of the members' woodland and market the timber produced. The success of the individual cooperative agency naturally depends on sound organization and good management.

A special study showed 15 cooperatives operating in the United States in 1943. Although they were organized for cooperative marketing, their existence is also due to the desire for good forest practice. Aid in logging has been a primary benefit.

The Otsego Forest Products Cooperative Association at Cooperstown, N.Y., is an example of a forest cooperative.[4] This association, which began in 1937, went through a great many difficulties during its early history. Many of these were due to the fact that the association attempted to acquire a modern processing plant before it had had sufficient time to determine how well the members would respond to the cooperative. It therefore borrowed large sums from the Farm Security Administration under conditions which made the efficient use of this money impossible. A heavy debt was incurred and a manufacturing plant erected that was considerably larger than the timber growth of member woodlands would support.

In 1946, as a result of a reappraisal of its operating possibilities, the cooperative worked out a settlement of its debt with the Federal government. This has permitted profitable operation. The association has provided its members with a good outlet for the products from their woodland, and it has established a good name for itself in the lumber trade for producing well-graded and well-manufactured lumber. Given good management in the years ahead, there is every reason to expect that it may continue to operate successfully for an indefinite period. As it proves itself to more and more timberland owners, membership should grow and operating profits should be maintained. These profits are not profits in the true sense of the word inasmuch as they are distributed to the member patrons in proportion to the value of the timber they have sold through the cooperative. The Otsego Forest Products Cooperative Association employed two foresters in 1951.

The Northeast Forestry Foundation of Boston offers forester-supervised timber management on a cost basis. As many as 10 to 12 foresters have been employed. Connwood, a cooperative timber harvesting and selling association, operates in Connecticut. Other cooperatives operate in New Hampshire, West Virginia, the Lake States, and elsewhere. There is a need for more such organizations. While cooperatives offer the advantage of pooling bargaining power, they are subject to certain weaknesses, however. Members may find it difficult to agree on policy and management. Managers may neglect member education, thereby permitting misunderstandings to arise. Men contemplating employment by cooperatives would do well to study the principles that have proved most successful in cooperative enterprise. They should also study carefully the specific cooperative with which they expect to work to be certain it has the physical and organizational means for success.

In Scandinavia and Finland, cooperatives have been outstandingly successful in bringing good forestry and good markets to owners of small woodland properties.

Other Private Employment. A broad and as yet largely undeveloped field for forest-land managers is with public service corporations. Extensive tracts are owned by railroads, water companies, mining companies, and industrial concerns in many fields. Most of these owners are likely eventually to turn to the opportunity of obtaining revenue from their forest lands through proper forestry measures and in so doing will provide additional positions for foresters. Also, private clubs or individuals who have acquired forest property for hunting and other recreational purposes provide opportunities for developing forest businesses.

United Nations

Forestry in the United Nations is handled in a division of the Food and Agriculture Organization. Its work covers the broad fields of forest economics, technical forestry, forest products, and forest legislation. Technical missions are sent to various lands upon request. In 1950 FAO employed about 20 foresters full time and many others for temporary missions. Effort is made to spread employment among nationals of the several member countries. Headquarters are in Rome, Italy, with working groups maintained for Latin America at Rio de Janeiro and for the Far East at

Bangkok. The United Nations FAO offers interesting and highly respected employment opportunities for mature foresters.

Foreign Employment

Many importer and exporter companies in the United States employ foresters for service outside the United States. Rubber companies, tropical dealers, and importers of chicle, cinchona bark, quebracho, tropical fruits and nuts are all potential sources of employment of foresters. So also are the tropical-fruit companies, which frequently own considerable areas of land in tropical countries and require managers for the fruit, agriculture, and forest land which they own. Jobs with well-established companies usually provide good salaries and living conditions and reasonable opportunities for vacations in the United States.

Men who go into tropical work, however, should give it careful thought as they cannot expect to make a high degree of success in a short period of time. Tropical forestry is far more complicated than forestry in the United States. Little research and experience are available to guide a man. It is truly a pioneering field. Men who are willing to devote their life to tropical service can look forward to good opportunities for splendid careers.

Aside from tropical importing and exporting companies, the United States government frequently sends technical missions to a foreign country. These may be sent to obtain information for our own country or they may be sent to advise a foreign government on forest policy, forest education, forest research, logging, lumber manufacture, pulp and paper manufacture, and related fields. As the Point 4 programs grow, these jobs will increase.

Opportunities also exist to obtain employment directly in a foreign corporation. Young foresters should review the working conditions offered in such a corporation very carefully before accepting employment. Otherwise, they are likely to find that the job they are called upon to perform is considerably different from the one they imagined while considering employment with the company. A careful understanding will save regrets, especially financial, on both sides. In general, American foresters have been well treated by foreign countries and foreign companies that have employed them. They have been provided good living quarters and opportunities for travel to the United States to keep abreast of developments in our own country. The opportunities are probably greater in South America than elsewhere.

Men contemplating service in other countries should, first of all, learn the language of the country in which they are to work. It is equally important to learn the customs of the people, particularly the social niceties that the country expects of them. In many cases they will find living conditions in foreign lands much more favorable for them than they are for foresters in our own country. They may be able to afford household service and various personal services that the forester in our country must forgo. They are likely also to find themselves occupying positions of considerable importance to the nation and to the companies for which they work. Consequently, a high degree of integrity and fairness is demanded of them.

Foresters who accept employment in Canada may expect about the same type of treatment they would receive if they worked in our own country. A forester considering foreign employment should give special thought to the limitations and advantages such employment may place before his family.

Self-employment in Forestry

Many experienced foresters have found that self-employment offered the widest opportunities for self-expression, the highest personal income, and the most fruitful over-all field of service. The widest field of self-employment is in consulting work.

Consulting Forestry. Consulting forestry was slow to develop until a number of foresters had had sufficient experience with industry to make them valuable timber appraisers, logging engineers, and experts on evaluation of timber-operating chances. That is, the consultant was hired when he was in a position to save money for his employer on a pending transaction or operation. Even as late as 1940 there were still relatively few firms of consulting foresters and the men who branched out into this field between 1930 and 1940 were considered daring individuals setting forth in a highly speculative field (Table 16).

TABLE 16. SOCIETY OF AMERICAN FORESTERS' LIST OF CONSULTING FORESTERS*

Date of publication	Number of independent firms listed	Estimated number of individuals employed
January, 1946	40	45
June, 1946	67	83
January, 1947	89	113
January, 1948	103	130
March, 1949	130	164
January, 1950	142	179
January, 1951	146	189

* These lists are published in the *Journal of Forestry* for the dates shown.

The Society of American Foresters lists consultants in 21 separate fields of specialization. These are:

1. Arboriculture and tree preservation
2. Cost and economic studies of forest operations
3. Forest and wood utilization
4. Forest management and silvicultural practice
5. Forest protection from disesae
6. Forest protection from fire
7. Forest protection from insects
8. Forest taxation
9. Game and wildlife management
10. Logging engineering
11. Market studies and trade promotion (forest products)
12. Naval-stores operations
13. Pulpwood operations
14. Range management
15. Surveying and mapping
16. Timber valuation and appraisal
17. Timber volume and quality estimates
18. Tree planting and reforestation; reproduction studies
19. Wood preservation
20. Wood seasoning
21. Wood technology

The individual consultants list these additional fields of specialization:

22. Imported and tropical woods and precious woods
23. Purchase and sale of forest properties
24. Management of municipal watersheds
25. Development of wildlife resource
26. Agent for absentee timberland owners
27. Park and recreation forest development
28. Farm forestry
29. Tropical forestry
30. Public-relations forestry
31. Forest litigation
32. Millsite locations
33. Negotiation of hunting, fishing, trapping, and water-storage rights
34. Technical editing
35. Highway location
36. Community forests and conservation education
37. Labor relations
38. Estate problems
39. Erosion and flood control
40. Seed collection
41. Timber-dispute arbitration
42. Preparing visual education material in forestry

Consultants specialize also by regions and territories as well as by activities.

It is expected that both the number of consulting foresters and the breadth of their activities will continue to increase in the years ahead. It is known that many of the consultants employ one or more assistants on a temporary basis. Hence, it is believed that the actual number of men engaged in consulting forestry for at least a part of the year probably exceeds 200. The Society of American Foresters' list does not include consultants in wood processing, such as experts on veneer, plywood, gluing, factory layout, design of wooden structures, and design of furniture and related activities. It is known that consultants also engage in these fields, though no published list is available at present of the number so employed.

CONSULTANT-CLIENT RELATIONSHIP. The relationship of a consultant to his client is highly confidential and must be conducted according to the highest ethical standards. The code of ethics of the Society of American Foresters includes several statements referring to this relationship. When an investor in forest industries or properties employs a consultant, he may of necessity divulge to this consultant a considerable amount of information about his personal affairs or the intimate affairs of the company he represents. The consultant must by all means keep this information confidential. He must also use his expert knowledge to gain for his client the full financial and technical advantage to which he is entitled by the employment of the consultant. Obviously, the consultant will from time to time learn from one client information that would be of utmost value to another. If he divulges this against the will of the first client, he violates a confidence and dishonors himself. It is for this reason that a consultant should always reveal to his prospective employer any other retainership he may hold that might handicap him in serving his new client effectively.

Another activity for which consultants are likely to be used extensively in the future is to advise an owner on the possibilities of augmenting income through multiple-use management. Public forests are managed for multiple use, but most private forest landowners obtain revenue from only one product, timber. The possibility of leasing hunting, fishing, trapping, summer home, and water rights are seldom explored. Wood-processing companies may have a considerable amount of waste either in the forest or from the mill that might be effectively processed through a subsidiary operation. Here is a special opportunity for the consultant to aid his client.

TRAINING AND EXPERIENCE NEEDED. A consultant should be one of the most highly trained foresters of all, for upon his advice others will locate plants, buy equipment, begin operations, and make other heavy financial commitments. Unless he can give highly valuable service, he will lose money for his client and diminish public confidence in foresters. The ideal training for a consultant would include a thorough knowledge of forestry with specialization in two or three fields, particularly in marketing, logging, and timber appraising. In addition, he should have an understanding of business law, trade practices, and markets. He should know about the reliability of timber operators and processors. He should be versatile and willing to undertake a wide variety of tasks, yet have sufficient strength of character to turn down a lucrative opportunity in a field in which he is unqualified. He must have the highest personal integrity. He should have experience working with other consultants or with government agencies or private corporations that will build up his knowledge of conventional practices in the timber-trade industry. He must be able to appraise logging chances and timber operations and have general knowledge of land and timber values.

Ordinarily a forester would not be prepared to take consulting work on an independent basis until after he had had a minimum of 10 years of practical experience. The actual amount will vary with the field of specialization.

COMPENSATION OF CONSULTING FORESTERS. The consulting forester must be prepared to build up his business over a period of years. Consequently, he can scarcely expect to earn more than $2,000 to $3,000 during his first year of operation. Thereafter he may expect his income to increase proportionately to the value of his services and his success in attracting clients with relatively large-sized jobs to be done. Many consultants make in excess of $10,000 a year. Relatively little capital is needed by the consultant. He requires a few simple instruments such as a staff compass, tape, an increment borer, diameter tape, tally boards, cruising ax, paint gun and similar equipment. His office equipment may include only a typewriter, slide rule, and essential books. The total expenditure for this purpose need not exceed $1,000. He may need some capital, however, to tide him over during the first two or three years while he is getting his business established.

FUTURE OUTLOOK FOR CONSULTING FORESTRY. Timber in the South grows at about the rate of 10 per cent a year, in the North at a rate of about 4 to 5 per cent, and in the West at intermediate rates. Any resource that can be counted on to appreciate in volume at the rate of 4 to 10 per cent compounded annually is certainly worth investigating from the financial standpoint. The cost of maintaining and servicing this property must be within reason. Value increments on forest properties can usually be made greater than volume increments through carefully planned forest operations. It is here that the consultant's true value comes in.

A great deal of education will be needed by the public and the consultants themselves to convince landowners that they can afford to employ a consultant before buying or selling timberland or selling stumpage from their land. The faster the professional foresters launch out on their own as consultants and convince clients of the financial advantage that will accrue to them through the use of consultants, the faster we can expect forestry to grow throughout the land.

Integrated Forest Operations. Pulp and paper companies operating in Maine are concerned primarily with processing spruce and balsam fir. The hardwoods from their lands must be sold to other operators or left standing until they rot. Operations on the West Coast may leave as much as 20,000 board feet per acre of sound material. These are two examples of waste through single-product use. Rare indeed is the timber property that boasts of a market for small-sized material, such as fuel wood, posts, poles, and pulpwood, and of saw logs for lumber and veneer, and at the same time realizes an income from hunting, fishing, trapping, and recreational resources that the property may provide. These potentialities of building up multiple-use forestry on private lands where only single-use forestry now exists have been explored only to a very limited extent. It seems highly probable that alert foresters can perform a valuable service to landowners, to wood-using industries, and to the general public through uncovering opportunities for multiple-use forestry.

Integrated utilization can be brought about by an individual company through expansion of its activities into new lines. It can also be brought about by organizing subsidiary companies or even by contracting for subsidiary operations. Many of the large business industries in the United States, such as automotive manufacture, shipbuilding, and a host of others, are primarily assembling operations. The manufacturer purchases spark plugs, tires, lights, frames, and other equipment and assembles them into the final product. This type of subcontracting has been only weakly developed in the forest industries. A survey of wood-processing plants revealed that in many cases a nearby plant could use the waste from another, if arrangements were made to do so. It is such opportunities that should be sought out and developed by men skilled in integrated utilization.

LANDOWNER AND TIMBER OPERATOR. Perhaps the highest ambition of many foresters would be to own, manage, and operate their own forest property. Considerable capital is required to build up a large property. Foresters have purchased small tracts of land, done their own logging, plowed back earnings, and gradually built up a property. A few now own several thousand acres.

An example is the Luther property near Saratoga Springs, N.Y. This was acquired some 40 years ago by the present owner's father, a hotel manager and sawmill operator. Much of the land was in poor-quality farms at the time of acquisition. These old farms were planted to trees and some timberland purchased for logging was held after cutting. The property is now in excess of 6,000 acres. It has returned a fair income on the investment since 1929, when the plantations first attained sufficient size for harvest of merchantable material. Most of the trees are still small and can be used only for fuel wood, pulpwood, or cooperage. The owner carries on his own timber-cutting operations. The plantations are thinned and only thinnings sold. A few plantations have returned to the owner all costs of establishment and a substantial profit above this, yet still are just entering their most rapidly growing period. Others have been far less remunerative, but the property as a whole returned a net interest to the owner in excess of 1.5 per cent during the 1930's, when business activities generally were at a low ebb. The owner feels that he has a good investment which will increase progressively in value during the years ahead.

A forester is likely to be more successful in building up and managing such a property if he does his own logging. For this reason it would be desirable to have logging experience before embarking on the venture. He should also make a thorough survey of markets and be prepared to engage in subsidiary activities to carry him through periods of slack income. He would be wise to plan on multiple use of his forest, deriving an income from as many uses as possible. Such a property started in a good location and added to progressively should yield a good livelihood during its acquisition and provide security for the future once it is well built up. During the early period the forester should plan on doing much of the physical work himself, particularly supervision of cruising and timber marking. Later on he may hire others to do the heavy work, concentrating most of his attention on marketing and planning the over-all operation.

To build up a good timber property is not an easy task. Well-stocked timberland sufficiently large in area may be difficult to find. Good business judgment must be used to become a successful private timberland owner. The capital required need not be excessive. Forest land throughout much of the United States could be acquired at prices of $15 an acre or less in 1950, and much of it could be acquired for the value of the stumpage alone. Judicious purchasing and certainly judicious operating once the land is acquired are the basis for success in private timberland management.

Other forms of private land management offer possibilities. The Christmas-tree industry has been estimated to amount to 20 million dollars wholesale or 50 million dollars retail each year.[6] When properly conducted, Christmas-tree growing involves a selective annual thinning of the stand with a continuous yearly crop as the objective. In poor years, when demand and prices are low, trees may be held on the land and later sold for pulpwood or for some other purpose.

TIMBER AND SAWMILL OPERATOR. A forester desiring to go into the timber-harvesting field can start as a logger and gradually build up his equipment for independent work. Contract felling and bucking with a chain saw may give him his start. Later a tractor for skidding, a truck for hauling, and a portable sawmill might be added to his equipment.

WOOD FABRICATION. The alert forester can oftentimes engage in his own wood-fabricating operations. Many markets exist for special products. Such prosaic things as bed slats, the scantlings used to hold up the spring of an ordinary bed, have been the basis for an independent enterprise. Others have engaged in manufacturing small turned products, novelties, small-dimension hardwoods, small-dimension softwoods, custom-built furniture, baskets, veneered products, and a host of other commodities. These industries remain typically small industries. Toy manufacture and many other relatively simple wood-processing operations can be done with a minimum of elaborate equipment. The forester needs to be handy with tools, have some skill in designing, and, in particular, know where he can find adequate markets, if he is to make a success as a wood processor and fabricator.

RESORT AND CAMP OPERATION. A few foresters operate children's camps and resort hotels for hunters, fishermen, and others. To succeed in this field, a forester must be enthusiastic, know how to provide reasonable comforts in outdoor surroundings, and, above all, have the capacity to make friends readily and to build up a permanent clientele.

A few camp operators with a deep feeling for the forest environment have been successful in building up splendid camps with a minimum of physical facilities and a minimum reliance on conventional camp programs. It is here especially that the man familiar with the forest and forest wildlife can expect to capitalize to good advantage on his knowledge. Swimming, boating, canoeing, hiking, mountain climbing, overnight camping, simple handicrafts, and woods lore can be drawn upon to popularize such a camp.

Invaluable experience for those desiring a camp-operating career can be obtained by serving as counselors in existing camps and by working with children in schools, Boy Scout and Girl Scout activities, and related programs.

LANDSCAPE AND NURSERY OPERATIONS. Many men with forestry training have become owners of small nurseries growing ornamental stock and have engaged in simple and even elaborate landscape and arboricultural activities. Competition in this field is well established. The business is stable and growing. Men wishing to embark in this field should be well trained in soils, silviculture, seeding, and planting and have some knowledge of landscape design as well as of business methods and operations.

A nursery requires a considerable investment in land, seed, equipment, and labor.

A forester may purchase stock at wholesale from nurseries and resell it to customers after making appropriate plans and designs and guaranteeing the success of his job. In this way a man can start in the business with little capital outlay other than that for his training and experience.

Other Fields for Employment

The foregoing discussions have dealt only with the major fields for forest employment, touching here and there on a few minor ones. Many other fields provide foresters employment. These include forest-land appraisers for banks; experts on analysis of forest accounts and records for income tax; organizers and guides of expeditions for special groups (the Trail Riders of the Wilderness organized by the American Forestry Association is an example); outfitters of expeditions for study and exploration in remote lands; photographers and lecturers on forestry, wildlife, and nature; specialists in operating winter-camping and winter-sports areas for adults; labor-relations foresters who work with corporations in attempting to improve working conditions and work output of labor, and those who work directly in the union attempting to improve the skill and productivity of labor and labor's economic position; specialists who work with societies interested in nature, bird protection, and protection of scenic and wilderness areas; and outdoor writers for sportsmen's and nature magazines, conservation departments, and periodicals that run feature stories on forests, wildlife, and remote lands. These offer potentialities for the imaginative forester.

LITERATURE CITED

1. Allen, Shirley W. 1947. "Handbook of Information on Entering Positions in Forestry," 2d ed., Society of American Foresters, Washington, D.C. 56 pp.
2. Anonymous. 1950. Employment Analysis Shows Industry Increase, *Journal of Forestry*, 48:444.
3. Commission on Organization of the Executive Branch of the Government. 1949. "Federal Personnel," Government Printing Office, Washington, D.C. 101 pp.
4. Rettie, James C., and Frank A. Ineson. 1950. "Otsego Forest Products Cooperative Association of Cooperstown, New York," U.S. Department of Agriculture Information Bulletin 17, U.S. Government Printing Office, Washington, D.C. 42 pp.
5. Spero, Sterling D. 1948. "Government as Employer," Remsen Press, New York. 497 pp.
6. Sowder, Arthur M. 1949. Christmas Trees—The Industry; The Farmer and Christmas Trees, "Trees, The Yearbook of Agriculture," pp. 248–254, U.S. Department of Agriculture, Washington, D.C.

CHAPTER 22. *Forest Employment by Regions and Fields*

EMPLOYMENT OPPORTUNITIES and working conditions for foresters are influenced by ownership of forest land, timber volume and growth rate, intensity of utilization, and general economic and employment conditions in the region. The North has intensive use of forests, the South rapidly growing forests, and the West the largest volume of standing timber. The West has unionized loggers and the highest wage. Workers here are also most productive and operations most highly mechanized. The East depends heavily upon portable milling operations.

Pulp and paper manufacture is restricted chiefly to the Northeast, Lake States, South, and Pacific Northwest. The shoe-last industry is centered in New York and Wisconsin, veneer and plywood on the West Coast. Even agencies are localized to some extent. The Forest Service and Bureau of Land Management have the majority of their lands in the West.

Trends in Regional Development

Employment opportunities tend to be more abundant and promotions more rapid in regions that are undergoing rapid increase in population and industry than in industrially mature regions. Population changes from 1940 to 1950 bring out some striking regional differences. Slowest growth occurred in the west north central region covering the states of Minnesota, Iowa, Missouri, North Dakota, South Dakota, Nebraska, and Kansas. These states together had a population increase of only 4 per cent during the decade. The east south central region, Kentucky, Tennessee, Alabama, and Mississippi had a 6.5 per cent increase. New England, the Middle Atlantic states, and west south central varied from 9.5 to 11.3 per cent increase. The east north central and South Atlantic showed 14.2 to 18.8 per cent respectively. The most rapid growth occurred in the West. The mountain states increased 22.3 per cent and the Pacific states 48.8 per cent.

Statistics on changes in numbers of employed foresters by regions during the past decade are not available. The Society of American Foresters has, however, prepared a tabulation of its society members by states for the years 1940 to 1949, which may be a fair index of changes in total foresters employed by states. Decreases in membership for the decade occurred in the following states: Colorado, Connecticut, Kansas, Massachusetts, Nebraska, Ohio, Oklahoma, Rhode Island, South Dakota, Wyoming, and the District of Columbia. Discontinuance of the Civilian Conservation Corps, Prairie-Plains Shelter-belt Program, and other emergency activities may have accounted in part for these decreases. None of these states ranks high as a timber producer. More than 100 foresters were employed in 1940 in Colorado, Connecticut, Massachusetts, and Ohio. Increases of 0 to 50 per cent occurred in Arkansas, California, Delaware, Idaho, Iowa, Kentucky, Michigan, Minnesota, Missouri, Montana, Nevada, New Hampshire, New Jersey, New Mexico, New York, North Dakota, Pennsylvania, Tennessee, Utah, Vermont, Virginia, and Wisconsin. This group includes the three Lake States, four of the Middle Atlantic states, and five mountain states. Increases of 51 to 100 per cent occurred in Alabama, Arizona, Indiana, Maine, Maryland, North Carolina, and Texas. Increases of 101 to 177 per cent occurred in Florida, Georgia, Illinois, Louisiana, Mississippi, Oregon, South Carolina, Washington, and West Virginia. Aside from Illinois and West Virginia, which had 86 and 85 foresters respectively in 1949, this group is made up entirely of southern and Pacific Coast states. The regions with rapid increase in foresters are regions with excellent opportunities in forestry.

Employment in the North. The North is characterized primarily by small woodland holdings, small timber operations, and relatively small wood-processing plants. Only a small amount of public forest land is found in the North. This region has a large number of the country's forestry schools and has had a longer period of forestry practice than the other regions. Forestry began early and has proceeded at a moderate pace.

Since the North is primarily an industrial and agricultural region, forest operations are secondary in the economy. The forest industries are relatively well developed in the North, though few of them are large in size. The North leads in pulp and paper manufacture, in furniture manufacture, in the manufacture of a wide variety of commodities that require wood of high technical quality, and in the use of lumber. In terms of potential employment it is obvious that the region offers substantial opportunities.

Potential opportunities for foresters to be employed in forest-land management in the North exist in that only 32 per cent of private lands operated in 1945 were under intensive or extensive management. Corresponding figures for the South and West were respectively 19 and 38 per cent.[5] Realization of this potential is handicapped by small and diversified ownership, depleted lands, and relatively slow growth of timber.

The opportunities for industrial employment in the North are promising because of the large number of industries and the intensity of use. Many manufacturers have small plants. Few realize the need for men with technical training. Perhaps the North can best be characterized by saying that since forestry is not a new activity and forest land ownership patterns, forest operating methods, and wood-using industry practices have been developed over a long period of time, timberland owners and forest industrialists are inclined to be conservative and to present a considerable amount of inertia to change. In the years ahead, it is almost certain that both forest management and industry practice will change a great deal and that this potentiality for change represents the real opportunity for the employment of foresters in both land management and forest utilization. Active interest exists throughout the North in improved methods of timber harvesting and utilization, and a great deal of interest has been taken in improved practices on the land. The handicap of small, scattered ownership will continue to plague forest practice in the North for many years to come.

Employment in the South. The South is likewise characterized by agricultural development alongside of forestry. Ownership of land is mostly in small holdings. Consequently, many of the lumber companies and wood-using industries are small. The pulp and paper industry of the South, on the other hand, is a big industry and is dependent upon the rapidly growing pine of the coastal-plain region. In the South also there is relatively little public land. Forests are mostly privately owned. The fact that timber grows rapidly in the South has encouraged many industries to engage in forestry on lands they own and to promote forestry throughout the territory from which they obtain timber. It is safe to say that nowhere else in the United States does the future for private forestry appear brighter than in the South.

Growing stock is inadequate over much of the South and timber yields correspondingly low. Wide opportunity exists for intensifying forest-land management and particularly for increasing the intensity of industrial use of southern woods. Although the South has long been underdeveloped industrially compared with the North, industrial development is proceeding rapidly and, consequently, opportunities are good. The forestry schools in the South, less numerous than in the North, do not turn out as many men as the region can absorb. The South probably offers the brightest opportunities for employment of foresters of any of the three major regions. A great deal of imaginative leadership is shown in the pulp and paper industry, the large lumber companies, and the furniture industries of the South. This interest is almost certain to be self-catalytic. Industrialization in the South will increase the need for forest products and particularly the need and opportunities for wood-using industries.

All phases of forest-land management and forest utilization offer opportunities for increased employment and intensification. Protection of the forests against fire is not yet universal. The educational program for owners still needs to be intensified. The development of cooperative management and trade associations lies in the future. Intensification of technology in wood utilization is badly needed. Timber is still being harvested faster than it grows in the South, but the potentialities for increasing growth rate have been barely touched. Forest land has been increasing rapidly in value and sustained-yield forestry has grown at a very rapid rate.[3]

Working conditions in the South Atlantic states are generally good and the companies tend to be progressive. Timber and the forest-products industries represent an important phase in the economy, which seems destined to increase as a moderate decline in the demand for agricultural products takes place. Though much of the land is in small holdings, large lumber and pulp companies are gradually accumulating lands for sustained-yield operation. This alone offers a good opportunity for the consulting forester and other men well versed in land values.

The climate in the Southeast is warm and consequently unattractive to many individuals accustomed to the North. The small size of many of the southern communities and the cultural lag in the South that followed the Civil War, and is just now being overcome, have caused many northern peoples to overlook the splendid developmental opportunities in the South. In the last two decades the southern people have taken vast strides to overcome local handicaps. Diversified farming has been introduced, and educational programs have been sponsored by local people in an effort to improve racial relations. Renewed efforts are being made to open up opportunities to develop the region industrially and to improve its use of resources. All these activities make the South a land of opportunity for the forester (Fig. 100). Private ownership characterizes the forest lands of this region. The Tennessee Valley Authority and the U.S. Forest Service are both trying to stimulate private forestry, and the forest industries themselves have taken outstanding leadership.

It is probable that in both the North and the South our forests have reached a low point in their condition and are destined to increase in area, in volume, and in importance with the coming years. The forest industries in these two regions, therefore, taken as a group, represent a growing rather than a declining economy. Individual industries, on the other hand, are still encountering difficulties, and local adjustments are to be expected.

FIG. 100. The open southern forests can be operated by trucks and wheel-mounted power saws. (*Florida Forest Service, and A.F.P.I. photo.*)

Employment in the West. The West differs radically from both the North and the South. Vast quantities of virgin timber still exist. The timber is huge in size, and special logging equipment and highly trained woods operators are needed to exploit it. As long as considerable accessible volumes in virgin timber exist, the market for second-growth will be depressed. Public ownership of forest is a major factor in the economy and is destined to become even more important in the years ahead as the volume of timber standing on private holdings is further depleted. The West has learned to cope with its rugged topography and the vast bulk of the trees harvested. This has required imagination, skill, and daring. The large companies are attempting to hold their forest lands and to expand the areas under management so that they may have a permanent timber supply for the future.

The Pacific Northwest remains the major lumber-producing area of the country and consequently offers good opportunities for employment. Large companies in this region employed over 500 foresters in 1950. The forest-practice laws in these states and in California tend to encourage the employment of foresters. The large size of operations, the large size of timber, and the high degree of organization and mechanization of the timber industries of the Pacific Northwest mean opportunities for trained loggers and men trained in wood utilization. The veneer mills, pulp and paper mills, furniture factories, and other wood processors of this region employ men trained in either forestry schools or other colleges where technical knowledge is gained of wood fabrication and use.

Industrial forestry has grown rapidly in the Pacific Northwest during the last two decades. Since the second-growth Douglas-fir forests have been found to grow very rapidly, timberland owners are encouraged to hold the land. The industries of the Pacific Northwest are large in size, well organized, and, on the whole, well financed. They appreciate the need for technical knowledge and, consequently, are financing forest-products research. This, in turn, will open up new avenues for use of wood and new opportunities for trained technologists in wood utilization and related fields. A number of new products financed by industry research are already being placed on the market. Most of the major companies are building up timberland holdings with sustained yield in mind, and at least two already have 100-year contracts with the Federal government to sustain the timber industry.

Integrated utilization is rapidly being developed in many communities. Examples are found in Everett, Bellingham, Tacoma, Longview, Shelton-McCleary, and Aberdeen-Hoquiam in Washington and in the Eugene-Springfield section and Coos Bay in Oregon. Effort is made to use all the products the forest provides. The very rapid development in the Pacific Northwest offers, for the present, excellent opportunities for foresters in the employ of private industry. The Forest Service, likewise, is engaging in research and in management of the national forests. Industrialization of the region has proceeded at a more rapid pace than in the country as a whole. Both industrial leaders and the people realize that timber is a major product and are determined to safeguard future supplies and future markets for forest products.

Although employment and living conditions are not always ideal, they are generally good. Much of the timber is still remote and many of the logging communities are small in size. The companies on the whole have adopted a progressive attitude toward their employees. They seek to erect good communities and good camps and to maintain good personnel practice. A forester who accepts a job with a Pacific Northwest company runs little risk of encountering undesirable living and working conditions.[2]

Dependence is still mainly on old-growth timber. The establishment of a forest industry based on second growth is largely for the future. Until this occurs, adjustments which may be critical for individual communities, companies, and workers will be in order.

Opportunities in Various Fields of Forestry

The changes in employment by fields of forestry have exceeded by far those occurring by regions. Unfortunately, there exists no nationwide breakdown of the employment by fields. Attempts to appraise this by membership in the several subject-matter divisions of the Society of American Foresters would certainly be misleading. The interests and even duties of many foresters embrace more than one field of work. Even less useful would be an appraisal based on articles published in the *Journal of Forestry*. Most writing is done by teachers and research workers, though they represent together less than 15 per cent of total membership.

Intensification of work has occurred in all fields. State foresters who formerly were concerned chiefly with protection of forests from fire are devoting increasing time to stimulating private forest management. Industrial foresters are becoming responsible for logging in addition to silviculture. Rangers on national forests may be responsible for forest protection and for managing the timber, range, water, recreation, and wildlife resources. Perhaps most spectacular of all has been the rapid development in use of technicians by wood-using industries. This is exemplified by the rapid growth of the Forest Products Research Society and the high enthusiasm that marks their annual meetings.

Forest Protection. Protecting forests against fire, insects, bacteria, fungi, animals, and other destructive agents is primarily the task of the forest-land manager. In this he may be assisted by specialists in the behavior of these forest enemies. The specialists find employment largely in Federal and state forestry organizations.

Control of fire is a major task of state forestry departments, requiring the greatest allocation of funds and much attention to organization and equipment. This work will continue to provide employment for foresters and forest guards for the indefinite future.

Entomologists and pathologists in the Bureau of Entomology and Plant Quarantine have the important task of keeping on the alert for incipient outbreaks of insects and disease-causing organisms and of controlling these before they reach large proportions. A limited number of opportunities exist for men of outstanding ability in these fields.

To bring protection against fire up to good standards would require intensive work on 121 million acres as of 1951. If completed in five years, this work might offer 16 new state fire-control jobs per year. Protection against insects and diseases at the level of highest years in the past would as of 1950 require 20 additional men per year for five or more years.

Forest-land Management. The broadest activities for foresters exist in some phase of land management. The forest-land manager is usually responsible for all resources and all activities that occur on the land under his jurisdiction.

The varied activities of the forest-land manager offer a wide range of job opportunities. The total number of openings is likely to be less, however, than in the broad field of forest utilization. It is largely through sale of forest products that money is obtained to finance manufacture of these products and growing and harvesting of the raw material. Forest-land management is largely an extensive operation. Because foresters have been attracted to it from the beginning, the field is more nearly saturated than that of forest utilization.

The annual replacement of foresters in public employment, most of whom are in land-management activities, is estimated at approximately 240. Replacements in industrial forestry will probably take at least 70 men and in consulting 10. This would make a total of some 320 jobs annually in forest-land management and related work.

Good reason exists for expecting both public and private forestry to expand in the years ahead. New legislation providing greater assistance to the private owners was enacted in 1951. Additional purchases of land for public ownership and increased areas brought under protection from fire through public cooperation are examples of the kinds of increased activity that will result in additional positions.

The Forest Service estimated the number of public foresters needed to render aid to the 4.2 million owners of forest properties below 5,000 acres in size. They assumed that only 131 million acres of the 261 million acres owned would be involved. The total number of foresters needed was 2,446. One hundred and seventy-three were employed in 1949. If this program expanded uniformly over the next 20 years, 114 men per year would be needed.

Private companies ultimately may need one forester for each 20,000 acres of intensively managed land. Assuming three-fourths or 62,000,000 acres of large private holdings suitable for such management, 3,100 foresters would be required. Industry in 1949 employed about 2,000 foresters in land management, leaving 1,100 for future employment. Assuming a 20-year period in which to employ these, there would be 55 new jobs a year. The total would then be 169 jobs per year in forest-land management. Adding the estimate of 320 replacements annually makes a total of 489 potential jobs.

Actually, during the period 1935 to 1950 new jobs in forestry increased at the rate of 300 annually, but many of these were provided by emergency work-relief programs. General forestry, therefore, can scarcely be expected to provide employment for as many as half the prospective graduates during the 1950's.

A realistic forecast of the number of new land-management positions cannot be made. A number of jobs could be developed if all the forestry organizations and forest landowners made a determined and concerted effort to increase the intensity of forest-management practices on their lands. The rapidity with which they do this will depend on economic outlook, intensity of demand for forest products, and the stimulus provided by public agencies. Public action likewise cannot be predicted, for it fluctuates widely with citizens' interest and in response to national and international affairs. The forestry student who prepares to meet keen competition will be acting realistically.

Managing Forests to Conserve Soil and Water. In the years ahead employment of protection foresters seems definitely destined to increase in the United States. Water for municipal and industrial use and for irrigation is becoming progressively more valuable year by year. It is also becoming scarcer. Floods tend to increase in frequency and in severity with the increase in use of our land. They also increase with the growth of our major cities, most of which are located in river valleys where important sections are subject to inundation. Our country is just beginning to recognize the truth of Raphael Zon's statement, "Floods begin on the land, and it is on the land where the major efforts toward their control should be directed."

It also seems inevitable that our people will continue to occupy and cultivate land subject to wind erosion in the prairie-plains region. A growing need for intensification of shelter-belt programs in this area will continue for many years ahead.

The U.S. Forest Service employs between 80 and 100 people on flood-control surveys and work programs. Others are employed on forest-influences research and other activities involving the protective value of forests. The Soil Conservation Service probably employs an equal number, though these men oftentimes have other than their forest duties to perform. A few jobs exist in other Federal agencies. Managers of municipal watersheds and others engaged in protective forestry might bring the total to some 250. This would indicate an annual replacement of 5 to 10 men. All these activities are gradually expanding so that employment opportunities probably exist for upwards of 15 foresters annually in the various phases of protection forestry. Since this is largely a specialist job, the men who enter the field should have training beyond that given in the ordinary forestry curriculum. In 1951 only one college of forestry had a professor of forest influences. Foresters will usually be wise to explore job opportunities in the field before investing too heavily in specialized training that would fit them primarily for work in protection forestry.

Wildlife Management. Wildlife management is still a young profession in the United States. It has already proved to be popular with young men. In 1949, 32 colleges and universities, mostly state institutions, were offering training leading to a degree in wildlife management. Thirty of these offered undergraduate training and two offered graduate training only. Of the 30 offering undergraduate training, 22 had men enrolled for the master's degree and 14 had men enrolled for the Ph.D. For the year ending in September, 1949, 279 undergraduate degrees were conferred, 71 master's and 19 doctorates.[1, 4] The total enrollment in the fall of 1949 for all schools was 1,463 undergraduates, 240 master's candidates, and 81 candidates for the Ph.D.

The conclusion of a special committee of the Society of Wildlife Management on employment outlook was that far more men were taking work in wildlife management than could be absorbed in the profession. They recommend two courses of action: that efforts be made to broaden employment opportunities, and that colleges restrict enrollment. They point to the possibilities of expanding the number of positions open to men trained in wildlife management by training these men as game protectors. They also point out, and rightly so, that it is the game protector or game warden who is familiar with conditions on the ground and that it is to him largely that the biologist must turn to get the results of his studies into practice on the ground.

In an earlier report a committee headed by Swanson pointed out that the United States would require the services of about 10,000 men and women trained in wildlife management to carry out all phases of the work. This committee definitely expected that college-trained men would become game protectors as well as biologists. Employment of such great numbers, however, is a long way off. Most of the students graduating during the 1950's will have to hunt hard to find good jobs. The annual replacement of men now in wildlife service is probably close to 110 per year, and new jobs may take up to 200 per year.[4]

During the first 10 years of operation of the Pittman-Robertson Act, employment in wildlife management grew rapidly. A total of 48 million dollars was collected from the excise tax on sporting arms and ammunition to support wildlife work. Other special taxes levied on hunting licenses have contributed to the funds of states for wildlife management. Federal and state expenditures under the Pittman-Robertson Act for the fiscal year 1950 amounted to 15.5 million dollars, enough to provide employment for over a thousand wildlife workers.

The Dingle-Johnson Act, a companion law to the Pittman-Robertson Act, is designed to provide funds for improving fish management. This should lead to wide expansion of work in stream improvement and other measures to increase the health and abundance of game fish. Employment prospects in this field are expected to be bright for a decade at least. The work will probably be administered along lines parallel to the Pittman-Robertson Act.

Foresters specializing in wildlife management find employment with the Fish and Wildlife Service, the land management agencies in the Departments of the Interior and Agriculture, with state conservation departments, schools, and a few private agencies.

Wildlife managers have some opportunity to go into business for themselves as trappers, fur dealers, or as leasers and managers of land for hunting, fishing, and wildlife production. The business at present is rather speculative but nevertheless supplies a comfortable income for a number of men who have gone into it. The potentialities are just being explored, especially on state and federally managed areas where production of wildlife is a primary or secondary aim. A good example is the production of muskrats from the migratory waterfowl refuge in Montezuma Marsh of central New York. Here the take of muskrats by trappers under permit has over an eight-year period totaled almost 35,000 animals and the pelts sold, about $103,000. The marsh is managed primarily for the protection and production of waterfowl, but muskrats play an important role in pond management. The two go hand in hand.

Range Management. The number of men estimated to be employed in range-management activities in 1951 was approximately 620. These carry out a variety of duties, including research, range administration, and range operation.

The Bureau of Land Management in the Department of the Interior and the U.S. Forest Service in the Department of Agriculture are the chief employers of range specialists. The Bureau of Land Management has about 140 million acres of land within the Taylor grazing districts and manages about 40 million acres outside of these districts. In 1949 the total number of districts was 58 and the average size 2.5 million acres.

The U.S. Forest Service administers about 140 million acres of land in the range region and makes extensive studies of range management in the regional forest and range experiment stations. The Bureau of Indian Affairs employs range managers to work on the Indian reservations.

The state agricultural experiment stations and agricultural extension services also employ specialists in range management. Their activities involve research in the nutritional needs of livestock, in livestock breeding, in livestock health, particularly the diseases to which range livestock are especially subject, in forage production, particularly the relative nutritional values of various types of forage, and in the over-all question of ranch management. This latter field needs stressing inasmuch as many of the ills of the western livestock industry at the present time are traceable primarily to uneconomic ranch units and ranch-management practices.

The state agricultural extension services perform for the rancher the types of service county agents perform for farmers. Men with training in range management are preferred for appointment as county agents where the range livestock business is the dominant agricultural activity.

A few large ranches in sections of the West are in a position to employ specialists in range management and do so. Others are in a position to employ consultants in range management, but this service is ordinarily provided free by the agricultural extension services and experiment stations.

A few men who have specialized in range management have acquired their own ranch outfit and become independent operators. Opportunities of this sort are many throughout the West. It is not always possible, however, for a man with limited income to get a start in the ranch business as considerable capital is required in animals, hay lands, and ranch equipment.

Job prospects in range management appear to be rather good for the long run. The stockmen of the West are insisting on range-improvement projects and other work that will build up the productivity of the range. This will require considerable numbers of men to carry out the technical phases of the work. The magnitude of the western livestock business is sufficient to justify use of many additional range-management specialists. The annual replacement is about 20 men and new jobs of about the same number can be anticipated.

Men who are to engage in range management need a fundamental understanding of range livestock and of ranch and livestock economics. They need to understand range ecology, how to detect signs of range deterioration through the invasion of weeds and other plants. They particularly need to understand the soils, their susceptibility to compaction and to erosion, and how to manage the soil so as to obtain a high yield of forage with a minimum of soil destruction.

Range-management men should possess good health and physical stamina as well as alert analytical minds. They should have a liking for the livestock business and a desire to meet people on friendly terms and to understand their problems. A courteous, sympathetic manner is an important quality for a man in this field.

Recreation and Park Management. Federal, state, county, and municipal parks together employ thousands of people. In 1949 state park personnel alone numbered 4,004 year-round employees and 6,245 seasonal. Of these, 344 were professional and technical men, 51 of them heads of agencies. The National Park Service in 1950 employed about 500 professional men of whom 91 were landscape architects.

The U.S. Forest Service employs about 21 landscape architects to assist in laying out administrative sites, picnic areas, and public campgrounds and in advising on the various other activities carried on by the U.S. Forest Service. Their advice is particularly sought on the location of roads and the treatment of road banks, scenic vistas, and special points of interest.

The United States government employed 16 landscape architects in 1949 in the Veterans Administration and some in other Federal agencies where the modification of the landscape for more effective human use and enjoyment is required.

County and municipal park authorities employ 150 professional men as well as many others and offer beginning opportunities for men trained in park planning and maintenance.

The American Institute of Park Executives, made up primarily of state, county, and municipal park men, now fully recognize the importance of properly trained men in their organization, and they have indicated a willingness to employ such men both for brief periods during summer seasons while they are still in school and upon graduation. Some municipalities employ city foresters to care for street and shade trees outside of their regular park organization. These men have duties which require a thorough understanding of vegetation and how it can be used to beautify cities.

State highway departments employ park designers to help with parkways and highway picnic and camp sites.

Various organizations employ men trained in recreation management to plan facilities for organized camps, such as those which the Y.M.C.A., Y.W.C.A., Boy and Girl Scouts, churches, service organizations, and other agencies operate to provide outdoor living experience for children.

Opportunities are available for private employment in resorts and on private estates. Large resort hotels that have various scenic and wildland attractions have particular need for the services of a landscape planner. The work would include the planning of the layout, buildings, and facilities, and golf courses and other recreational areas so as to make the best possible use of the natural attractions of the setting. Some resorts feature hunting, fishing, and horseback riding as well as golf and other outdoor activities. If the over-all planning and maintenance of trails and facilities is done with an eye toward preserving and enhancing natural beauty, the enjoyment of these activities is greatly enhanced.

Finally, men trained in the landscape field may find employment with nurserymen, who sell stock to cities for park purposes and for suburban developments, and even to individual home owners who wish to landscape their property. Many develop their own business as landscape specialists, and some operate tree-expert companies that service shade and ornamental trees for the public and for home owners.

Of the 190 graduates of the State University of New York, College of Forestry, specializing in Landscape Engineering, 60 to 70 per cent were engaged, as of 1950, in the work for which they were trained. For the long run it can be anticipated that approximately 85 jobs a year will be open to replace men now in park work, and up to some 700 jobs a year in other phases of park work, especially maintenance. Many of these can provide a satisfactory beginning in the park-management field.

Timber Operating. However tight employment opportunities for foresters in land management may be, plenty of openings in logging and sawmill operation are likely to exist for decades to come. Such employment is remunerative and offers excellent training and opportunities for advancement. It is the foresters who understand how to harvest and process timber efficiently who will occupy the key jobs in the profession in the future, just as such men are the key industrial foresters of today. Logging and milling are the two basic forest industries without which others cannot operate, and they offer experience the apprentice can ignore only at jeopardy to his professional future. It is hard, hazardous work, but for these very reasons needs the freshness of viewpoint, technical training, energy, and agility that forestry graduates can bring it.

Logging is the first major operation in manufacturing useful commodities from timber. It is common practice among large timber-holding and operating companies to have all woods activities in charge of a woodlands manager. The woodlands manager is responsible for purchasing wood from outside owners, for planning and carrying out all operations on company lands, and for supervising all activities of the forestry division. He may also supervise an accounting division, a supply division, an engineering division, and the general staff offices. A woods manager in a large company will have district managers responsible for logging in the various districts, who may also be responsible for the forestry activities.

In Maine and to a limited extent in the Pacific Northwest, a considerable volume of pulpwood is transported by river driving or floating. Timber floating requires special skills. Stream improvements and various engineering works, such as dams, sluice gates, storage ponds, and equipment for lifting pulpwood from the water into stacks where it is available for winter use, must be provided. River driving is a highly seasonal operation. When skillfully carried out, it is exceedingly economical, but it oftentimes requires a huge investment in dams and stream improvements before it can be successfully undertaken.

Foresters also may be called upon to operate the pulpwood yards at the mill. This job includes receiving pulpwood from suppliers and contractors, scaling the volume, and directing where the pulp is to be piled in the yard. A progressive program of use is necessary to avoid spoilage.

The over-all job of regularly supplying adequate volume of pulpwood to feed some of the huge modern mills requires a high degree of managerial skill. Foresters have attained many such positions. These are among the most responsible and highest paid forestry positions in our country.

Initial employment on either the professional or nonprofessional basis is usually of temporary nature. Work is performed and paid for on an hourly basis. Heavy snowfall in winter and periods of extreme fire danger in the summer may force the interruption of logging operations.

Work in logging must often be done in unfavorable weather. In the Douglas-fir region the rainy period sometimes extends from October to June and unless it is physically impossible, work continues during this period. In the western pine region on the eastern slopes less rainfall is experienced, but low temperatures and heavy snowfall in the winter accentuate seasonal fluctuations of the industry.

Accident rates in logging are among the highest in the country. Work in the woods requires a person who is mentally alert. Eastern timber, being small, is less hazardous to log than the heavy timber of the West. However, because operations are small, widely scattered, and often inadequately supervised, occurrence of accidents per 1 million man-hours of labor is higher in the East than in the West. The power saws used in felling and bucking in the South add to accident hazard, and tractors introduce a greater hazard than teams.

Although remuneration is rather low for the beginning jobs, wage increases are likely to be more rapid than for public forestry jobs. The minimum wage for the industry in western Washington and Oregon in 1950 was $1.65 per hour based on a 40-hour week. Time and one-half was paid for Saturday work and double time for Sunday work. The rate is slightly less in the pine region of Idaho, Montana, eastern Washington, and Oregon. From these beginning figures, pay may climb to the $10,000 to $15,000 annual salary of foresters employed by industry in managerial jobs.

The beginner entering the field of logging must accept the fact that not until he has acquired experience and demonstrated his worth to management may he expect a job with responsibilities. This apprenticeship period may last for several years.

Most beginning professional jobs will involve surveying and may include anything from axman on a survey crew to instrument man. It is more apt to be the first-mentioned than the latter. While performing these simple tasks, the forester begins to establish his reputation with his

FIG. 101. A portable bunkhouse in an Adirondack logging camp.

employer. Advancement will depend on demonstrated initiative, intelligence, ambition, and personality.

The forester may enter the field by accepting subprofessional or labor employment, for this type of employment is almost always open. Here, work would be directly associated with logging operations and road construction.

Persons accepting labor employment may be required to join a labor union. Membership in a union and direct personal participation in work with union employees may give the forester valuable insight into workers' viewpoints that will pay high dividends should he later be responsible for negotiating labor contracts (Fig. 101).

A prospective logger-forester will need to be basically well grounded in mathematics, surveying, and physics. As advancement is made in the industry, knowledge of forest pathology, mensuration, logging techniques, accounting, and business methods will assume importance. Ability to express oneself both in speaking and in written reports is of paramount importance.

College training will help the forester to compete successfully with the practical logger. It is to be expected that the forester will rarely achieve the skill on a particular job possessed by the practically trained man of long experience, but the practical man often lacks the imagination and ability to conceive and execute planning and executive jobs.

Because entering jobs in logging usually are at the labor level, no attempt to estimate opportunities is advisable. Jobs of some sort that will provide the basic experience needed for advancement will always be open whenever logging operations are in process.

Lumber Manufacture and Sale. Employment opportunities in sawmills parallel to some extent those in logging. Throughout the United States are some 40,000 to 60,000 operating mills. Almost half of the lumber is sawed by fewer than 1,000 mills. The many scattered small mills, like the small loggers, operate intermittently and have high rates of business failure.

The big mills cutting 100 million board feet or more per year are all in the West and South. Each one saws a few species into a limited number of products. They are usually located in small communities and are affected by the same general conditions that apply to logging. Work is highly mechanized.

Working conditions in a small sawmill may be primitive. This is particularly true in the North and South where small mills predominate. The small sawmill is typically located in an open, unheated shed and has a minimum of safety devices. Hazard in sawmill operations is high, second only to logging among American industries. The lumber industry as a whole fluctuates widely in activity and in parts of the country operations are seasonal.

Unionized labor characterizes the large sawmills of the West and some of those in the Lake States. By 1951 organized labor had made only limited headway in the Northeast and in the South. Living conditions in small sawmill towns are likely to be far from ideal, but this is not true of many western mill communities, some of which are among the most attractive western villages. The sawmill town, though, is typically small with restricted social and cultural life.

BEGINNING JOBS. The forestry-college graduate has to make his own job in the mill. The jobs depend more upon experience than technical training and positions at the top are normally beyond the ability of a young fellow lacking experience.

Fortunately, the picture is not discouraging. There is likely to be increasing opportunity among the small mills where the top and bottom are not far apart. As the manufacturing processes go farther beyond the sawing of logs into lumber, the technical problems become greater and the need for technically trained men increases. Various phases of remanufacture, fabrication, and merchandizing offer good opportunity to college men.

In many mills the job of dry-kiln operator offers an entering wedge several rungs above the bottom of the ladder. Scaling, tallying, and assisting in inspection are good ways to gain a quick knowledge of a mill's operation and such jobs may be available to a man young in the profession. Time studies, cost analyses, production control, and quality control require abilities not ordinarily developed by even the above-average worker. Here the college man enjoys a distinct advantage. Some may enter the field by way of general office and accounting routine. College background along these lines enables the trained man to by-pass much of the step-by-step advancement up the labor ladder. A college-trained man should be better able than others to appreciate the many problems of personnel and labor relations. The ability to read and make use of such material as is issued both by government agencies and trade associations is something that is not learned by handling lumber. Starting salary range for these jobs of from $2,500 to $3,000 in 1950 might still fall short of the semiskilled labor wage, but the potential range is much higher. The average hourly wage for sawmill labor was something over $1 in 1951. Generally, high rates prevail in the West, low ones in the South.

The small sawmill also offers the forester with some experience opportunity to go into business for himself. Capital requirements are modest and financing can often be arranged through concentration-yard operators. There are no large blocks of virgin timber, a situation which makes the small sawmill the dominant harvesting machine in the East. The operator of a small mill should employ efficient methods to succeed. A forester with business ability who applies good forest practices to his operation, maintains an adequate timber supply, uses efficient handling methods, and reduces waste to a minimum has a good chance to succeed.

One feature above all must not be neglected—marketing. The small mill owner who wishes to realize full value from his labor must learn to grade his lumber, season it, and hold it for truck or carload shipments. Otherwise he will spend his time working for concentration-yard operators. Time spent in finding profitable market outlets for his product is likely to be highly remunerative.

As in the case of logging, no numerical estimate of job opportunities is feasible. Sawmill jobs offering worth-while experience far exceed forestry-school enrollments.

LUMBER SELLING AND CONSTRUCTION. Forestry colleges in 1951 numbered among their graduates several presidents of lumber companies and many successful wholesalers and retailers. Beginning jobs in lumber selling will be mostly of the apprentice type or common labor. Advanced jobs include price estimating and materials handling and expediting for the contractor. Planning and designing homes are important jobs which must be well done. Some may wish to train for such work, which leads to partnerships in the business and trains the worker for the establishment of an independent contracting business. Business ability and security for borrowing the necessary working capital are essential for independent contracting, but the capital outlay need not be large.

Although the typical American dwelling is a custom-built product, more and more houses are "built to sell" rather than on the customer's order. In either case, a beginning contractor needs only sufficient capital to erect one home at a time. Usually he can borrow money on land and on a partially constructed building. Financial backing is not difficult to obtain when tangible security in land and buildings can be offered.

Many small building contractors enter the business from the journeyman route. They are carpenters, masons, plumbers, or electricians who branch out as builders. To begin with, they usually do much of their own work, hiring one or two assistants. This keeps capital requirements low and almost eliminates overhead. As experience and capital are accumulated, they increase their business. Few college men have entered the field. Hence, competition is largely from men who have had limited opportunities to learn the fine points of the business.

As in milling, lumber-sales jobs are many and widely dispersed. One estimate indicated the need for over 100,000 trained men in the light-construction field. Opportunities are plentiful for those desiring careers in house construction.

Wood Processing. The wood-processing industries are numbered in the thousands. Over 1,600 operate in New York State alone. Their products are legion, their basic processes limited. These include seasoning, machining, gluing, and finishing, in all of which technical knowledge is valuable (Fig. 102).

A thorough understanding of basic physics and chemistry and some knowledge of economics and business administration are needed. Knowledge of wood technology, including wood structure and identification, timber physics, wood seasoning, wood machining, glues, and finishing, and sufficient advanced chemistry to understand the chemical treatments of wood so as to modify its properties will enable the forester to make a real contribution. Ingenuity is especially valuable in designing new tests to meet special requirements of industrial research in wood processing.

The forestry graduate will ordinarily have to find his own job. Frequently, he will be obliged to accept a job at labor wages to get a start. If he is alert, he can expect, in time, to prove his worth to his employer so that he will be given responsibility and opportunities to introduce important improvements in plant operation. Patience, good practical sense, and willingness to undertake any task that needs to be done are attributes that should lead to success in the long run (Fig. 103).

As factory conditions in America go, working conditions are likely to be far from ideal. This is partly because many of the plants are small

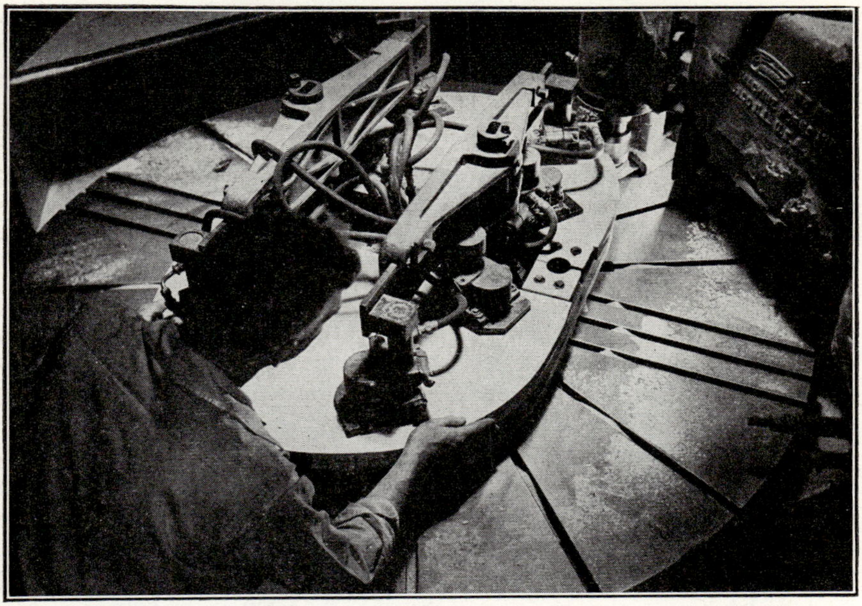

FIG. 102. Cutting a table top to contour. (*Mengel Company, A.F.P.I. photo.*)

FIG. 103. Fine furniture requires the hands of a craftsman. (*A.F.P.I. photo.*)

and partly because the workers in many plants are unorganized and have not demanded the improvements found in the highly organized fields.

It is possible to get positions where a minimum opportunity exists to demonstrate the value of technical knowledge. If such opportunity does not open up or cannot be opened up after a reasonable period of employment, it may be desirable to change jobs. Before leaving a job, a forester should have a discussion with his employer making clear his reason for desiring a change. It is far better to leave a friendly rather than a hostile employer behind.

Gradually, men entering the wood-processing industries may be expected to work into managerial positions. They may start as technicians, eventually become heads of technical departments, and then rise into administrative and executive positions. Because the wood-using industries are characteristically small and because relatively few foresters or wood technologists have entered these industries, opportunities for advancement to high posts are good. Foresters, considering the number entering, have probably made a better record of achievement in these industries than have graduates in business administration or engineering.

Salaries for most beginning jobs are usually in accord with those of unskilled labor in the plants, ranging from $2,000 to $3,000 per year. Ultimate salaries are largely dependent upon size of the organization, type of product, and the man himself. Executive positions in the larger plants range from $10,000 up. Other positions such as superintendencies and plant managerships are from $5,000 up, depending on size of plant.

Employment in the wood-processing industry will depend a great deal upon the prosperity of the industry at the time the graduate applies. During the period 1946 to 1951 the wood-processing industries enjoyed prosperity and expanded their technical and labor forces. Additional expansion on a somewhat less rapid scale can be anticipated. Graduates willing to take whatever jobs may be open in the company of their choice should encounter little difficulty in getting a start in the industry during normal times. When the industry is depressed, employment may be exceedingly difficult.

The long-term employment outlook is very favorable for the reason that so few wood technologists are now working in the wood-processing industries. The industries taken together bulk large and important and can absorb ultimately a large number of technical men. The rate at which technical men are employed will depend to a considerable extent upon how rapidly those now working for the industry demonstrate their worth and how rapidly they rise to positions of authority within the companies, where they themselves can employ recent graduates. It is anticipated that over the next two decades employment in the wood-processing industries will grow at a geometric rate. Leveling off should not be reached until such time as wood technologists are performing all functions they effectively can in the modern wood-processing industry. This will require technicians numbered in the thousands.

Because the wood-processing industry is made up of a great many small establishments, it is possible for a man, after having accumulated a modest amount of capital, to enter into the business for himself as part owner or full owner of a small wood-processing plant. Before starting out on his own, he will be wise to get some experience in the industry and to establish contacts in the markets for the products he plans to manufacture.

Pulp, Paper, and Wood-chemical Industries. The pulp and paper industry is characterized by relatively large operations. A progressive, though fluctuating, increase in output has occurred over the years. Small mills have generally been at a disadvantage in competition with large ones unless they devoted their efforts primarily to specialty papers. During periods of business depression the pulp and paper industry suffers along with the others. During periods of high business activity it has difficulty in keeping up with the demands. The long-term outlook for the industry continues to be favorable. Year by year new uses are found for paper products and new skills are developed in manufacturing, which make it possible to maintain a strong competitive position.

The pulp and paper industry, particularly the pulp industry, is highly dependent upon its raw-material supply, which it uses in great abundance. The decline of many pulp and paper mills in the past has been due in part to shortage of local sources of suitable wood for manufacture of the pulp. The industry generally recognizes this shortcoming and many companies have undertaken progressive steps to establish a dependable pulpwood supply. This is being brought about by purchase of land, planting of trees, and encouragement of good forest practice on the lands of private owners tributary to their mills.

Other things being equal, an integrated pulp and paper mill with a substantial area of timberland from which to draw its material is a more stable company for the long pull than those that manufacture pulp only or paper only. During the period 1943 to 1949 a large number of pulp and paper mills were bought up by publishing concerns seeking to ensure a supply of paper to meet their needs. The industry on the whole, therefore, is becoming more and more characterized by a few large companies. Such control, however, does not mean that independents are frozen out. The more the large manufacturing plants are devoted to a few major products, the greater the opportunities for the small concerns with versatile equipment to produce a wide variety of specialty papers and thus maintain their position in the field.

Entering positions in the paper industry are as control chemists, assistant research chemists, assistant foremen, or as assistants to the superintendent. Some of the larger companies have special training programs by which new men are moved from department to department until they are finally placed in positions in the manufacturing or sales department. Starting salaries in 1951 were $3,000 to $4,200 a year. After 10 years' experience many men have become assistant superintendents, superintendents, assistant chief chemists, or chief chemists with salaries ranging from $5,000 to $12,000 a year. Further advancement leads to positions in management with salaries going considerably above this level.

Entering positions in the chemical-manufacturing companies are generally as research chemists in the technical-sales department. Starting salaries are about the same as those in the paper industry. After 10 years' experience men are generally employed as salesmen, special technical representatives, or as head chemists with salaries ranging from $7,000 to $10,000 a year. These positions often require considerable traveling.

On the whole, employment opportunities in the paper industry have been good. Of 410 living graduates of the Pulp and Paper Department of the State University of New York, College of Forestry, over 91 per cent were employed in the paper or related industries in 1951.

Opportunities have been particularly good since 1945. The industry is still expanding and technical problems are becoming increasingly acute; so the employment outlook for the future is bright.

Students trained in pulp and paper manufacture or in special courses based on pulping with instruction in high polymers added are equipped for technical positions in the cellulose-plastics field. Working conditions and salaries are similar to those in the pulp and paper industry. Employment may be in large or small companies that manufacture consumers' goods.

Because of the large investment required for pulping and manufacturing paper, little opportunity exists for a man without large financial resources to become an independent manufacturer; however, he may become a consultant or broker.

Industries engaged in extracting and refining naval stores, tanning materials, essential oils, and other wood-chemical products also offer employment opportunities. Forestry schools offer only limited training in these specialities. Most of the technique must be learned on the job. Men with a good background in pulp and paper manufacture might, however, seek out these fields should employment in paper mills be slack.

LITERATURE CITED

1. Graham, E. H., C. R. Gutermuth, Lloyd W. Swift, Clarence M. Tarzwell, and R. M. Rutherford. 1950. It's Time to Take Stock, *Journal of Wildlife Management,* 14:237–242.
2. Moir, Stuart, DeWitt Nelson, Michael Bigley, William H. Price, and W. D. Hagenstein. 1949. What is Private Forestry Doing in the West?, "Proceedings, Society of American Foresters' Meeting, Boston, 1948," pp. 169–181, Society of American Foresters, Washington, D.C.
3. Seigworth, Kenneth J., Ray F. Bower, Harley Langdale, Jr., and Paul M. Garrison. 1949. What is Private Forestry Doing in the South?, "Proceedings, Society of American Foresters' Meeting, Boston, 1948," pp. 159–169, Society of American Foresters, Washington, D.C.
4. Turner, David B. 1948. "Professional Opportunities in the Wildlife Field," Wildlife Management Institute, Washington, D.C. 208 pp.
5. U.S. Department of Agriculture, Forest Service. 1948. "Forests and National Prosperity—A Reappraisal of the Forest Situation in the United States," Miscellaneous Publication 668, Washington, D.C. 99 pp.

CHAPTER 23. *Education in Forestry*

FORMAL AND INFORMAL courses in forestry were offered as early as 1870.[5] Professors of agriculture, botany, or horticulture at Michigan, Michigan State College, Minnesota, Nebraska, Iowa State College, Cornell, and Yale were among the first to begin collegiate-grade instruction in forestry. Education at the professional level began in 1898 with the establishment of the New York State College of Forestry at Cornell University and the Biltmore Forest School at Biltmore, North Carolina. After a period of interruption the former was moved to Syracuse; the latter has been discontinued. The Yale School of Forestry, established in 1900, is the oldest operating college of forestry in the United States. The first two schools were headed by European foresters, the third by an American who took his training in Europe. To begin with, all three schools and the others that were established during the 15 years immediately following the turn of the century were obliged to depend heavily upon European forest technique for subject matter and teaching methods. It was not until after some 15 years of forest practice in our own country and after many forest investigators had studied American forests that a truly American forestry began to emerge. Teachers now rely heavily upon the results of practice and research that have proved their worth in American forests.

Between 1903 and 1914, 21 additional schools of forestry were established, most of which are still functioning.[5] The number of schools has continued to grow. As of 1951, 34 colleges and universities were offering four-year training in forestry. The rapid increase in colleges of forestry during the past 50 years has been paralleled by the greatly enlarged scope of the profession. The original concept of forestry embraced mainly protecting forest lands, regulating the timber harvest, and starting new timber crops on their way. Today forestry includes all phases of forest-land management, the harvesting of timber and other forest crops, and the processing of forest products for human use. In other words, foresters serve where trees serve and where wood serves (Fig. 104).

FIG. 104. Many properties and uses of wood are determined largely by its microscopic structure.

Requirements for Forestry

The technical requirements and working conditions in forestry impose certain standards that men who desire to enter the profession must meet. The education offered in colleges of forestry demands high intellectual capacity and serious devotion to study.

Leadership in forestry requires men of the highest intellectual capacity and with great maturity of judgment, for their jobs require them to perform very important functions in our national economy. Forestry also has jobs in which men of good intelligence, high character, and devotion to duty can expect to achieve a satisfying life of service. Colleges of forestry admit and graduate only men who the faculties believe have the capacity to perform useful service in forestry.

A forester's work must be done largely on his own initiative and through his own ingenuity and sometimes in opposition to pressures from special interests. Foresters should, therefore, possess high standards of character, responsibility, and personality. They must be able to endure isolation without losing intellectual alertness. They must often persist in the face of discouragement. Emotional stability, facility in meeting and dealing with people, self-denial necessary to place the requirements of the job above physical comfort, and frequently even above the wishes of friends, are tests of character every successful forester must meet. A forester assumes a professional obligation to serve the public whether he works in public employment or private employment. To the extent that he unselfishly performs a public service, he will gain public confidence and esteem.

The Training of a Forester

A forester requires a broad education in the natural and social sciences as well as intensive training in his own special field. Entrance to a professional career depends upon such preparation. To become a successful forester a man must be able to express himself clearly and forcefully in written and spoken English, in simple mathematical language, and by means of sketches and drawings. Failure to acquire facility in expression is almost certain to handicap his future. Written reports inform his superior of the work accomplished and supply the information upon which future plans can be based.

Oral expression is indispensable to clear understanding and smooth relationships among people. Mathematical formulas are in almost constant use by foresters in their day-to-day jobs. Maps, graphs, pictorial charts, sketches of simple layouts for headquarter sites, recreation areas, and roads, plans for buildings, and designs for new equipment all require some facility in graphic expression.

A forester needs to understand fundamental biologic laws. Forests are far more than land covered by trees. They involve the complete biologic, hydrologic, and edaphic complex that occupies the land. Unless a forester knows how earthworms, insects, other arthropods, fungi, and bacteria break down the forest litter and convert it into a life-supporting soil rich in humus, he lacks understanding of the key to forest productivity. Insects of both harmful and beneficial nature abound in the forest. Small mammals help keep insects in check but may also destroy tree seed and even girdle small trees. To manage this exceedingly complex biological community so as to harvest from it products useful to man with a minimum expenditure of money and human effort requires understanding of the interrelationships between the various plants and animals that make up the forest community.

A forester needs basic knowledge of chemistry, physics, geology, and engineering. This helps him to understand this biologic community and to survey land, lay out and build roads, and erect buildings and fire-lookout towers. It also helps him to understand the principles of telephones, radios, internal-combustion engines, and the many types of equipment that are used in developing the forest and harvesting and processing the products the forest yields.

Understanding of fundamental economics and business methods, government, and simple business law are the basis for operating a forest property economically. The major purpose of forestry is to provide for human needs. This must be guided by basic economic and business principles. A forester needs to understand elements of cost and to make appraisals of timber to be offered by sale. He should be able to understand simple balance sheets and earning statements and know how to interpret and prepare budgets. He should have broad enough grasp of basic economics to understand the way in which forestry is related to the national economy as a whole. The management of land, the harvesting of timber, and the conversion of forest products into useful commodities involve operations affecting the economic welfare of communities as well as of individuals and companies. The forester is obliged to conduct his operations in conformity with Federal and state laws governing use of land and employment of labor. To be successful, he must do more than stay within the law. He will seek to conduct his operations so as to strengthen his community and enlist the good will of citizens and government officials. By setting objectives acceptable to all men of good will and making these understood by all, he can help strengthen human cooperation in productive enterprise and free government.

Finally, a forester must understand how to live and work with people. Forestry is to a large extent a public activity and almost every forester must deal with timberland owners, timber operators, employees, customers, and the citizen. Each forester should know something about administration, including the selection and training of personnel, the enlistment of public support for essential forestry measures, and how to draw forth collective opinion on community problems so as to enlist wholehearted cooperation in their solution. He should be aware of civic responsibilities and learn to live a happy and useful life.

The foregoing may be summed up by saying that a forester should have a good general education in the broad meaning of this term. It is especially important that he understand and deal effectively with himself, his fellowmen, the society of which he is a part, and the physical universe in which he lives. Thereby he is prepared to live a productive life in harmony with his associates and environment.

Curriculums in Forestry. Colleges of forestry have sought to establish curriculums that would prepare men for service throughout the United States. They also seek to offer specialized training by fields of forestry and by regions. A standardized curriculum for all colleges has few advocates. Feeling is widespread, however, that all curriculums in forestry should include certain basic subjects, which have been set forth in the 1950 revision of Bylaws of the Society of American Foresters under requirements for junior members. Bylaw 5 reads in part:

Graduates of approved curricula in institutions accredited by the Council of the Society of American Foresters shall be eligible for election to the Junior Member grade on the endorsement of three voting members. For a curriculum to be approved it must include the following professional courses: Silviculture—3 semester units (based on ecology—3 semester units; botany, plant physiology, and dendrology—6 semester units, and soils—3 semester units); Protection (fire, insects, or disease)—3 semester units; Management—3 semester units (based on mensuration—3 semester units, surveying—3 semester units, and mathematics through trigonometry); Economics—3 semester units (based on general economics—3 semester units); and Utilization—3 semester units.[6]

Major curriculums offered by colleges of forestry in 1951 covered the following subject-matter fields:

General forestry
  General forestry with science option
  General forestry with management option
  General forestry with industrial option
Range management
Wood utilization
  Utilization with the chemical-technology option
  Utilization with engineering option
Wildlife management
Wood technology
Recreation management
Lumber merchandising and light construction
Furniture industry
Logging engineering
Conservation of natural resources
Pulp and paper manufacture
Wood chemistry and plastics

Additional curriculums have been urged upon the schools by various industries. When the society's Bylaw 5 was applied to these curriculums, it was found that many failed to meet its requirements.

Debate is vigorous between those who would limit the title "forester" to men who have been trained to manage land for production of timber crops and those who would broaden the title to include men with forestry background who specialize in wildlife, range, watershed, and park management, in wood technology, and in forest utilization.

As forestry grows in complexity, specialization becomes inescapable. The question remains as to where the line between foresters and nonforesters should be drawn and whether academic training is a sound basis for making the distinction (Fig. 105).

A study of the tasks performed by foresters graduated from the State University of New York, College of Forestry reveals some interesting

FIG. 105. Scale models help students to visualize construction details which can only be indicated in blueprints.

facts. Supervision and administration occupy 41 per cent of their time, research 13 per cent, sales and promotion 11 per cent, and technical forestry work only 8 per cent. Of course, forestry knowledge is used in many of their supervisory, administrative, and other duties. Still it may appropriately be questioned whether training to perform the tasks of these advanced jobs should not be considered as desirable as technical knowledge.

Detailed descriptions of various curriculums in forestry may be found in catalogues of the several colleges of forestry. Almost all colleges are studying their curriculums with thought toward improvement and changes are introduced yearly. Critical analyses of curriculums and other features of colleges of forestry are given by Graves and Guise.[2]

Crowded curriculums are characteristic of almost all colleges of forestry. To encompass the subjects desirable in the training of a forester, some colleges have required a fifth year. Others require 150 or more semester hours of work for graduation, thereby requiring the equivalent of five years' work at normal undergraduate loads. A few are seeking to condense subject matter so that an undergraduate professional curriculum can be completed in four years without being overloaded with courses.

Graduate Study in Forestry. Two colleges of forestry, Yale and Duke, accept only men who hold a bachelor's degree and who have completed basic prerequisites in natural and social science. Graduate training has a two-fold purpose: to increase breadth of understanding and to permit concentrated effort in a special field. Independent scholarship and research are usually required. The master's degree is awarded for 30 or more semester hours of work beyond the bachelor's and the doctorate for 90 or more semester hours of work beyond the bachelor's. A thesis based on independent research may be a part of the master's program and is always a requirement for the doctorate.

Men studying at the graduate level may specialize in any one of the several subject-matter fields of forestry or in basic sciences underlying forestry. Undergraduate students who have definite plans for graduate study would usually be wise to use elective hours to broaden their knowledge of the basic sciences and to study one or more languages. They will then be in a strong position to pursue work at the graduate level. Men graduating in the upper half of their class are generally considered capable of profiting from graduate study (Fig. 106).

A graduate degree is considered a requisite for most college-teaching and research positions. Men planning to enter these fields should look forward to studying for the doctor's degree.

Colleges of Forestry. In 1951 the Society of American Foresters reported 25 accredited schools and 8 nonaccredited schools. In addition,

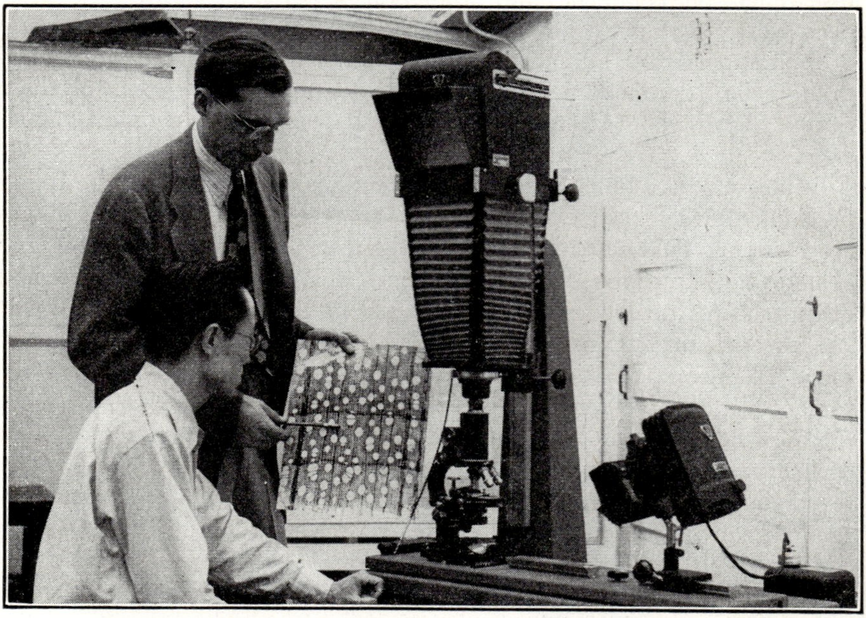

Fig. 106. Photographing the microscopic structure of wood.

24 schools were giving introductory or special courses, and 6 ranger schools were operating. The names of the accredited institutions are listed below, together with the number of students enrolled in undergraduate and graduate work in 1950.[1, 3]

Alabama Polytechnic Institute, Department of Forestry, Auburn, Ala.	113
University of California, School of Forestry, Berkeley 4, Calif.	253
Colorado State College of Agriculture and Mechanical Arts, Division of Forestry and Range Management, Fort Collins, Colo.	191
Duke University, School of Forestry, Durham, N.C.	76
University of Florida, School of Forestry, Gainesville, Fla.	195
University of Georgia, School of Forestry, Athens, Ga.	174
University of Idaho, School of Forestry, Moscow, Idaho	235
Iowa State College of Agriculture and Mechanical Arts, Department of Forestry, Ames, Iowa	243
Louisiana State University, Department of Forestry, Baton Rouge, La.	129
University of Maine, Department of Forestry, Orono, Me.	240
University of Massachusetts, Department of Forestry and Wildlife Management, Amherst, Mass.	41
Michigan State College of Agriculture and Applied Science, Department of Forestry, East Lansing, Mich.	358
University of Michigan, School of Forestry and Conservation, Ann Arbor, Mich.	315
University of Missouri, Department of Forestry, Columbia, Mo.	118
University of Minnesota, Division of Forestry, St. Paul, Minn.	298
Montana State University, School of Forestry, Missoula, Mont.	253
State University of New York, College of Forestry, Syracuse, N.Y.	796
North Carolina State College of Agriculture and Engineering, Division of Forestry, Raleigh, N.C.	207
Oregon State College, School of Forestry, Corvallis, Ore.	318
The Pennsylvania State College, Department of Forestry, State College, Pa.	374
Purdue University, Department of Forestry, Lafayette, Ind.	205
Utah State Agricultural College, School of Forest, Range and Wildlife Management, Logan, Utah	249
University of Washington, College of Forestry, Seattle, Wash.	264
West Virginia University, Division of Forestry, Morgantown, W.Va.	178
Yale University, School of Forestry, New Haven, Conn.	55
Total enrollment in 25 accredited schools	5,878
Enrollment in 9 nonaccredited schools	656
Total enrollment	6,534

Men pursuing work for a master's degree totaled 483 and doctoral candidates 118. Graduate students therefore accounted for 9 per cent of total enrollment in 1950. Enrollment by undergraduate classes was: freshmen, 1,442; sophomores, 1,379; juniors, 1,384; and seniors, 1,730.

Total enrollment in 1949 was 8,017. Together these figures presage a decline in the number of forestry graduates up to the year 1954.

Scarcely an annual meeting of the Society of American Foresters passes without some debate on how many men should be trained in forestry. Opinion reflects the speaker's conception of the breadth of forestry. Those advocating a broad interpretation foresaw no surplus of trained men even at the peak rate that prevailed in 1948.[4] Those with the narrower concept have stated many times in the past that too many foresters were being trained. Yet the profession has absorbed most of these. Little loss to society results, whatever a man's training, provided it prove as useful to him in later life as any other he might have pursued. Certainly this is true of foresters who later become wildlife managers, loggers, or sawmill operators.

Employment in Forestry Education

Estimates show that 600 men are employed in education in forestry. Probably at least 200 additional foresters are employed in extension, public relations, and teaching in ranger and secondary schools. Together this indicates an annual replacement of 25 to 30 men.

College Teaching. Working conditions for teachers in colleges of forestry are among the best the profession affords. Academic life provides many delightful amenities. Association with students, alumni, and faculty presents constant and stimulating challenges.

Colleges and universities recruit their faculty with great care. Records are scrutinized and personal interviews often precede appointment.

Training for teaching in forestry is of two types, academic and practical. Schools prefer men who hold the master's and Ph.D. degrees, and who have practiced their profession in the field. The ideal recruit is a young man with five or more years' practical experience who has just been awarded the Ph.D. degree. It is advisable for men contemplating college teaching to gain teaching experience either while pursuing work for the Ph.D. or by carrying on special public relations or educational work in the job in which they get their practical experience.

A man planning to enter college teaching should select as his specialty one of the major fields of forestry, such as silviculture, mensuration, forest protection, management, or wood technology. Opportunities appear to be brightest for men trained to teach forest economics, forest utilization, and perhaps the new field of forest work studies, which is now being rapidly developed in Europe. So far this has received no formal attention from the colleges of forestry in the United States. It deals with methods of increasing worker output and worker satisfactions from logging and other forestry activities.

Lucid writing and speech are requisites of good teaching. So also is skill in the art of teaching. The latter may be acquired best by training under a good teacher who knows how to grip students' attention and awaken a zest for learning.

Training on the job at colleges and universities is the accepted method for acquiring excellence in teaching. Few colleges require applicants for teaching positions to study the theory and philosophy of education, but many young teachers have found it profitable to do so. College teachers are chosen primarily for their knowledge of the subject matter, personality, and scholarly ability.

ATTRIBUTES OF A GOOD TEACHER. The successful teacher should be patient and should understand youth. He should enjoy working with young people and helping them with personal as well as academic problems.

To keep abreast of latest developments in the profession and related fields, teachers must be lifelong scholars. They must have the background and critical ability to evaluate new developments and to give them proper perspective in the over-all training program. Constant refreshing is necessary to keep abreast of rapidly developing techniques in field, factory, and office.

The teacher usually is expected to carry on research. It is through research by their staff that individual schools of forestry gain special distinction. Students, especially graduate students, select their college on the basis of scholarly eminence of faculties.

College authorities realize that education has a very important role to play if world peace and democracy are to live. Skimpy training in such broad fields as citizenship and the art of successful living has been often charged against technical colleges. A vigorous debate goes on yearly as to the proper distribution of effort between specialized subjects that prepare the students to earn a living and general education that fits them to participate effectively in a free society. No simple solution to this difficulty is apt to be forthcoming in the near future. No course or group of courses will likely be found to provide the specialist with a complete general education. Instead, it will be up to the teachers of professional subjects to inculcate in their students' minds their obligations to society as a whole and to set before them desirable standards of successful living.

Forestry curriculums, in common with those of other broad professions, fortunately provide a great deal of general education. Forestry deals with human beings and the resources they use. A student planning to enter forestry education should understand the broad implications of forestry and its obligations to modern society.

WORKING CONDITIONS. Salaries for college teachers are likely to be below those in private and Federal work and to match those in state work.

Teachers may have limited opportunities to augment their regular salaries by gainful vacation employment, part-time consulting work, lecturing, and writing. Promotions may be slow. Frequently they depend upon suitable vacancies, and as turnover in college employment is low, vacancies occur infrequently.

Vacations for college teachers are ordinarily generous—usually a month and sometimes as much as two or even three months. When faculty members are free of academic duties, they are expected to travel and do other work for professional advancement. Sabbatical leave is available in many institutions for professional improvement, serious writing, and research. Retirements from college faculties are usually provided under either a state retirement system or a special retirement contract with an annuity company. Retirement age in most institutions is at 65, though it varies from age 60 to age 70.

College teachers have wide opportunity for self-expression. They are expected not only to teach, but to develop their subject matter, carry research, and engage in technical writing and professional activities. They are encouraged to serve their profession through membership on committees and through holding offices in the professional societies.

College teachers have wide latitude in hours of employment. They themselves, in the last analysis, are the only judge of their own working hours and standards. Wide difference exists in the services that individuals actually perform, with consequent difference in promotion rates, salaries, and the prestige teachers enjoy among their fellow foresters. Teaching is not easy work. A regular teaching schedule is likely to take up 30 to 40 hours a week in actual instruction, preparation for class work, laboratory work, and correction of papers. In addition, teachers are called upon for committee work, to aid in student activities, and for university-wide services. Scholarly and research activities are also expected and encouraged. To perform these tasks and at the same time to remain an inspiring teacher with fresh subject matter require a high degree of personal application, imagination, and energy.

It is very easy for a teacher to dissipate his time in activities of secondary importance and never to achieve national recognition for expertness in his field and for outstanding teaching skill. But for men of imagination, energy, devotion to professional achievement, interest in research, and, above all, real understanding and sympathy with the problems of students, teaching offers satisfactions rarely found in any other type of employment. Many forestry teachers become national figures and enjoy high prestige on their campus, in their profession, and as citizens.

Other Teaching Opportunities. RANGER SCHOOLS. Teachers in ranger and other subprofessional schools require somewhat different preparation from college teachers. The curriculum is short, must be highly practical, and is ordinarily concentrated in presentation. Their work is vocational in nature. The ranger-school teachers, if outstanding in their fields, may have opportunities for advancement equal to those of college teachers. More depends upon the individual and the working conditions that surround his own job than on the type of institution in which he works.

JUNIOR COLLEGES. Men with forestry training have found satisfactory employment as teachers in community and junior colleges. Such institutions are growing rapidly and a number of them are introducing preforestry training. The forestry work offered at such institutions is often parallel with that offered in regular colleges and universities that have a few courses in forestry but do not provide professional training.

SECONDARY SCHOOLS. High schools, secondary schools, and trade schools also offer opportunities for teachers with forestry background. Teachers in such institutions must generally be trained in education as well as in forestry. One weakness of the forest industry in America is lack of men with subprofessional training.

The jobs of harvesting timber, manufacturing lumber, processing wood, and handling the various other technical jobs that lie between the forest and the consumer of wood products yearly become more technical and difficult. The amount and types of power equipment used in woods operations have grown rapidly during the last decade. Efficient use of power chain saws, mechanical loaders, special skidding devices and new portable sawmills, places new demands on workmen. Specifications for particular products, such as handle, lathe, and bowling-pin stock, aircraft veneer logs, and blanks for ball bats, require a degree of knowledge for the selection of material that hitherto was not required of the timber operator and sawmill worker. Operating head saws, supervising lumber piling, operating dry kilns, and grading lumber are jobs that require training and devotion to detail.

Safety alone in logging operations is of sufficient moment to justify special training for woods foremen and others who supervise logging and sawmill operation. Accident rates are high, compensation insurance expensive, and evasion of responsibility by financially weak operators, a constant source of annoyance to state administrators.

Extension Forestry. Forestry extension is administered by the Federal Extension Service in cooperation with the state agricultural extension services and thus constitutes a part of the over-all agricultural extension program of the land-grant colleges. The major objective of the farm-forestry extension program is to maintain a system of management for farm woodlands which will enable these areas to contribute fully and regularly to farm income and a more satisfying farm life. Chief responsibility for the accomplishment of this objective rests upon the extension foresters, who work through some 7,000 county extension agents, local farm leaders, and numerous county and community committees representative of the farm people. The work of extension foresters includes a certain amount of assistance to individual owners of farm woodlands. The bulk of their activity, however, is directed toward reaching large numbers of people and thus involves the use of such media as public talks, film strips, movies, radio scripts, news releases, bulletins, circulars, and method demonstrations.

In 1948, 45 states and the Territory of Puerto Rico were cooperating with the Department of Agriculture in farm-forestry extension work. The number of extension foresters employed on the staffs of the state extension services totaled approximately 65 in 1950. In addition, two extension foresters were employed by the Federal extension service to serve in a liaison capacity between the Department of Agriculture and the state extension services.

Farm Forestry. The work of farm foresters is authorized under the Cooperative Forest Management Act of 1950. While this program is similar in many respects to farm-forestry extension work, it is considerably more intensive in its approach. It concentrates upon providing in-the-woods technical advice and assistance to individual owners of small woodland areas. In most states the farm foresters are employed by state foresters, with the U.S. Forest Service cooperating in supervising and financing the program. The foresters are organized on a project basis within the state, the territory covered by each averaging from three to four counties. The farm forester will prepare a simple management plan for a woodland, covering such operations as timber cutting, planting, thinning, and protection from fire, insects, disease, and grazing. When the farm forester finds the owner has timber ready for harvest, he recommends a method of cutting. He marks trees for cutting, estimates volume, and suggests timber buyers. The forester uses his technical knowledge to determine the best management policy for a particular forest property and counsels the landowner in the proper methods of carrying out the management practices required. Farm foresters under the cooperative forest-management program do not serve as agents for the farmers in making sales.

As of 1948, 40 states were cooperating under the Norris-Doxey farm-woodland-management program which preceded the Cooperative Forest Management Act. The number of projects established totaled in excess of 170. The U.S. Forest Service estimates that the nation ultimately should have upwards of 2,000 such projects if at least a minimum amount of technical assistance is to be provided for each of 2 million or more owners of small woodlands. Farm forestry represents an expanding field and may be expected to provide a number of jobs in the years ahead.

Public-relations Foresters. Railway companies, power companies, chambers of commerce, and pulp and paper and lumber companies employ foresters to promote forestry in the territory they serve and to explain the policy of the companies to landowners and others. Programs of forest-fire control, planting, and forest management have been sponsored by various railways in the South. Other railways in the Lake States and elsewhere have promoted tree planting, care with fire, and conservative methods of timber harvesting that will promote future timber growth. The major objective of all of these companies is to build up the forest resources of the territory they serve. These jobs require special training and aptitude for meeting the public and convincing individuals that conservation is their business, as well as that of timber companies.

Public-relations foresters are employed by the American Forest Products Industries and by various regional associations. Federal land banks oftentimes employ foresters to promote good forestry among their clients. All public forestry agencies such as the U.S. Forest Service and the Soil Conservation Service must get their messages across to the public if they are to be effective. The need for developing specialists becomes progressively more acute.

Special training in public-relations work is now offered at more than one college of forestry. Success in this field requires training in the techniques of radio, motion pictures, television, news writing, and other means of mass education. But success also depends upon the public-relations forester's having a good background of field experience upon which he can draw to get his message to his public.

Trade-association and Conservation-association Foresters. Many of the national and regional trade associations of lumbermen, pulp and paper producers, and other forest-industry groups employ technical foresters, who serve as staff advisors, liaison men to instruct and assist association members, and promotional personnel. The American Pulpwood Association is an outstanding example. It has undertaken to help test new mechanical equipment developed to facilitate pulpwood cutting and to disseminate the results to its members. It also has a program for training woods workers, promoting safety in woods operations, and preparing films and other instructional material to bring its message home to timber workers. The need for this type of service is growing constantly.

Conservation associations, sportsmen's organizations, and organizations such as the Boy Scouts of America, mountain clubs, and other organizations dealing with the out-of-doors employ some executive officers. Men with forestry training may find in these desirable employment. Opportunities and employment conditions vary widely from one organization to another. Such jobs provide opportunities for wide contacts and satisfying service.

LITERATURE CITED

1. Anonymous. 1951. Colleges and Universities in the United States Offering Instruction in Forestry, *Journal of Forestry,* 49:672.
2. Graves, Henry S., and Cedric H. Guise. 1932. "Forest Education," Yale University Press, New Haven, Conn. 421 pp.
3. Guise, Cedric H. 1951. Statistics from Schools of Forestry for 1950: Degrees Granted and Enrollments, *Journal of Forestry,* 49:7–12.
4. Herbert, Paul A. 1949. Are We Set Up to Train Too Many Professional Foresters?, "Proceedings, Society of American Foresters, 1948," pp. 21–25, Society of American Foresters, Washington, D.C.
5. Hosmer, Ralph S. 1950. Education in Professional Forestry, "Fifty Years of Forestry in the U.S.A.," pp. 299–315, Society of American Foresters, Washington, D.C.
6. Hostetter, Robert D. 1951. Vote Establishes Student Grade, *Journal of Forestry,* 49:56–57.

CHAPTER 24. *Research in Forestry*

RESEARCH, APPLIED TECHNOLOGY, and human ingenuity have contributed to the American way of life fully as much as an abundant supply of natural resources. The American tradition implies high wages, short working weeks, and low prices of manufactured goods. These, in turn, require high output per worker and mass production and distribution. America has seen tremendous progress in agriculture during the past two centuries. In colonial times 80 per cent of our workers were engaged in agriculture to support the other 20 per cent engaged in trade and manufacture. In 1950 only 13 per cent of our nation's workers were engaged in agriculture, yet we were better fed and clothed than at any preceding time in the nation's history. Applied research and technology were largely responsible for this change.

During the period 1900 to 1932 output per worker in the lumber industry remained essentially stationary, but that of steelworkers increased 24 per cent; stoneworkers, clay workers, and glassworkers, 58 per cent; workers in paper and printing, 62 per cent; in nonferrous metals, 120 per cent; and in motorcars, 1,000 per cent. The increase in output per worker has been almost directly related to the amount of research carried on by these several industries. Today the chemical, electrical, and automotive industries are leaders in productivity, wages, and value of product. They also are leaders in research. Coal mining, lumber manufacture, and several others that have sponsored little research have lagged in output per worker and in prices of products.

Development of Research

Research in forestry began long before extensive practice of forestry by either public or private agencies. Early investigations were concerned mainly with descriptions of species and general descriptions of forest types. John Bartram in colonial times and his son William in post-Revolutionary days made extensive studies in Georgia, Florida, and other states. Johann David Schöpf, a German surgeon with forestry training, made critical observations on the forests of Pennsylvania, Maryland, and

Virginia during the days of the Confederation. In 1796 André Michaux and his son François-André Michaux published the first authoritative studies of American trees. The latter's "North American Silva" in 1857 was the first comprehensive work. Nuttall, Emerson, Hough, and Sargent brought the study of American trees to a high level. The Arnold Arboretum founded in 1872 has been the outstanding center for study of woody plants.[3]

The first Federal recognition of forestry was the authorization in 1876 for the Department of Agriculture to investigate forest production, consumption, and influences.[4] Research was the chief activity of the first three Federal foresters, Hough, Egleston, and Fernow. Before the year 1900 substantial knowledge of forest trees, timber resources, and the behavior of timber types was ready for use by the first practicing foresters.

In 1908 research was formally organized in the U.S. Forest Service according to a plan prepared by Raphael Zon. Experiment stations were established on the national forests where long-term experiments could be carried out. At the beginning the value of these stations as training centers and demonstration forests was envisioned. The Forest Products Laboratory at Madison, Wis., was established in 1910. In 1921 research was extended to eastern forests. The McSweeney-McNary Act of 1928 provided organic legislation for nationwide coverage by forest experiment stations, for the nationwide survey of timber resources, and for enlarged research appropriations.[4] This followed an extensive study of research needs by Clapp.[2]

From the beginning, other bureaus of the Department of Agriculture assisted in forest research. As of 1951, Federal research in forestry was centered in the Forest Products Laboratory, the eleven regional and two Territorial forest and range experiment stations, in the Divisions of Forest Insect Investigations and Forest Pathology of the Agricultural Research Administration, and in other cooperating agencies, including the Soil Conservation Service, Bureau of Animal Industry, Fish and Wildlife Service, Weather Bureau, Bureau of the Census, Tennessee Valley Authority, and Bureau of Foreign and Domestic Commerce.[3]

Agricultural colleges and experiment stations and universities began forest research before the year 1900. This work has now largely been centered in the colleges of forestry. Many of these have substantial programs of research in forest-land management, game management, and use of forest products.[3, 6] In the years since 1945, activities have expanded rapidly. In 1951 one school reported a budget for research totaling close to $400,000. Another had approximately half this sum. Many others likewise had substantial research budgets.

State conservation departments have carried on research since 1900. Maine has been outstanding for its investigations on forest insects. Pennsylvania, New Hampshire, New Jersey, Ohio, New York, Texas, Michigan, and California have done important research in forestry.

Research in forestry by industries is less well known but is rapidly approaching in cost that by all other agencies. Much of it is directed toward improving forest products or processing techniques. The Timber Engineering Company has its own product development research department and laboratory. The American Pulpwood Association and National Lumber Manufacturers Association also carry important programs. Altogether, some 900 manufacturers, processors, and consumers of forest products and their associations participate to some extent in forest-products research.[3] A few private research institutes have carried projects of importance to forestry.

The total expenditure for forest research in 1949 was estimated by the U.S. Forest Service to be $16,894,000, with corporations responsible for 32 per cent, Federal agencies 29 per cent, states 24 per cent, private institutions and colleges 15 per cent.[8] The several defense agencies of our country help finance a substantial portion of government and college research.

Importance of Forest Research. Our nation would not be spending annually almost 17 million dollars for forest research unless those responsible for these appropriations were convinced that they produced returns commensurate with costs. Compared with expenditures for agricultural research, which have yielded rich returns, those spent for forestry are still modest. Yet forestry exceeds agriculture in complexity. The forester must work with natural and economic environments he can modify but little. He can rarely afford to weed, spray, fertilize, or cultivate his crops. Nor can strains of trees resistant to insects and fungi be quickly developed by breeding. The forester must look largely to natural forces, which he can manipulate only to a limited extent, to maintain desirable soil structure and fertility, to control the composition of the stand, to minimize damage by windstorm, sleet, ice, destructive insects and fungi, and to provide desirable parents for future stands. Trees are a long-term crop, yet the forest is ever changing. With its changes go changes in soil and changes in economic factors that determine what can and cannot be done in a practical way.

Objectives of Forest Research. The broad objective sought through forest research is to increase human benefits from forest land by discovering and applying scientific knowledge to forests and forest products. Timber growing needs to be made a more certain and profitable enterprise. Output and safety of workers in timber harvesting and processing needs to be increased. Properties of wood must be well understood in order that timber stands, various kinds of wood, logs and boards may be put to their highest uses. If we are to maintain our forests and per-

FIG. 107. Infiltrometer used to measure rate at which soil will absorb artificial rainfall. Tent removed. (*U.S. Forest Service photo.*)

petuate their products and services, we must understand how forests retard erosion, promote water penetration into soil, and shelter wildlife; how forests can best be protected against destructive insects, diseases, fire, animals, and windstorm, and how timber can be economically harvested (Fig. 107). Fundamental as well as applied research is needed.

Organization of Forest Research

Much forest research involves experiments that are continued for several years or resources beyond the means of the individual worker. He must have the backing of an organization such as a forestry school, a forest experiment station, or a large industrial corporation if long-term work he starts is to be carried to logical completion. This means that careful working plans must be prepared and instructions issued so that anyone properly trained can carry on an experiment once it is started. Organized, long-term research is expensive. Much forest research by its very nature requires fairly large-sized operations. Cutting methods must be tested on plots of 40 acres or larger. New pulping techniques must undergo pilot-plant tests. Properties of wood vary with the growth rate of the tree, with the territory from which it comes, and with other factors. Consequently, wood from many trees and different locations must be tested before valid data representative of the species in question are obtained.

Federal Research. Federal forest and range experiment stations are organized on both territorial and functional lines. Each station director is responsible for all research activities of the Forest Service in his respective territory. The directors, in turn, report to an assistant chief in charge of forest research. The assistant chief has staff officers assigned to dendrology and range-forage investigations, forest-management research, fire-control research, range research, forest-products research, forest-economics research, and forest-influence research. Each experiment station has one to five functional activities in these several fields with a division chief for each. A division chief has reporting to him two or more project leaders. The projects are the broad lines of inquiry. An example is spruce-fir management in the Northeast. The project leader is responsible for all work on his project. He prepares a problem analysis to determine the relative importance of problems, assigns priorities, and recommends a definite work program. A working plan is prepared, which sets forth in detail the types of individual studies to be conducted and outlines the methods by which they will be carried out and by which the data are to be summarized and analyzed. Individual studies of the project are usually assigned to assistants to the project leader, who collect information on the ground, analyze results, and prepare office reports. The end result of a well-conducted project is one or a series of bulletins that set forth proper management techniques for the particular type.

Much collection of data and sample plot work is done at work centers established by the station. A large station may have six or more work centers, each in charge of a well-qualified research worker. These are centers for field work and for demonstrating good forest practice to forest owners.

The Divisions of Forest Pathology, of Forest Insect Investigations, and of Forest Biology—the latter under the Fish and Wildlife Service in the Department of the Interior—maintain workers at many of the forest experiment stations. These function as a division in the station and coordinate their work with the station program.

The Forest Products Laboratory at Madison, Wis., has its own director and a series of division chiefs. The research divisions are timber physics, timber mechanics, silvicultural relations, industrial investigations, material containers, derived products, pulp and paper, information and education, wood preservation, and forest pathology.

Research in Forestry Schools and Agricultural Experiment Stations. Formerly, most research in forestry schools was organized on an individual basis. Each instructor was free to pursue research in whatever field of forestry he found most interesting and productive. The work took the form of building up school forests, establishing permanent sample plots, and doing detailed work in dendrology, silviculture, wood technology, and other specific phases of forestry. Forest mensuration, management, and economics have been given particular attention by workers in forestry schools (Figs. 108 and 109).

Within the last decade schools of forestry have also been entering into organized research. An outstanding example is the establishment of the Wood Products Laboratory operated by Oregon State College at Corvallis, Ore. Financed by a severance tax on stumpage, this laboratory employed 20 or more workers in 1950. It is organized on a division basis very much as is the Forest Products Laboratory at Madison. A similar laboratory is projected for the University of California and one is being planned in the South. The research program of the State University of New York, College of Forestry is organized under a research director, who works with the several department heads in the college. The program at this college required 35 research men in 1951. Important research in forest products and forest management is being done at other colleges of forestry.

Where forestry colleges are part of a land-grant college system, they normally participate in the research activities of the agricultural experiment station. This helps to finance projects and lends stability to long-term experiments. Agricultural experiment stations have made some important advances in research in silviculture, forest soils, forest entomology, forest pathology, and utilization of forest products.

Private Agencies in Forest Research. Private research organizations vary widely from company to company. Generally, they are headed by a director of research, who may have one to several project leaders and assistants reporting to him. Much of the research carried on in industrial laboratories is kept confidential as it is expected to improve the company's processes and products and thereby enhance the competitive position of its sponsor.

Industrial corporations also sponsor research projects at the Mellon Institute and at various colleges and universities. Sometimes this may be merely sponsorship of an industrial fellowship, where wide latitude is left to the institution and fellow to select the problem and the methods of study. In other cases, special, confidential studies are undertaken in which the industry may participate directly. Work at the Mellon Institute and at the industry-supported Institute of Paper Chemistry at Appleton, Wis., may be pursued by individual companies on a confidential basis.

The forest-products industries support the Timber Engineering Company Research Laboratory in Washington, D.C. It is organized by divisions to carry out important work in timber engineering, timber mechanics, wood chemistry, and related fields. Their program of work included

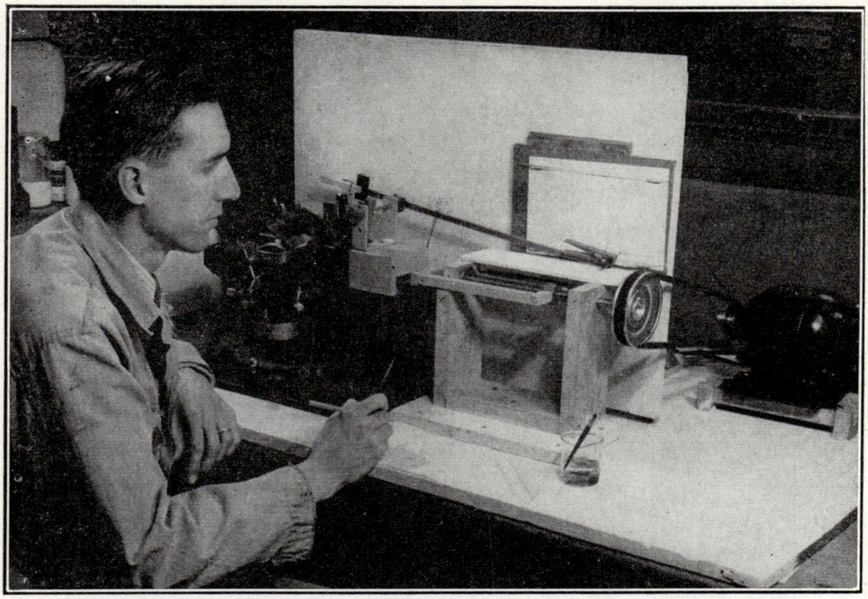

Fig. 108. Research in wood properties requires skill in devising laboratory testing apparatus.

Fig. 109. Determining operation data for applying electrical high-frequency heating to cure glue lines in fabricating wooden articles.

in 1950 chemical treatments to improve bending, wood as a raw material for the chemical industry, joining, gluing, laminating, coating, and flameproofing wood, house construction, lignin research, marketing surveys, pulping, seasoning, preserving, timber testing, plywood and veneer, wood-working machinery, and use of waste (Fig. 110).

Cooperation in Research

Cooperation between research agencies is a common practice. In 1949 the Forest Service had 644 cooperative agreements for conduct of research work.[5] These were with industries, state agencies, agricultural experiment stations, and private institutions and colleges. The Society of American Foresters, the Forest Products Research Society, and other scientific organizations serve to bring research workers together for ex-

FIG. 110. Testing a bolted timber connection. (*Timber Engineering Company photo.*)

change of ideas and coordination of effort. These many formal and informal exchanges facilitate over-all efficiency without restricting freedom of action by workers.

Expenditures for Forest Research

Funds devoted to forest research by Federal and non-Federal agencies have tripled since 1940.[8] Expenditures totaled about $5,700,000 in 1940. By 1949 these had increased to approximately $16,900,000. U.S. Forest Service expenditures more than doubled, yet did not keep pace with those of non-Federal agencies.

The expenditures for forest- and range-management research in 1949 totaled approximately $5,740,000, of which some $2,940,000 was spent by the U.S. Forest Service and $2,800,000 by non-Federal agencies. The over-all expenditure on forest-resource investigations approximated $1,322,000. Some $870,000 of this was contributed by the U.S. Forest Service. The expenditure on forest-products research, $9,833,000, greatly exceeds that devoted to all other types of research. About $1,200,000 of this was spent by the U.S. Forest Service with non-Federal, mostly private, agencies contributing something in excess of $8,650,000.

Of particular significance is the rapid expansion that has taken place in recent years in forest-products research. In 1946 the Forest Products Research Society was organized. By 1951 its membership had grown to approximately 3,000, including several industrial memberships. This organization comprises not only wood technologists but also engineers and others concerned with the use of wood.

Achievements in Forest Research

Forest research has many outstanding achievements to its credit. Almost revolutionary developments have occurred in the use of wood. These vary from such prosaic work as how to design and nail crates for packaging material to how to extract vanillin from lignin (Fig. 111). Modern timber connectors make possible the erection of huge unsupported spans; laminated timber and plywood avoid natural weaknesses; modern adhesives make possible better joining of wood. Impregnated wood is hard, fire-resistant, and dimensionally stable. "Compregnated" wood has extreme density, high strength, and resistance to abrasion and can be made with variable density. Striking advances have occurred in the manufacture of pulp and paper from pines and hardwoods and particularly in the manufacture of dissolving pulps for rayon and a host of plastic materials.

Important also are the achievements in silviculture, forest surveys, and forest management. Little publicized but highly useful has been the research to determine proper methods of collecting, extracting, storing, germinating, and sowing forest tree seeds. The huge reforestation program of the 1930's created great need for such information. Research was stimulated and a nationwide effort put forth to assemble, evaluate, and organize the results in usable form. The "Woody-plant Seed Manual," published by the U.S. Department of Agriculture in 1948,[7] is the outcome of this work. It sets a very high standard for research and compilation of results in an important forestry field.

Forest nursery problems have also yielded to systematic research. Proper nutrition in nursery beds, weed control, protection against insects and damping off fungi, hardening stock to resist frost and drought, proper field planting techniques—all of these problems were successfully attacked by forest-research workers.

FIG. 111. Research in wood chemistry and plastics often requires involved laboratory techniques.

Factors affecting forest-fire occurrence and behavior have been studied under various conditions of weather and fuels. Systematic classification and integration of these factors has greatly narrowed unknown elements in fire control. Good judgment thereby becomes the easier to apply to this highly dynamic forest enemy.

The classification of trees by rate of growth, value increment, and susceptibility to insect and fungal attack has simplified the process of selecting appropriate trees for harvesting. New hybrids have been developed through systematic tree breeding that offer promise of rapid growth.

During the war period forest-research workers discovered that output per worker in the naval-stores industry could be increased from 60 to 80 per cent through the use of a new method of bark chipping and the application of acid to the freshly cut streak. Output per worker in hardwood-pulp cutting could in certain timber stands be increased from 40 to 80 per cent through restricting cutting to trees 9 to 16 inches in diameter. Yield of spruce-fir lands in Maine could be almost doubled by reducing the cutting cycle from 60 to 20 years. These and other findings are rapidly reducing forestry from a highly speculative financial venture to a normal-risk business enterprise.

Special Aptitudes Required of Forest-research Workers

Forest research requires high scholarship, lively imagination, persistent endeavor, critical and objective thinking, self-discipline to work hard, patience to await results, and determination to acquire the necessary training and background through study and search of literature before embarking on a new undertaking. Forest research requires elaborate training.[1] Men entering research in the biological phases of forestry need to be well grounded in mathematics, physics, chemistry, and fundamental biological sciences and to have a reading knowledge of one or more foreign languages. Those entering forest economics need to understand basic economics, political science, and business administration as well as forestry and the forest-products industries. They should be able to appraise the effect that proposed public measures are likely to have on forest practice throughout the country. They need to be well informed on the total forest resources, the potential growth rates, and the present condition of our forests.

Forest-products research requires comparable skills in all phases of wood technology, especially in timber physics and wood chemistry and their related engineering applications.

Employment and Outlook

At least 750 men were employed in forest research in 1951. This figure includes those with forestry training only. Probably even more men were engaged in research on forest products of a type requiring chemical or engineering backgrounds. Research was still expanding in 1951. As the boundaries of the known expand, our awareness of fields still to be explored grows.

Opportunities probably exist for 30 to 50 openings yearly in forest and forest-products research. Expansion will not be steady. Federal expenditures fluctuate widely from year to year with the congressional appropriation, private expenditures with the prosperity of industry. Both are influenced by business cycles and military activities.

The long-term outlook remains good. The only means by which our country's needs for forest products can be continually met and expanded with a dwindling old-growth timber supply is through applied research.

Opportunities in Tropical-forest Research

The tropical forests of the Western Hemisphere represent the world's greatest virtually untapped forest resource. Intensive use of this resource depends upon detailed and extensive research into the properties of the various woods, tree-growth rates and ecology, methods of exploiting the forest, types of transportation, and methods of making the tropical lands healthful places in which workers can live and work. A number of tropical countries are interested in developing this great resource.

Very little research has been carried out in tropical forestry in the Americas. A tropical forest experiment station is maintained by the U.S. Forest Service at Rio Piedras in Puerto Rico. Other research in tropical forestry is being carried on at Turrialba, Costa Rica, Tinga Maria, Peru, and São Paulo, Brazil, and on the island of Trinidad. The countries of Latin America understandably wish to maintain their own research institutes, receiving some help and guidance from the United States. Opportunities for North Americans to participate in this research are not clearly defined. The countries naturally favor their own citizens over outsiders for important public jobs.

LITERATURE CITED

1. Bailey, I. W., and H. A. Spoehr. 1929. "The Rôle of Research in the Development of Forestry in North America," The Macmillan Company, New York. 118 pp.
2. Clapp, Earle H. 1926. "A National Program of Forest Research," American Tree Association, Washington, D.C. 232 pp.

3. Dana, Samuel T. 1950. Forest Research, "Fifty Years of Forestry in the U.S.A.," pp. 316–339, Society of American Foresters, Washington, D.C.
4. Graves, Henry S. 1947. "Problems and Progress of Forestry in the United States," Society of American Foresters, Washington, D.C. 112 pp.
5. U.S. Department of Agriculture. 1950. "Report of the Chief of the Forest Service," Washington, D.C. 69 pp.
6. U.S. Department of Agriculture, Lake States Forest Experiment Station. 1938. "Forest Research in the United States." 138 pp.
7. U.S. Department of Agriculture, Forest Service. 1948. "Woody-plant Seed Manual," Miscellaneous Publication 654, Washington, D.C. 416 pp.
8. U.S. Senate. 1950. The Forest Research Program, Excerpt from Hearings before the Subcommittee on Appropriations, United States Senate, Eighty-first Congress, Second Session, on the Agricultural Appropriation Bill for 1951, pp. 469–502, Government Printing Office, Washington, D.C.

CHAPTER 25. *Administration in Forestry*

MOST FORESTERS sooner or later find themselves engaged in administrative work at some level. In response to a questionnaire 41 per cent of the graduates of the State University of New York, College of Forestry including those out of college only one year indicated that they were engaged chiefly in supervisory or administrative work. Supervisory and administrative skills therefore weigh heavily in determining promotion rate, salary, and ultimate position. No forester can begin too early to cultivate those attitudes and traits that fit him for administrative responsibility. Opportunities to exercise administrative traits can be found readily almost everywhere. A survey of this important field of work, therefore, is a desirable feature of an introduction to forestry.

The responsibility of administrators is to attain policy goals, utilizing the personnel and physical resources placed at their disposal. Administrators leave their impress on an organization by determining to a large extent its objectives, ethics, and the amount, kind, and standards of work performed. Forestry has been fortunate in having been influenced by wise and competent administrators. The degree to which the profession satisfies its future obligation will depend largely upon the intelligence, foresight, judgment, and ability of the administrators who are responsible for directing the work. To the extent that wood becomes scarce or too high-priced, that wild lands produce floods, that natural resources are needlessly destroyed, forestry fails to attain its goal. Such failure is the failure of forest administrators to accomplish their task, although it may not be the direct fault of any one or any group of administrators.

Every type of work that is organized requires administrators. Large organizations are usually headed by skilled administrators. Small organizations may tolerate certain administrative weaknesses, provided they are not sufficiently outstanding to wreck the organization. The techniques of administration employed by governmental and private organizations are much the same. The conditions under which the public administrator works, however, differ markedly from those of the private executive. The man in governmental service must always strive to put the public interest before that of any individual or special-interest group. If he fails to do this, he fails to fulfill his responsibility as a public servant and leaves himself open for public criticism. Moreover, the public administrator is obliged to act within the laws and regulations the public sets. This may cause him at times to appear indecisive or lacking in initiative. But what he does must be well considered and capable of withstanding detailed public scrutiny. Administrators in private enterprise generally are more free to act than the public servant. They are answerable for their actions only to the head of the organization—in small enterprises, possibly to but one individual, or in large corporations, to a board of directors. Moreover, the private administrator is primarily concerned with advancing the welfare of his organization and weighs his actions in that light. Men in private enterprise are, of course, conscious of the public interest, but they make their decisions without undue fear of being rebuked by individuals outside of their organization.

The Nature of Administrative Work

The American public enjoys caricaturing persons in high offices. In this way those otherwise unapproachable may have their own shortcomings revealed. But caricatures of busy administrators are likely to be far from typical or complete. For example, there is the administrator caricatured as a grown-up playboy spending his time on yacht cruises, theatre parties, golf, and entertaining important customers. At the opposite extreme is the irreproachable man who sits behind a huge desk at the end of a long room and glowers ominously at those who dare encroach on his sacred time. Then there is the bossy type who calls in his lieutenants, one by one, hears their reports, arbitrarily reverses their decisions, and dismisses them curtly. And finally, there is the high-powered superexecutive whose office desk is piled high with important papers, whose telephone is constantly jingling with long-distance calls, yet who decides affairs of great moment with machine-gun rapidity.

An administrator may enjoy expensive pleasures and pleasant companions, and his time is certainly limited. He does have the task of reviewing and sometimes of correcting the work of his subordinates. Oftentimes he works under heavy pressure. But instead of being aloof, he is a friendly man who makes visitors feel at ease, listens intently to what they have to say, lets them know that he considers their opinion important, and points out to them others' points of view that he must consider before reaching a decision. He makes sure he hears all sides and appears unhurried in reaching decisions so that everyone will consider him judicious, fair, and considerate of their viewpoint. He may make few decisions himself, and these mostly only after having all evidence brought to his attention through personal investigation and competent staff work. As many decisions as possible are delegated to those near the scene of action. The administrator is the one who sets broad policies and objectives, who delegates authority to others, and backs them up in its use. He recruits personnel, divides territories and responsibilities, and sets working standards that others can follow.

The Functions of Administrators

An administrator is a person who is responsible for the efforts of others, both as to policy and practice, and exercises authority in seeing that decisions are carried out.

The man in an administrative post is not primarily engaged in doing the detailed work of his organization, although he functions prominently in developing over-all plans and in setting up general policies. The administrator is chiefly concerned with providing leadership and with doing the specialized work associated with maintaining the organization in operation.

Luther Gulick [2] defines the duties of an administrator as including these functions:

1. Planning the work to be done and how to do it.
2. Recruiting personnel, training them, and establishing favorable working conditions.
3. Organizing the functions and defining the responsibilities of various subdivisions and the channels through which their activities are to be reported and coordinated.
4. Budgeting funds for the several divisions and activities and maintaining fiscal control.
5. Coordinating the various activities and divisions for smooth and effective operation.
6. Reporting to those to whom he is responsible on the work of the entire organization and maintaining free two-way communication between himself and his subordinates through suggestions, records, inspection, and research. He also interprets his organization to his employees and the general public.
7. Directing the work by making decisions and transmitting these through directives and instructions. Serving as leader of the organization.

The extent to which an administrator performs each of the major duties depends, of course, entirely upon the size and complexity of his organization. In small private enterprises the owner usually acts as the policy maker and performs the bulk, if not all, of the administrative functions himself. In large enterprises the above functions are organized in separate departments.

Perhaps a clearer picture may be gained by describing the functions of a high administrative officer, the Chief of the Forest Service, and those of a typical foreman in industry. The average forester is likely to find his responsibilities lying somewhere in between these two levels of administration.

Functions of the Chief of the U.S. Forest Service. PLANNING. The Forest Service is the government's chief agency for forestry. It therefore is responsible for advising the President and Congress on all forestry questions as well as for performing those tasks specifically authorized by Congress. This position is analogous to that of the defense agencies. Not only must they be prepared to repel invasion, but they must advise the President on military situations throughout the world that threaten the security of the United States.

The Chief directs the Forest Service to make periodic surveys of our country's needs for forest products and services and of the capacity of the nation's forests to meet these needs. He represents our country in international organizations dealing with forestry. He sees to it that our government is accurately informed on world forest resources and problems. He causes studies to be made and proposals to be drawn up for action by the President and Congress to safeguard our country's forest needs.

In legislative as well as administrative actions the desires of one group of forest users may sometimes run counter to that of another or to the general national welfare. It thereby becomes his responsibility to reconcile conflicting interests, if possible, and when impossible, to propose fair adjustments. He may not knowingly sacrifice the national good under special-interest pressure without violating the trust imposed in him.

It is his responsibility to make sure that all plans and suggestions from congressmen, the military establishments, other government agencies, the general public, and members of his own organization are carefully reviewed and appropriate actions taken.

The Chief must also plan the major functions and special projects of the Service. In an organization as large as the Forest Service the Chief's function is not so much to originate proposals as to decide which ones among the many presented to him are to be adopted. Detailed study of proposals may be delegated to others. In the end, after appropriate staff consideration, decision is his responsibility.

STAFFING. The Chief is responsible for final decisions on all major appointments in the Service. He appoints the assistant chief, the regional foresters, and the experiment-station directors as well as other key personnel. He establishes over-all working standards and conditions. He provides for training forest officers. He directs that appropriate actions be taken to discover young men of talent within the organization and that special effort be made to broaden their training and experience.

ORGANIZING. Organization becomes a major task in a highly decentralized agency such as the Forest Service. The Chief is responsible for the regional and functional distribution of activities. This is not a task done once for all. As new responsibilities arise, he has the constant task of channeling these to the divisions he deems most competent for each assignment. Although established channels are used when appropriate, he does not hesitate to go outside of these when effective operation indicates the necessity for so doing.

BUDGETING. Annually the President indicates to the Director of the Bureau of Budget the total expenditures he is prepared to recommend to Congress and the new activities he wishes financed. The director provisionally allocates this total fund among departments and independent agencies. The Secretary of Agriculture makes a similar allocation to bureaus. The Chief requests estimates from the regional foresters, experiment-station directors, and staff agencies of the costs of operation for the next fiscal year. All proposals for new expenditures must be carefully justified. Since this process begins about 14 months before the fiscal year opens, advanced budget planning must be made for 26 months ahead.

The proposals of the several directors and regional foresters are scrutinized carefully by the assistant chiefs. New activities lacking strong justification or timeliness are eliminated. Finally the strongest possible budget statement is drawn up for consideration by the Secretary. In due time the Chief is called upon to defend his budget before officers of the Secretary. After department eliminations and approval it is then passed on to the Bureau of the Budget. Again the Chief is called on for explanation and defense of his proposals. Further eliminations are made. The final recommendations of the budget director, if approved by the President, become a part of the President's budget and as such must remain strictly confidential until presented to the Congress early in January of each year.

The Forest Service budget is assigned to the subcommittee on agriculture of the House of Representatives Committee on Appropriations. It is through control over authorizations and expenditures that Congress exerts powerful influence over the government administration. The Chief and his assistant chiefs are called before the subcommittee for hearings that may last several days. It is common practice for individual committee members to inquire into minute details of proposed expenditures, especially those for new items. The fact that a previous Congress has authorized a given activity is no guarantee that appropriations adequate to finance it will be made. Other congressmen, individual citizens, and representatives of organizations may appear before appropriation subcommittees to argue for or against a given appropriation. Support from outside often determines whether or not a particular new activity will be started or an existing one continued.

After final vote in the subcommittee the budget goes to the full committee and then to the floor of the House. Here further cuts may be made or new items inserted. When passed by the House of Representatives, the appropriation bill goes to the Senate, where committee hearings are repeated. What occurs in House and Senate hearings to a large extent determines the appropriations and hence the effectiveness of the Forest Service. This matter is one of the Chief's heaviest responsibilities.

Eventually the Senate also passes the bill, conferees iron out differences between Senate and House versions, and the bill is finally approved by both Houses. Following signature by the President, the funds are quickly allocated to departments and bureaus and within the Forest Service to the various stations and regions. Usually this is done within two weeks after the bill passes in final form. The Chief has thereby given his approval and committed government funds for on-the-ground operations. He still has the further task of reviewing expenditures for conformity to allocations. This is carried out by staff officers who refer questionable cases to him.

The budget is further complicated by grants-in-aid to states for cooperative fire control and forest management.

The budgeting functions of the Chief are covered in some detail, though by no means exhaustively, because they require great effort, grasp of immense detail, and the utmost in patience and persuasive powers to convince an economy-minded congressman that taxpayers' dollars will yield returns commensurate with costs.

COORDINATING. Distributing activities, especially new ones, is a constant function of the Chief. Keeping the several activities in proper balance and appropriately interrelated is another. This is particularly difficult where decentralization as practiced in the Forest Service exists. Coordination is affected by having clear-cut, carefully stated objectives.

The most effective tool is the annual conference of directors and regional foresters in which policies and programs are forged by thorough discussion. Only after such discussions are major policies and programs adopted. Integrated inspections are another tool of coordination. High officers, usually an assistant chief and a top-ranking field officer, inspect all operations within a region, covering administration, research, and cooperative activities with states and other agencies. The chief gives careful consideration to their reports and takes action where needed.

REPORTING. Annually each director and regional forester submits reports to the Chief covering operations during the past year and plans for the following year. Many special reports are also submitted. The chief in turn prepares his official report to the Secretary of Agriculture. He also reports to Congress as a part of the hearings on budgets.

FIG. 112. Forest and range administrators testifying before a congressional committee. (*Bureau of Land Management photo.*)

In recent years Congress has tended to lay down definite standards of legislative policy and to leave the detailed application of these standards to the administrative agencies. Consequently, the Chief and his staff take great pains to maintain good relationships with the Congress and to suggest desirable changes in forest policy as expressed in existing law.

The Chief has the further task of directing the flow of information from the Service to the public. He personally addresses important gatherings and may conduct or participate in hearings on controversial subjects. He is responsible for public expressions of other forest officers which deal with Service policy. He directs the general flow of information from the Service to the public through signs, folders, pamphlets, bulletins, and professional journals. Obviously this can be done only through delegating much responsibility to others and establishing clear directives to guide their actions (Fig. 112).

DIRECTING. Day-to-day decisions covering general operation are made largely by field and staff officers. Only strictly top-level decisions reach the Chief for action. These involve mainly items already discussed, top personnel, policy, public relations, Service morale, and questions concerning the Secretary, President, and Congress.

The Chief Forester's influence over the personnel as a whole is exercised not so much through orders and directives, important as these are, as through the tone he sets for the organization. By seeing as many officers and other workers as possible both in Washington and in the field, he convinces them of their importance, of his interest in their welfare, and of the broad objectives of the Service. Through timely Service messages, he keeps them informed and knits the organization together. By his own example of friendliness, consideration, fairness, and industry, he builds devotion to the Service and *esprit de corps.* When unpopular decisions must be made, he explains the reason for them.

Most important of all, he directs through well-understood objectives, by maintaining a free flow of ideas up and down the administrative channels, and by delegating responsibility for decision and action as fully as possible to the man in the field. In large private organizations the duties of top-level executives are not unlike those of high administrators in the government service. They too are concerned with planning, organizing, staffing, directing, coordinating, reporting, budgeting, as well as with maintaining good relationships for their organization with the public.

The Chief Forester, assistant chief, regional forester, national forest supervisor, and ranger are all concerned with administration, that is, with carrying out the seven functions described. Important objectives of administration are to get trees grown and harvested, fires prevented and extinguished, range protected and utilized, and roads built and maintained. The last men in the administrative hierarchy are the assistant ranger and foreman.

Duties of a Foreman in Industry. Riegel generalizes the duties of a typical foreman in industry as follows: [4]

I. To administer his own department
    1. To plan and schedule operations in his department in cooperation with any central planning unit
    2. To execute plans
        a. To organize operations in terms of personnel, materials, and machines
        b. To direct operations to attain standards of quality, quantity, economy, and safety
        c. To keep department clean, orderly, and safe
    3. To develop subordinates
    4. To improve the efficiency of operations by the development of methods and processes
    5. To develop and maintain desirable personnel relations

II. To cooperate with superiors, associates, and other departments in furthering general aims of the enterprise

III. To represent his management to his employees and his employees to his management

IV. To increase his fitness to meet present and new situations by improving his abilities and personality traits

The assistant rangers and foremen on national forests have analogous responsibilities. An excellent discussion of supervising work in forestry is given by McCulloch.[3]

Aptitudes for Success in Administration

In "The Art of Leadership" Ordway Tead [6] has listed the most important qualities demanded of the administrator as (1) abundant physical and nervous energy, (2) a sense of purpose and direction, (3) enthusiasm, (4) friendliness and affection, (5) integrity, (6) technical mastery, (7) decisiveness, (8) intelligence, (9) teaching skill, and (10) faith. Howard Smith in his book "Developing Your Executive Ability" [5] states that the dominant urges of the good executive must be (1) to deal with his fellow men on the basis of concern for their welfare as well as his own and (2) to climb to leadership on the basis of demonstrated worth and with the approval and respect of his fellows. The necessary administrative qualities are consequently those of character, personality, and ability.

Mason [1] in an attempt to single out traits of good administrators compared successful executives with successful salesmen, production men, and technical men. Unclassified college students were used as a control group. No significant differences were found among the five groups in these abilities: (1) reasoning in arithmetic, (2) judgment in estimating, (3) symbolic relationships, or (4) word comparisons, though all these abilities must be exercised by administration.

Mason next selected 33 executive traits as a basis for comparison. He grouped these into four categories for convenience and to obtain greater objectivity. The categories used were called (1) health and drive, (2) judgment of fact, (3) reaction to human qualities, and (4) leadership. The scores on this test are shown below:

	College students	Sales group	Executive group	Technical group
Health and drive	0	−4	+16	+2
Judgment of fact	0	−3	+10	+4
Reaction to human qualities	0	+6	+10	−2
Leadership	0	+1	+14	−6

The administrative group received markedly higher scores in each category than any of the other groups. Obviously, the executive is a well-rounded individual. Many traits and abilities are important for success.

No wholly reliable test for executive ability has yet been devised. A successful administrator must draw the best from the personnel in his organization. He must at the same time give each person a sense of personal accomplishment. He must have the ability to make an organization more than the sum of its parts and to make the teamwork of the group far more effective than the separate accomplishments of the individual workmen.

Not All Men Should Aspire to Become Administrators. Many men will find their greatest usefulness, and therein their greatest satisfaction, by serving in staff positions with responsibility for the development of certain technical phases of forestry. Others will find satisfying achievement in subordinate administrative positions, finding outlet for extra service in professional societies and institutions and in general publicity activities directed toward sponsoring improved management of forest resources. Those whose energy is limited or who dislike working under pressure and assuming heavy responsibility under trying conditions will find administration highly irksome.

Even men who aspire to leadership should realize that administrative posts are few and competition keen. Many fortuitous circumstances may arise to stop their career short of their life's ambition. They should, therefore, be willing to adjust their lives to whatever level of administrative activities they can achieve and not consider themselves failures and become embittered if they fall short of the goal they set for themselves. In both public and private careers men in advanced administrative positions may be called upon at times to take a courageous position which may later on, however, disqualify them for advancement. Such occurrences are most frequent in the field of politics, but they occur in private business and public administration as well.

Training Required

The student who wishes to prepare himself for forest administration may select the curriculum which includes the field of his primary interest. Attention should be paid to the obvious fact that graduates from certain curriculums find employment with larger groups of men than do others. As the number of men doing the same kind of work within an organization increases, the need for administrative direction increases. It is generally true, however, that a considerable degree of success in any field will bring with it administrative opportunity and responsibilities.

The election of courses in history, policy, administration, public administration, business administration, economics, and psychology is desirable in preparation for administrative work. Also, participation in campus extracurricular activities is helpful. Holding class offices, serving on committees, participating in the work of local churches, and being active in social and professional societies provide useful experience. The contacts formed are valuable in developing poise and self-confidence. Moreover, they offer an opportunity to listen to varying points of view, to weigh evidence, to suggest logical conclusions, and to carry out activities that build up the qualities of leadership.

Work in voluntary organizations has the special training value of requiring the individual who directs such work to elicit wholehearted cooperation without the possibility of exercising disciplinary control. It is characteristic of American institutions that activities are directed with little use of disciplinary measures. When such measures become necessary, they are, to some extent, a reflection of breakdown in administration as well as breakdown in function of the subordinate. The skillful administrator anticipates weaknesses and corrects them long before disciplinary measures need to be taken.

Many men receive valuable administrative training as commissioned or noncommissioned officers in the armed forces.

Graduate study is of increasing importance in providing specialized training for men who have demonstrated their aptitude for administrative work. In the past most administrators have been selected through an unplanned struggle, on the assumption that the best men would gradually thrust their way upward. As a result sheer aggressiveness has often been weighted more heavily than other qualities of leadership demanded by executive duties. Much of our labor unrest in America can be attributed to the fact that both labor unions and many industries have promoted men to leadership for their ability to get things done rather than for their skill in diplomacy, their broad understanding of objectives and policy, their genuine friendliness and interest in their fellow workers, and their willingness to adopt a judicious attitude toward points of view in conflict with their own. Struggles for administrative leadership have also brought forth many talented administrators, but in the process young men have wasted considerable time learning skills that could have been acquired more rapidly through systematic study. Personality traits can be evaluated, intelligence recognized, and, to some extent, talent for leadership predicted. When men selected for these basic qualities are given graduate training in administrative functions and activities, they normally advance rapidly toward important executive posts.

Many employers provide training in administration on the job or in special classes. A number of colleges are now giving short courses in administration. On-the-job training is frequently provided by special assignments for short periods of time or by frequent transfers. Most employers believe that young men who show promise as administrators should be given opportunity to develop as rapidly as possible. They assign such men to positions of increasing responsibility as the men demonstrate that they are ready for them. To the man impatient to get ahead, such training programs may seem onerous, but short cuts have their pitfalls. Political preferment, nepotism, favoritism because of fraternal affiliations, common alma mater, personal liking, and other reasons exist to a limited extent perhaps in all types of organizations and particularly in small ones. Men whose position depends on such support may find their tenure fleeting.

All forms of favoritism lower employee morale. Men will not do their best work unless assured of fair treatment and opportunity to advance. The administrator who holds his position as a result of favoritism therefore has a doubly difficult task in maintaining the satisfactory performance of his organization.

Career Ladders

Few men will become important forest administrators soon after finishing college. For the first few years technical competence is of major importance. After obtaining practical experience, some men will find themselves gradually working into positions of more and more administrative responsibility; others will either fail to advance or develop particular ability in a specialized field of work.

It is not possible to describe the jobs taken by young men interested in forest administration, because any job that is satisfactory in other respects will serve as a starting point. The rapidity of advancement and the speed with which administrative opportunity will come depend on the ability of the man and on the opportunities within the organization.

Within each organization there is a theoretical career ladder, that is, steps through which a man is expected to pass as he advances from one level of administrative responsibility to another.

Simplified career ladders are shown in Figs. 113 and 114 for a private and a public forestry organization. Men may follow separate career ladders until they reach a certain plateau where both are performing primarily high administrative functions. At this level they may interchange posts with a minimum of difficulty. It is not uncommon for men who have attained high administrative leadership in the U.S. Forest Service to transfer to similar posts of high authority in business, in association work, and in other government agencies. The many facets of a forester's job demand the type of thinking and acting that is required of administrators.

Opportunities for advancement in administrative work are greater in organizations that are growing rapidly, in organizations that handle a wide variety of tasks rather than routine work, and those that are concerned primarily with action programs rather than with research or plan-

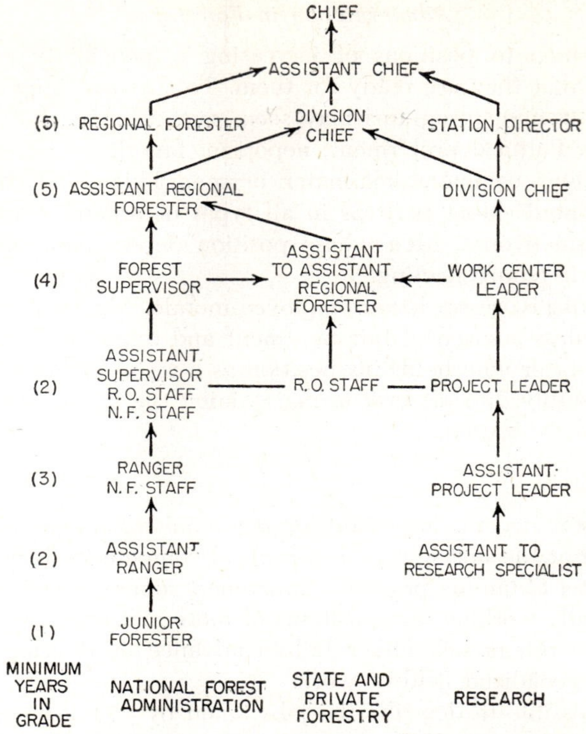

Fig. 113. Career ladders in the U.S. Forest Service (simplified).

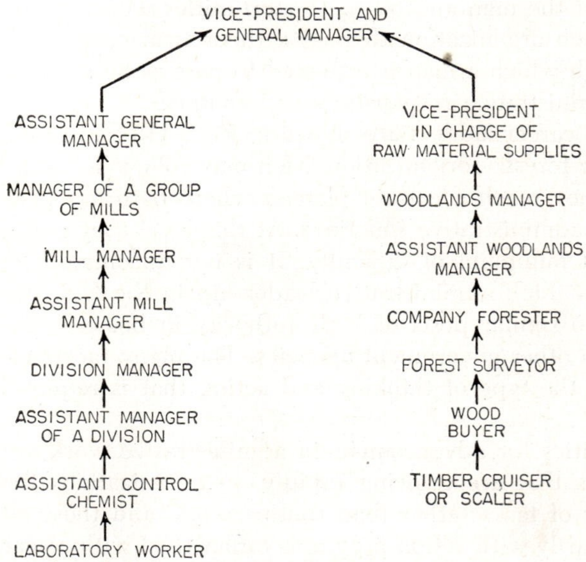

Fig. 114. Career ladder in a pulp and paper company (simplified).

ning. A man may develop his administrative skill by shifting from one employer to another among small organizations. It is advisable to move to another employer soon after he has reached his ceiling within any one company. Another method is to seek employment with a large organization and gradually work up within it.

Young men desiring to advance in administrative work should follow recognized rules of effective conduct. Loyalty to the organization and to superiors and subordinates is indispensable to advancement. Friendly consideration for coworkers and readiness to accept responsibility and to discharge it fully will gain a reputation with superiors. Participation in nonofficial tasks, if important, helps to develop leadership. Performing extra jobs the superiors will appreciate attracts favorable attention.

Helpful Rules for Administrative and Supervisory Work

As a man works in the field of administration he develops and is guided by general rules. Among rules of good administration are the following:

Have clear-cut objectives to which all workers can wholeheartedly subscribe.

Interpret these objectives in terms of the job of every man in the organization.

Have clear-cut lines of responsibility and delegate authority commensurate with responsibility.

Except in emergencies, never act without facts, even though these may require time to collect.

Anticipate contingencies by keeping constantly alert to needs and morale of organization.

Dispel unfounded rumors by tracing them to their source.

Be accessible to everyone in the organization who feels that he must see you personally, but do not encourage men to bring things directly to the top that should go through channels.

Be clear and explicit in instructions and patient to answer questions.

Never criticize a man before others but offer praise before others.

When criticism or correction is needed, do it promptly, but first put the man at his ease and praise his good points.

Most men want to please their superiors. Rare is the individual who deliberately causes mischief. The latter should be promptly taught to mend his ways or be discharged.

Constant carping, criticism, and faultfinding will have one of two effects: break the employee's self-confidence and morale; cause him to become indifferent because nothing he does pleases.

Be genuinely concerned about the personal welfare of the workers and show it.

Through conferences and otherwise, encourage free discussion of future plans and policies; but once these are decided, pursue them with vigor.

Keep physically and mentally alert. When ill, stay at home. Take necessary vacations.

Have systematic measures for checking performance and systematic methods for keeping informed on the performance of your entire organization. Use staff conferences to build teamwork and to encourage reliable staff work.

Have confidence and faith in yourself, your staff and organization.

Appoint some competent assistant to act for you when you are away; check his work upon return, praising where deserved, suggesting better courses of action where he made mistakes.

Frankly and promptly admit your own mistakes. This will encourage others to do likewise rather than to cover up mistakes or shift blame. Nothing is more destructive to morale.

Have a definite training program for each new employee and for each man eligible for promotion.

Make it a practice at least once a year to review each man's performance and to discuss with him his future career in the organization. If the door to his advancement is closed because all higher posts are filled by young men or because he has reached his own ceiling, frankly tell him so. Do not try to hold a man in your organization provided brighter opportunities exist elsewhere. He will be discontented.

Avoid gossip. Hold in strict confidence such of the essentials concerning the organization and its employees as deserve to be kept confidential. Neither seek nor repel the personal confidences of associates and employees, but when they are offered, hold them inviolate.

Become a clear writer and an effective public speaker (Fig. 115).

FIG. 115. Foresters in advanced positions give frequent public addresses. Michigan Conservation Department. (*A.F.P.I. photo.*)

Give careful thought and effort to building up and maintaining favorable public relations and public attitude toward your organization. Appropriate news releases and speeches are far more effective than advertising. Convince the public of the sincerity of your desire to serve them.

Place service to the objectives of your organization ahead of profits or salary, but recognize that operations must be fiscally sound, compensation fair, and, if possible, generous.

Always be generous in giving credit to others.

Organize the work so that a minimum of routine or detailed matters come to the top for decision. Plan constantly for improving the organization and its performance.

Men who cannot or will not accept and carry heavy responsibilities are unfitted for forest administration. However, for those who are willing to devote themselves to the objectives for which their organization is working and who are temperamentally fitted to lead and to carry the load, it is the field of endeavor which offers largest monetary compensation and rich opportunity for worth-while service.

LITERATURE CITED

1. Cleeton, Glenn W., and Charles W. Mason. 1946. "Executive Ability, Its Discovery and Development," Antioch Press, Yellow Springs, Ohio. 540 pp.
2. Gulick, Luther. 1937. Notes on the Theory of Organization, "Papers on the Science of Administration," Institute of Public Administration, Columbia University, New York. 195 pp.
3. McCulloch, Walter F. 1950. "The Forester on the Job," Oregon State College Cooperative Association, Corvallis, Ore. 105 pp.
4. Riegel, John W. 1941. "The Selection and Development of Prospective Foremen," Bulletin 11, Bureau of Industrial Relations, University of Michigan, Ann Arbor, Mich. 69 pp.
5. Smith, Howard. 1946. "Developing Your Executive Ability," McGraw-Hill Book Company, Inc., New York. 225 pp.
6. Tead, Ordway. 1935. "The Art of Leadership," McGraw-Hill Book Company, Inc., New York. 308 pp.

CHAPTER 26. *Trends and Outlook in Forestry*

IN THIS CHAPTER the author gives his own interpretation of trends and outlook in forestry. Obviously this is largely a personal view and should be considered as such. National and international events not now foreseen can completely upset the predictions made. Moreover, the forecast is limited by the author's own knowledge of events and his skill in interpreting them.

National and international policies influencing the economy of the United States had by 1951 clarified greatly the uncertainty that prevailed immediately at the end of the Second World War. The anticipated postwar depression had failed to materialize in substantial form. The action of Communists in eastern Europe convinced the Western world that military and economic imperialism had not died with the Nazi leaders. The rise of Communists to power in China with their subsequent aggressive action in Korea, Tibet, and Indo-China convinced the world that aid to tottering and unpopular governments would neither maintain peace nor the *status quo*. Promises, however fleeting, to lighten the burden of oppressed and poverty-ridden people had overpowering appeal. The starving man makes quick choice between promises of food and freedom.

The United States was forced unwillingly to choose between assuming a dominant role in world leadership backed by adequate military and economic force and surrendering this leadership to communism together with the freedom of political and economic choice this entailed.

Our country has chosen to lead rather than surrender.

The outcome of this decision is gradually becoming clear to our people. Industrial strength is the basis for military strength and for economic power to help other peoples. The United States, powerful as we may become, can never become the major producer that supplies goods without recompense to all underprivileged peoples and nations. Only by helping other nations to develop their own physical, human, and spiritual resources can we develop freedom-loving allies with the hope, faith, and courage to resist aggression.

If the United States is to succeed in its role of world leadership, we must have a vigorous economy at home, and we must take necessary steps to aid other nations to improve human economic productivity.

Natural resources, particularly those that are renewable, form the base for wealth. Skill in developing and using these resources is the base for increasing worker output.

For more than half the world's people, land is pitifully scarce. As living comes from the land, effective use of land becomes an important feature in our country's foreign and domestic policy. The programs we sponsor in other lands through the Food and Agriculture Organization of the United Nations, the Economic Cooperation Administration, and other Point 4 programs deal with improved agriculture and forestry as well as industrial output.

For the peoples of the world living above the subsistence level depends upon wise use of all land resources. In many regions, of which Israel is an example, land rehabilitation must go hand in hand with more intensive land use.

Future Need for Forest Products

The stimulus to general economic activity that arose from consumer demands unfilled during the war period kept total production at a high level from 1946 to 1950. The outbreak of war in Korea in 1950 thrust a huge military program on top of an already heavily burdened industrial economy. Expansion of productive plant became essential. Since no early decline in military tension can be foreseen, high industrial activity with its demands for new plants, new homes for workers, and new products seems assured for at least a decade.

Lumber production apparently will remain between 35 and 40 billion board feet annually. Already our country is importing Paraná pine from Brazil, prefabricated houses from Finland, and birch lumber and veneer from Sweden. In part, this is a result of need for dollar exchange in these countries and, in part, of the high costs of domestic production.

Need for pulpwood, veneer, plywood, piling, and poles is expected to continue increasing for the next two decades. Demand for hewn ties, cooperage, and fuel wood probably will decline. Other wood products, such as fence posts, mine timbers, and distillation wood, are likely to show little increase from present consumption.

As high industrial output for military and civilian use places increasingly heavy demands on iron, nonferrous metals, and other exhaustible resources, wood, paper, and cellulose-based plastics seem destined to grow in use.

Over-all decline in demand for forest products for domestic use seems improbable (Fig. 116).

Export of forest products to Europe, Japan, and other lands probably will continue, but as programs to stimulate the development of resources in other lands becomes effective, more of the world's forest resources will be put to use. This may mean increased imports by our country and decreased export markets. A growing international trade in forest products would be a natural outcome of improved world economy.

Future use of lumber, veneer, and other forest products as well as the position American wood products occupy in international trade will depend upon their cost and utility to the ultimate consumer. Vigorous effort will be required to halt the rapid increase in prices of lumber in comparison with other prices. The cost of lumber in use, that is, in houses, furniture, and other consumers' products especially, must be competitive with other materials.[3] Lumber and other forest products

FIG. 116. Good truck roads are required for heavy log loads. Edward Hines Lumber Company operation. (*A.F.P.I. photo.*)

have been widely used in the past because they were cheap, and their future usefulness will depend in no small measure on a competitive price. Efforts in forestry should be directed toward low price of product.

Water Use

As cities grow, industry expands, and agriculture spreads to arid lands, water becomes an increasingly important commodity. No longer can abundant supplies for domestic and industrial use be taken for granted.[2] Conflicting uses become questions for serious concern. Water supplies determine use of land and use of land in turn affects water supply. The nation is showing increasing interest in river-basin planning for most effective over-all use of land and water resources. This is no new development but has been under study for decades. River-basin planning committees have recently been appointed by the President. Their purpose is set forth in the following excerpt from Mr. Truman's directive establishing such an agency for New England and New York:

> The comprehensive study of land and water resources of this area should include, among other matters, coverage of electric power generation and transmission, forest management, fish and wildlife conservation, flood control, mineral development, municipal and industrial water supply, navigation, pollution control, recreation, and soil conservation. The necessary first step in such a study is an inventory of the land, water, and all of the related natural resources available for utilization, together with a survey of the projected regional and national requirements which might be met through more effective utilization of the natural resources of the region. When these basic facts on resources and needs have been collected and analyzed, the committee should then proceed to determine what development and conservation projects are feasible and desirable, and to prepare recommendations for specific action to carry them out.*

Forests have important influence on interception of rain and snow, infiltration of water into soil, use of water from the soil, protection of soil from erosion, detention of water during periods of heavy rainfall, accumulation of snow, snow melt, and other factors influencing water yields and flood crests.[1] These influences increase in human significance with increased importance of water and land.

Forests for Recreation and Wildlife

Increasing urbanization makes a growing percentage of our people dependent upon playgrounds, parks, and public and private forests for recreation. Both public and private recreational developments will therefore be needed. Integrated use of public and private lands is likely to

* Quoted from the Annual Report of the Northeastern Forest Experiment Station, 1950, U.S. Department of Agriculture, Forest Service, Upper Darby, Pa. 102 pp.

Fig. 117. The forest in winter has especial appeal to hardy recreationists. Green Mountain National Forest. (*U.S. Forest Service photo.*)

develop following the pattern of trails along which public shelters and camping facilities are provided. Examples are the canoe routes of the Superior National Forest and the Adirondacks, the Appalachian Trail, Long Trail, and Horseshoe Trail. Other systems are features of many national and state forests and parks (Fig. 117).

Wildlife is an important recreational feature of forest and park lands. Because of money collected through taxes on sporting arms and ammunition and on fishing gear, efforts to improve conditions for wildlife are increasing in importance.

Economic Use of Land

Counties, communities, and states are becoming increasingly concerned with the proper use of land. Improper use may place people in remote land areas where they must be provided public services at heavy cost and yet have only a meager livelihood because of the limited resources available on the land. A proper classification of lands such as has been carried out in Wisconsin, New York, and other states is the foundation upon which public-land policy and zoning of land use can be established. Zoning may not become widespread as a public measure; nevertheless, it seems that the public must exercise some limitations on the extent to which it will go in providing services to dwellers in remote areas.

Equally important is concern that lands be well used. The livelihood of people in any community depends to a large extent upon the resources available for their use. If these resources are mainly forest lands that have been badly depleted of growing stock, the people earn a poor living. Rehabilitation of forests is an essential to raising general standards of living in the forested regions.

Role of the Federal Government

The Federal government may be expected to continue its important work of land management on national forests, national parks, public grazing districts, public-domain lands, Indian lands, national wildlife refuges, and other public lands. Some increase in these lands may occur. Greater coordination in management seems destined soon to be effected. Widespread expansion of national forests appears unlikely for the next decade or more.

As timber from more accessible areas is depleted, products from national forests will become increasingly important.

Federal forestry activities will be focused especially on the over-all national and international programs. National goals and objectives are the proper sphere for the Federal government. Broad programs designed to meet these goals will be worked out with detailed administration frequently channeled through state foresters.

Water use, because it involves navigation and often affects more than one state, will get renewed attention by the Federal government, with strong effort made to coordinate state and local efforts in unified land and water management.

Research in forestry also seems destined to remain an important Federal function, though research by private and other agencies is growing rapidly. The Federal government will be obliged to strengthen its research on those broad economic problems that affect national and international policy, regional economy, and land use. It will continue to play the leading role in research on forest influences. Forest, range, wildlife and park-management research will require continued Federal support, but state and school activities in these fields are growing in importance.

Role of State Agencies

The states are now recognized as the main action agencies in forest-fire control and cooperative forest management on private lands. Some are assuming responsibility for services to loggers, lumbermen, and wood processors. Several states engage in control of destructive forest insects and disease organisms.

State forestry organizations are assuming responsibility for the major forestry action programs dealing with private owners. Wildlife-management and research programs likewise channel through state departments.

Expansion of state parks, game refuges, and public shooting areas probably will occur.

Actual forest-land management by states can be expected to increase, but not rapidly. Neither state nor Federal agencies are likely to engage in extensive land-purchase programs when private purchasers stand ready to buy all forest land offered for sale.

Research by state-supported and other colleges of forestry is growing rapidly in importance. Education in forestry is everyone's business with the colleges of forestry taking complete responsibility for professional training. The colleges share with state conservation departments, extension foresters, Federal agencies, and industry-supported agencies the broad task of informing the general public on forest conservation.

Role of County and Local Governments

County, city, and town governments have major responsibility for local parks and the special responsibility of providing water supplies. Many own and manage watersheds on which they maintain forest cover. Watershed management will become increasingly important to local governments.

County, community, and school forests may also be expected to expand, but it remains questionable how far such expansion may go. If well managed, such properties have high value for local recreation, education in conservation, and demonstration. As the cooperative forest management and comparable state programs expand, public foresters with local experience become available for expert service to small community forests.

Public Programs for Private Landowners

Public aids have gradually been extended to private timberland owners. These have taken the form of special tax measures, education, technical assistance in planning, management, and marketing, forest research, distribution of tree planting stock, aids in tree planting, benefit payments for forestry measures, control of fires, insects, and diseases, and extending public credit to timberland owners. Public aids to private owners may be expected to expand in the years ahead. The tax problem is by no means settled. Inheritance taxes as now administered can force the liquidation of a managed forest property, thereby destroying much of its value as a permanent timber-producing area. Property taxes can be so high as to take all the income the land produces in timber growth. Various social-security taxes sometimes weigh heavily on the forest landowner. In New Hampshire a yield tax has already been substituted for the general property tax, and special rebates are provided to those who practice good forestry on their lands.

Public aids, though now substantial, still fall short of providing financial aid to the owner of badly depleted forest property which will require investment of capital over a period of 20 or more years before income therefrom can be realized. If such lands are to remain in private ownership, substantial additional public aids appear to be needed. Furthermore, many of these properties are so small in size and widely scattered that public ownership would be uneconomic. Public aid seems appropriate for owners of such properties who are willing to build up the forest capital but can now scarcely expect to realize any personal benefits from so doing.

Regulation of Private Forests

Already some 15 states have adopted laws to regulate private forestry. Some of these are merely educational laws carrying no penalties. Others require the licensing of timber operators and the meeting of certain minimum cutting-practice standards on their operations.

Extension of regulation laws by the several states and strengthening of these laws to make them more effective seem assured. As public aids are extended to private timberland owners, it is only reasonable that the public require minimum practices of private owners.

Regulation, however, is only one measure toward improving forest production in the United States. A more important task is to improve the level of practice on all forest lands. It is not minimum practices that are of greatest importance but standard and average practices. Gradually the public can be expected to concentrate effort to ensure that every forest owner understands and adopts measures leading to high forest productivity on lands economically suited to intensive management.

Role of Private Forestry

From 1940 to 1950 there occurred substantial growth in land holdings by pulp and paper, lumber, and other timber-using industries. Land has been acquired primarily to protect future raw-material supply. Rapid progress has been made in increasing the intensity of forest practice on private lands. This has occurred both on the lands of large timberland holders and on the lands of small holders from whom the large companies buy their raw material. The progress has been neither uniform nor universal but is definitely in the right direction. It is realized that levels of forest practice must vary widely from region to region and even from company to company. What is important is the recognition by companies that they are in the landowning business from now on and that they must grow the timber that they use. This fact has caused a substantial increase in the value of land, particularly in the value of land that supports healthy young growth. Already speculative values have been attached to some forest land beyond the potential yield from such properties. We can expect private timberland owners to employ additional foresters in the management of their lands and to apply forest-management practices that will build up the forest for the future while yielding returns for today.

Many problems remain to be solved. Forest-land ownership as a business still has few precedents to guide it. Little is available from the records of successful operators to determine the size of property needed for successful sustained-yield forestry under various conditions of growth, accessibility, and markets. The extent to which the owner should attempt to carry his timber operations also has not been thoroughly investigated. Generally speaking, those owners who log their own timber realize higher returns in the form of stumpage than those who rely on outside loggers. Frequently, those who carry the timber processing to the form of rough lumber or other primary products realize higher returns than those who sell only logs.

Many private forest-land holdings at present are questionable economic units. They are too broken up by holdings of others, inadequately stocked, or contain too much low-value land to make them potentially productive forest properties for an individual interested in permanent income from forestry. The public in extending aids to private timberland owners will doubtless give consideration in the future to whether or not the owner has, or is in the process of acquiring, a forest property that is economic to manage.

The Lumber Industry. It seems likely that the per capita consumption of lumber in the United States will continue to decline in the years ahead. The decline is affected by the rapid increase in lumber prices and the costs of using lumber on the job.[3] Fortunately, progress has been made in recent years in the mechanization of woods operations that will lead to an increase in output per worker. Similar achievements have occurred in increasing efficiency in use. It seems almost inevitable that increased use will be made of lumber in small sizes and short lengths that can be produced from small and defective logs.

Lumber probably will remain the chief outlet for products from the forest for a number of years to come. It also seems likely that for the indefinite future timber that can produce high-quality saw logs will command a price well above the price it would bring for pulpwood, chemical wood, or other conversion products.

Veneer Industry. We can expect the veneer and plywood industry to continue to develop rapidly. The shorter our timber supply becomes, the more necessary it is that each log be used in the most effective manner possible. Logs of sufficiently high quality to make veneer will increasingly be converted into this product.

Expansion of veneer use for laminated furniture appears probable and will offer new potentialities for design. Research in wood technology aimed to increase the utility of wood can be expected to advance rapidly in the years ahead and to bring into the general market many products in which wood is used in laminated and plywood form. Along with the research in wood technology will come new and better products that can be fabricated from wood.

Pulp and Paper Industry. The outlook in pulp and paper is for continued increase in the use of paper and other wood-fiber products. The utility of paper and paperboard for packaging has caused them to pervade the merchandizing field. High-strength laminated paperboard and paper plastics probably will occupy an important position ultimately in the manufacture of products that require high strength and good electrical- and heat-insulating properties. The manufacture of heavy board such as wallboard will also continue to increase.

In the chemical-pulping field great increase in the use of hardwoods for pulp seems to be the next major step in broadening raw-material supply. Already hardwoods are used widely in the soda and sulfate cooking processes. Hardwood semichemical pulp is just entering into general use. It is this field that is likely to expand most rapidly. The sulfite-pulping process is due for a modification soon to avoid dumping vast quantities of waste liquor into streams. The states of Pennsylvania and New York have already taken vigorous steps to curtail stream pollution and other states may be expected to follow their lead.

Wood Chemistry and Wood-based Plastics. Research in wood chemistry is extending the uses of wood. Solving the secret of lignin use lies still ahead. Already, however, many valuable by-products of lignin have been isolated. It seems highly probable that lignin chemistry will ultimately give rise to new wood-based plastics and other valuable chemical products. The cellulose-plastic field is already well developed, yet it is rapidly growing and seems destined to continue this rapid growth for a number of years to come.

Integrated Industries. Several companies owning timberland have realized the economic advantages of having diversified products. Integration of industry makes it possible to use the various products the forest produces and to use in one mill the products that may be waste in another. Slabs and edgings, if bark-free, can readily be used in the pulp mills. Small-sized waste from one wood-fabricating plant may supply just the material needed for another.

Good use of all the timber produced in the forest and of the small-sized waste products from the various wood-fabricating plants requires planning, development of markets, and a reasonable amount of capital to collect and hold sufficient waste material to justify its further processing. Just how far waste utilization can go is still uncertain. Integration of industry and markets seems to be one of the most fruitful ways in which this burdensome problem can best be solved.

Public-Private Cooperation

Great opportunities for high service lie in developing cooperative programs among industries for integrated use and between industry and government to promote good forestry throughout the land. Trees For Tomorrow, Inc., originally an industry-sponsored program, now enlists the aid of the U.S. Forest Service, University of Wisconsin, Wisconsin Conservation Department, and Wisconsin Department of Public Instruction in a broad program of public education and demonstration.

The American Forest Products Industries sponsors in cooperation with state foresters and schools nationwide education in forestry directed toward forest landowners, timber operators, school children, and the general public. The Tree Farm and Keep Green programs are especially noteworthy.

Less spectacular but perhaps of equal long-term significance are the citizen and industry advisory committees that work with various Federal and state agencies. These bring citizen viewpoints to administrators of national forests, forest experiment stations, grazing districts, public parks, wildlife programs, water use, and public aids to timberland owners.

Other associations and cooperating agencies have a large role to play in leading our country toward wise use of all forest lands. Only through cooperative effort by citizens and government can conflicting uses best be reconciled (Fig. 118).

It is characteristic of the American people to respond with enthusiasm and devotion to the general public good once issues are made clear and all sides presented. Statesmanship is required of Federal and state forest officers to build such public support behind the major forest programs.

Already citizens' organizations such as the American Forestry Association and Friends of the Land have done much to promote conservation. Fear that fully representative bodies or voluntary associations would become dominated by special-interest groups has probably inhibited their growth. It also may have caused public foresters to go slowly in delegating authority to citizens' councils and advisory groups. Growth of such coun-

Fig. 118. Future forestry programs must be developed with young trees and young people. (*U.S. Forest Service photo.*)

cils in strength and authority seems likely in the years ahead. Public foresters should learn how they can be used effectively and the proper sphere for their action. In this way strong democratic support will be gained for our nation's forestry effort.

LITERATURE CITED

1. Frank, Bernard, and Anthony Netboy. 1950. "Water, Land and People," Alfred A. Knopf, Inc., New York. 331 pp.
2. Report of the President's Water Resources Policy Commission, "A Water Policy for the American People (General Report)," Vol. 1, U.S. Government Printing Office, Washington, D.C. 445 pp.
3. Zivnuska, John A. 1951. Future Wood Markets and Forest Management, *Journal of Forestry*, 49:326–330.

Index

A

Acetic acid, 202
Acre, definition of, 110
Activated carbon, 203
Adhesives, animal, vegetable, synthetic, 164
Adirondack Forest Preserve, 22, 36, 460
  use of, 237
Adirondacks, private-estate employment, 375
Administrative work (*see* Employment, in forest administration)
Aerial photographs, use in timber estimating, 115, 116
Africa, 295
Agencies employing foresters (*see* Employment)
Agricultural economist, 284, 285
Agricultural Extension Service, 38, 230, 362, 363
Agriculture land, depletion of, 10, 289
Aircraft, use for fire control, 76, 77
Alabama, 58
  national forests area, 311
Alabama Polytechnic Institute, 419
Alaska, forests of, 66, 67, 313, 315, 366
  national, area, 311
Alaska cedar, 66
Alderfer, R. B., 252*n.*
Allen, E. T., 30
Allen, Shirley W., 22*n.*, 341*n.*, 347, 359*n.*
Allen, Zachariah, 39
Alpine fir, 60, 67
Ambrosia beetles, 80
American Association for the Advancement of Science, 23, 24, 341, 342
American Bankers Association, 322
American elm, 53
American Forest Products Industries, Inc., 30, 41
  cooperation, 466
  programs, 321, 322
  public-relations work with, 425
American Forestry Association, 24, 322, 333*n.*–335*n.*
  conservation promotion, 466, 467
  forest-resources estimate, 297
American Forestry Congress, 1905, 24, 26
American Forests, 11
American Geophysical Union, 342
American Institute of Biological Sciences, 342
American Institute of Park Executives, 247, 400
American Legion, 322
American National Cattlemen's Association, 226, 227
American Paper and Pulp Association, 191*n.*, 192*n.*, 322
American Pulpwood Association, 333, 425, 429
American Society of Mechanical Engineers, 342
American Turpentine Farmers Association, 322
American Viscose Corporation, 67
Anderson, M. L., 337*n.*
Andes forests, 12
Anobiids, 80
Aphids, 78
Appalachian Hardwood Manufacturers Association, 325
Appalachian Trail, 460
Arbor Day, 22
Argentina, 289, 294, 295
Arizona, 61
  national forests area, 311
Arkansas, 59
Arnold Arboretum, 428
Ash, occurrence, 5
  properties, 160
  use, 160, 195
Aspen, growth rate, 48
  occurrence, 46, 47, 60, 67
  use, 48, 51, 196, 224, 282
Assyria forestry program, 13

Atlantic white cedar, 198
Australia, 295, 296
  hardwoods of, 291, 296
Austria, comprehensive forestry ordinance, 14
  forest resources, 291
Ausubo, 68
Avalanches, 93, 253
Ayres, Phillip, 30
Aztecs, forestry measures, 13

B

Bailey, I. W., 437*n.*
Bailey, Vernon, 222
Baker, F. S., 98*n.*
Baker, Hugh P., 30
Baker, Richard St. Barbe, 289*n.*, 293*n.*
Baker, W. J., 162*n.*, 164*n.*, 169*n.*, 196*n.*
Bald cypress, 54
Ballinger, Richard, 28
Ballou, Robert O., 11*n.*
Balsa wood, occurrence, 290
  properties, 158
Balsam fir, 81
  cutting cycle, 118
  occurrence, 44, 46
  use, 48, 206, 383
Bark beetles, 62, 78–80
Barrett Committee Report, 227*n.*
Bartram, John, 427
Bartram, William, 427
Basswood, occurrence, 44
  properties, 160
  seed, 96, 100
  use, 47, 196
Beaver, 89, 224
Bedwell, Jess L., 82*n.*
Beech, growth rate, 48
  occurrence, 5, 44, 46, 52
  use, 195
Beech scale, 78
Belgium, 337
Bennett, Hugh Hammond, 251*n.*
Bigley, Michael, 64*n.*, 393*n.*
Bigtree, 62
Biltmore forest, 39
Biltmore Forest School, 412
Birch, growth rate, 48
  occurrence, 46
Birth rates, declining, 288
Black, S. R., 333*n.*–335*n.*
Black cherry, growth rate, 48
  use, 47
Black gum, 54
Black locust, 53
Black oak, 52
Black spruce, occurrence, 46, 67
  use, 48, 206
Black walnut, occurrence, 52
  use, 53
Blue spruce occurrence, 60
Board foot, definition of, 111
Borers, round-head, 80
Bower, Ray F., 58*n.*, 391*n.*
Boy Scouts of America, 320, 322, 425
Boyce, John Shaw, 84*n.*
Brauns, F. E., 181*n.*
Brazil, 289, 293, 294
  Paraná pine imported from, 457
Brewer, E. G., 82*n.*
Bristlecone pine, 60
Broome County, N.Y., forestry in, 276–279
Brown, A. A., 73*n.*
Brown, H. P., 155*n.*, 157*n.*, 159*n.*
Brown, N. C., 120*n.*, 126*n.*, 134*n.*, 140*n.*, 142*n.*, 143*n.*, 149*n.*, 162*n.*, 166*n.*, 195*n.*–197*n.*, 264*n.*, 269*n.*
Brown-spot needle disease, 74
Brown-tailed moth, 79
Bruce, Donald, 113*n.*
Bryant, Ralph C., 126*n.*
Buckeye, 53
Buell, Jesse H., 26
Building cycles, 266–268
Bur oak, 60
Bureau, of Animal Industry, 428
  of Biological Survey, 364
  of Census, 428
  of Entomology and Plant Quarantine, 356, 361, 362, 394
  of Fisheries, 364
  of Foreign and Domestic Commerce, 428
  of Forestry, 26
    changed to Forest Service, 27
  of Indian Affairs, 356, 366, 367, 398
  of Internal Revenue, 368
  of Land Management, 314
    forest practices in, 323
    grazing work in, 226, 227, 398
      (*See also* Employment)
  of Plant Industry, Soils and Agricultural Engineering, 356, 362
Burma, 295
Business acumen in forestry, 265, 266
Business cycles, 266–268
Butler, Ovid, 24*n.*, 30
Buttrick, P. L., 262*n.*, 269*n.*

C

Caldwell, David H., 13*n.*
California, forests area, 64
national, 311
state-owned, 36, 319
furniture manufacturing, 169
regulation, of private forestry, 335
of timber cutting, 318
shelter-belt planting, 258
Cambium, 154
Camp Manufacturing Company, 40
Canada, forests of, 296
Capper, Arthur, 29
Career ladders, 451–453
Caribbean Research Council, Committee on Forestry, 305
Carpenter ants, 80
Cass Lake, Minn., 281
Ceanothus, 65
Cedar, properties, 160
use, 198
Cedars of Lebanon, 11
Cellophane, 180
Cellulose, chemical nature, 178–181
empirical formula for, 179
properties and derivatives, 180
Cellulose and plastics industries, 193, 194
Central forest formations, 52–54
description, 52
growth rate, 49, 54
uses, 52–54
Central and South America, 293–295
Centre International de Silviculture, 304
Chain, definition of, 110
Champion Paper and Fibre Company, 40
Chapman, Herman H., 113*n.*, 262*n.*
Chapman, Oscar L., 67*n.*, 219*n.*, 220*n.*, 246*n.*, 314*n.*
Charcoal, 202, 203
Chemical debarking, 197
Chemigroundwood, 184
Cherry, 5, 206
Chestnut, occurrence, 52
use, 202
Chestnut oak occurrence, 52
Chicago, 320
Chicle, 208
Chile, forests area, 295
China, 295
forest destruction, 12
timber shortage, 11
Chinard, Gilbert, 18*n.*, 21*n.*
Chippewa national forest, 281, 282
Chips, use of, 199
Christmas tree and greenery industry, commercial value of, 204–206, 385
Cicada, 79
Cinchona, 208, 295
Civil Service Commission, 357
Civil Works Administration, 31
Civilian Conservation Corps, 31, 355, 369
Clapp, Earle H., 33, 335*n.*, 428*n.*
Clawson, Marion, 227*n.*–229*n.*
Clearwater Forest Protective Association, 41
Cleeton, Glenn W., 448*n.*
Cleveland, Grover, 25
Cleveland Cliffs Mining Company, 325
Climate, 2, 258, 259
Climax forest, 5, 104, 105
definition of, 104
Coconut palm, 67
Code of ethics, for foresters, 342–344
Society of American Foresters, 382
Coe and Pingree estate, 39
Colleges of forestry, 319
Collins Almanor Forest, 40
Colombia, 295
Colorado, 60, 61
national forests area, 311
regulation of timber cutting, 319
state forestry, 36
Colorado Agricultural and Mechanical College, 419
Comité International du Bois, 303
Commercial forest areas and growth rates by regions, 49
Commission on Organization of the Executive Branch of the Government, 359*n.*, 360*n.*
Community forests, 319, 320
in France and Switzerland, 14
Compreg, 177, 275
Compton, Wilson, 30
Comstock, J. H., 222
Coniferous stem structure, 156
Conklin, Howard E., 277*n.*
Connaughton, C. A., 82*n.*
Connecticut, regulation of timber cutting, 38
state forestry, 36
Connwood, 378
Consolidated Water Power and Paper Company, 40
Construction of dwellings, 146
prefabricated, 147
sandwich and box, 166, 167
Consulting foresters, Society of American Foresters' list, 380

Consulting forestry, 380–383
  compensation for, 383
  experience needed, 382, 383
  future outlook for, 383
  (*See also* Employment)
Controlled burning, 74
Cook County forest preserve, 237
Cooperage industry, 195, 196
Cooperatives, 377, 378
Cope, J. A., 204*n.*
Copeland Report, 31
Cord, definition of, 111
Cork oak, 208
Cornell University, 412
Costa Rica, 208, 294
Cotta Master School, 15
Cottonwood, occurrence, 52, 54, 60, 315
  seed, 100
  use, 53, 196
County forests, 319, 320
Craddock, George W., 253*n.*
Crafts, Edward C., 30*n.*
Craig, James B., 323*n.*
Craighead, F. C., 80*n.*
Crocker, Clayton S., 73*n.*, 77*n.*
Crossett Lumber Company, 40, 152, 269
Crossties, 20, 195
Crown-Zellerbach Company, 40
Cunningham, R. N., 51*n.*, 52*n.*, 280*n.*
Cutting methods, 107–109
  improvement, 107
  selection system, 108
  unrestricted, social dislocations of, 281
Cylinder machines, 190
Cypress, use of, 196, 198

D

Dambach, Charles A., 87*n.*
Dana, Samuel T., 428*n.*, 429*n.*
Danish forest laws, 15
Darwin, Charles, 222
Davis, Arthur A., 219
Davis, Kingsley, 288*n.*
Day, Albert M., 213*n.*, 219*n.*
DDT, 81, 84
Dead River Company, 40
Deer, 88
  range essentials for, 216
  white-tailed, 215
Definitions, acre, 110
  board foot, 111
  climax forest, 104, 105
  cord, 111
  forest, 3
  forest land, 2
  forest management, 110
Definitions, forestry, 6
  form class, 113, 114
  hectare, 111
  hemicelluloses, 179
  saw kerf, 112
  section of land, 110
  silvics, 109
  township, 111
  tree, 3
  virgin forest, 104, 105
  wildlife, 210
Defoliators, 78, 79
Delaware, 50, 54
Denmark, 340
  forest practices, 338
  forests, growth rate, 302
  regulation of timber cutting, 15
  stand yield per acre, 118
Department, of the Interior, 34, 313–315
  of Labor, 368
Depression, economic, 31
Dietz, Martha A., 297*n.*
Direct seeding, 100
Disease losses, 78
Division, of Agriculture and Conservation, 304
  of Forest Biology, 431
  of Forest Insect Investigations, 362, 428, 431
  of Forest Pathology, 362, 428, 431
  of Forestry, 24
Douglas-fir, forests, employment in, 393
  logging, 402
  growth, 65
  occurrence, 60, 62, 65, 104, 105
  use, 195, 202, 206, 296
Dowling, Ellen C., 52*n.*, 313*n.*, 319*n.*–321*n.*
Doyle rule, 112
Doyle-Scribner rule, 112
Drought injury, 90
Dry kiln, 138, 140, 144
Duke University, 417–419
Dutch elm disease, 83, 84
Duthie, George A., 320*n.*
Dwellings, construction of, 146, 147
Dyestuffs, 202, 208

E

Eastern hemlock, 44
Eastern redcedar, 53, 198
Ebony, 13, 290
Economic Cooperative Administration, 369, 457
Economic Stabilization Agency, 368

Education in forestry, beginning of, 15, 16
(*See also* Employment)
Egleston, Nathaniel H., 24, 25, 428
Egypt, 8
forestry programs of, 13
Elm, occurrence of, 52, 160
seed of, 100
Elm bark beetles, 84
Emerson, George B., 428
Employment, advanced jobs, 347, 348
with American Institute of Park Executives, 247, 400
beginning jobs, 346, 347, 404
with Boy Scouts of America, 425
career ladders, 452
city service, 370
college teaching, 420–423
construction, 405, 406
consulting forestry, 380–383
education, 420–425
extension forestry, 423, 424
farm forestry, 424
Federal service, advantages and shortcomings of, 356–369
Bureau, of Entomology and Plant Quarantine, 356, 361, 362, 394
of Indian Affairs, 356, 398
of Internal Revenue, 368
of Land Management, 356, 365, 366, 398
of Plant Industry, Soils and Agricultural Engineering, 356, 362
Fish and Wildlife Service, 222, 356, 357, 398
(*See also* Wildlife)
National Park Service, 247, 356, 363, 364
salaries and promotions, 358
Soil Conservation Service, 222, 356, 362, 396
on-the-job training, 357
public-relations training for, 425
stability of, 358
Tennessee Valley Authority, 356, 367, 368
U.S. Forest Service (*see* U.S. Forest Service)
foreign, 379, 380
in forest administration, 441, 442
aptitudes, 448, 449
career ladders, 451–453
functions, 442–448
nature of, 441, 442
responsibilities, 440, 441
rules, 453–455
training required, 449–451
Employment, forest-land management, 394–396
forest protection, 356, 361, 362, 394
forest research, 438
forestry, 275, 346, 352, 354
earnings, 349, 350
with Girl Scouts of America, 386
intermediate jobs, 347
in junior colleges, 423
landscape architect, 399, 400
logging, entering jobs, 404
lumber companies, 372, 373
lumber manufacture, 142–144, 372, 373, 404–406
lumber sales, 405, 406
National Conference of State Parks, technical jobs, 247
in North, 389, 390
park management, 237
private estates, 375–377
private industries, 371–378
public-relations forester, 425
pulp, paper, and wood-chemical industries, 372, 373, 409, 410
recreation and park management, 399–401
regional, trends in, 388–393
in secondary schools, 423
self- (*see* Self-employment)
soil and water conservation, 396
in South, 390, 391
state, 369
conservation departments, wildlife posts, 222
fish, game, and highway departments, 370
park services, 370
teaching opportunities, 420–423
timber operations, 401–404
top jobs, 348, 349
trade-association and conservation-association forester, 425
tropical forestry, 379, 380
with United Nations, 348, 378
various fields, opportunities in, 388, 394–410
in West, 392, 393
wood-processing industries, 373–375, 406–409
wood-using industries, suppliers to, 375
working conditions, 350–352
Engelmann spruce, 60
Engelmann spruce bark beetle, 80
Engineer regiments, First World War, 29
Entailed estate, 15
Erosion, 289

Erosion, in India and China, 295
  (*See also* Shelter-belt program)
Erythrinas, 68
Europe, democracies, forestry programs, 337–339
  forestry, 290–293
  regulation of timber cutting, 15
Evergreen magnolia, 54
Excelsior, 195, 196
Explosives industry, 177
Extension forestry, 38
  (*See also* Employment)
Extractive industries, 177

F

Fabre, Henri, 222
Farm Credit Administration, 362
Farm land, abandoned, 308
Farm Security Administration, 377
Federal cooperative programs, 316
Federal Emergency Relief Administration, 31
Federal Income-tax Law, 270
Federal Resettlement Administration, 37
Federal-state cooperative regulation plan, 335
Fernow, Bernhard E., 24, 25, 13*n.*, 428
Field planting, tree-planting machines used in, 102
  tree seedlings, 101, 102
Finch, Pruyn and Company, 40
Finland, 305, 336
  cooperatives, 378
  forests, 291
    growth rate, 302
  prefabricated houses from, 293, 457
  regulation of private forestry, 338, 339
Fir growth from seed, 100
  (*See also* Douglas-fir)
Fire, attitude of man toward, 73
  control of, 72–78
    aircraft use, 76, 77
    expenditures, 73
    grants-in-aid, 28, 29
    organization, 77, 78
  control pact, northeastern states, 77
  damage by, 73
  early use of, 8
  kinds of, 76
  presuppression and suppression activities, 74, 75
  prevention of, 74
  review of, 78
  unprotected areas, 77
Fire-control equipment, 376
  aircraft, 76, 77
Fire-control equipment, bulldozer, 375, 376
  Byram haze cutter, 375
  fire-danger meters, 74, 75
  Osborne fire finder, 375
  tomcat logging tractor, 125, 375
Fire-control personnel, 75, 76
  fire boss, 75, 76
  fire dispatcher, 75
  smoke chaser, 75
  smoke jumper, 77
Fires, in Idaho, Michigan, Minnesota, and Wisconsin, 72
  in Maine, and Montana, 73
  in Oregon, 72, 73
Fish and Wildlife Service, responsibilities of, 219, 315, 364
  (*See also* Employment)
Flinn, Edwin S., 202*n.*
Flint, Elizabeth Canfield, 352*n.*
Flood-control surveys, 255
Flood damage, 254
Florida, early studies in, 427
  forests, 54, 58
    tropical, 67
  state forestry receipts, 319
Foley, John, 30
Fontanna, Stanley G., 37*n.*
Food and Agriculture Organization (*see* United Nations, Food and Agriculture Organization)
Forbes, Robert H., 73*n.*
Foreign employment, 379, 380
Foreman, duties of, 447, 448
Forest credit, 265
Forest dweller, primitive, 9
Forest formations, Alaskan, description, 66, 67
    use, 66, 67
  central, 52–54
    description, 52
    growth rate, 49, 54
    uses, 52–54
  northern, 44–52
    description, 44, 46
    growth rate, 48, 49
    uses, 46–48
  Pacific Coast, 62–66
    description, 62
    growth rate, 49, 65
    logging methods, 63, 65
    uses, 63–65
  Rocky Mountain, 59–62
    description, 59, 60
    growth rate, 49
    uses, 60–62

Forest formations, southern, 54–59
description, 54, 55
growth rate, 49, 54–57
uses, 56, 58, 59
tropical, 67–70
description, 67, 68
growth rate, 68
uses, 68–70
Forest insects, ambrosia beetles, 80
aphids, 78
bark beetles, 62, 78–80
borers, round-head, 80
brown-tail moth, 79
cicada, 79
defoliators, 78, 79
elm bark beetles, 84
Engelmann spruce bark beetle, 80
gypsy moth, 79, 81
investigations of, Michigan and New York, 428, 429
New Hampshire, 424
New Jersey, Ohio, and Pennsylvania, 428
Texas, 428, 429
ladybird and May beetle, 78
larch and Leconte sawfly, 79
pine shoot moth and white pine weevil, 78
scale insects, 78
seven-year locust, 79
spruce budworm, 79, 81
tussock moth, 81
(*See also* Wood-destroying insects)
Forest insurance, 265, 332
Forest labor, 263, 264
Forest land, definition of, 2
ownership, 324
farmers, 324
lumber companies, 324
national, 324
miscellaneous private, 324
pulp companies, 324
state and local governments, 324
world, 289, 290
taxes on, 331
Forest laws and other legislation, Capper and Snell bills, 30
Carey Act, 21
Clarke-McNary law, 29, 30, 37, 38, 317, 330
Cooperative Forest Management Act, 316–318, 424
Cooperative Sustained Yield Units Act, 312, 314
Danish, 15
Desert Land Act, 21
Dingle-Johnson law, 364, 397
Forest laws and other legislation, Forest Lieu Act, 21
Forest Management Act, 38
Forest Pest Control Act, 330
Fulmer Act, 317
General Exchange Act, 30
Hewitt law, 37
Homestead Act, 21, 36, 365
McNary-Woodruff law, 30
McSweeney-McNary law, 30, 428
National Industrial Recovery Act, 32
Norris-Doxey Farm Forestry Act, 38
Pittman-Robertson Act, 220, 221, 364, 397
Preemption Act, 21
Smith-Lever Act, 38
Swampland Grant Act, 21
Swedish, 339
Taylor Grazing Act, 34, 226, 365
Timber Culture Act, 21, 22
Timber and Stone Act, 21, 24
Transfer Act, 26
Weeks law, 28, 29, 37
yield-tax laws, 336
Forest management, definition of, 110
(*See also* Employment)
Forest manager, job of, 262–266
(*See also* Employment)
Forest nurseries, 100
Forest policy, controversial, nature of, 329
role in, of citizens and foresters, 340
struggle for, 42
Forest practice, character of, on farm, national, private, and state forests, 324
on private holdings, large, 323–326
medium-sized, 324–326
on public forests, 323
on small holdings, 326, 327
Forest productive area, continents and countries exclusive of U.S.S.R., 291
U.S.S.R., 291–293
Forest products, 6
export of, 458
future need for, 457–459
international trade, 458
markets, 264
in national defense, 273–275
Forest Products Laboratory, establishment of, 29, 428
organization of, 431
Forest Products Research Society, 1, 394, 434, 435
Forest regulation (*see* Regulation)
Forest research, achievements in, 435–437

Forest research, in agricultural experiment stations, 431, 432
aptitudes for, special, 437
cooperation, 434, 435
development, 427–430
employment, 438
expenditures, 429, 435
Federal, 431, 461
in forestry schools, 431, 432
importance of, 332, 429
by industries, 429
objectives, 429, 430
organization, 430–434
private agencies, 432, 434
tropical, 438
Forest reserves, administration, 25–27
authorized, 24, 25
transfer to agriculture, 25, 26
Forest resources, estimates of, by American Forestry Association and U.S. Forest Service, 297
Forest stands, age of, 72
reproducing, 109
Forest Survey Staff, 51*n.*, 52*n.*, 280*n.*
Forest tent caterpillar, 79
Forest tree seed, 100
Forest vegetation in United States, 45
Forest work studies, 336
Foresters, agencies employing (*see* Employment)
code of ethics for, 342–344
earnings, 349
farm, 326, 327
obligations, 345
training, 414–420
Forestry, administration, 440–455
beginning of education in, 15, 16
in Broome County, N.Y., 279
business of, information on, 260–272
favorable factors, 270
policy, 271
unfavorable factors, 269, 270
business acumen in, 265, 266
colleges of, 319
in colonial times, 17–20
consulting, 380–383
curriculums, 415–417
definition of, 6
degrees granted, 346
early measures, 15, 16
emergency programs, 31
employment in (*see* Employment)
in Europe, 290–293
extension, 38
farm, 37, 38
Forestry, Federal, advantages and shortcomings of employment in, 358–360
(*See also* Employment)
beginnings of, 23–29
chiefs, Clapp, Earle (acting), 33
Egleston, Nathaniel H., 24, 25
Fernow, Bernhard E., 25
Graves, Henry S., 28, 29
Greeley, William B., 28, 29
Hough, Franklin B., 24
Silcox, Ferdinand, 32, 33
Stuart, Robert Y., 30
Watts, Lyle, 34
in Department of the Interior, 313–315
expansion of, 360
organizations promoting (*see* Organizations promoting forestry)
programs, 311–316
reorganization of, 336, 337
research, 431
responsibilities, 461
sustained-yield units, 312
in First World War, 29
graduate study in, 417, 418
integrated operations in, 383, 384
miscellaneous revenue from, 270, 271
multiple-use, 6, 384
as opportunity for America, 306–308
private, in Adirondacks and Maine, 39
history, 38–41
regulation, 34, 335, 336, 342
returns from, 39, 40
role of, 463–466
as profession, age and size, 345, 346
growth, 1
satisfactions offered, 352, 353
programs, 310–327
in European democracies, 337–339
generally supported, 330–333
proposed, 329–340
requirements for, 413, 414
responsibilities, Federal government, 461
local governments, 462
private landowners, 462, 463
states, 461
scope of, 1
in Second World War, 34, 35
social benefits of, 273–275, 332
state, 35–38
management practices, 323
programs, 316–319
promotion, 1911–1932, 41
support for, 331
in various states, 36–37, 319
unattained goals, 330

Forestry, working conditions, 350–352
world organizations, 303–306
(*See also* Forests)
Forestry commission, Federal forest reserves, suspended creation of, 25
Forests, care of, 96
climate influence on, 258
community, 103
community concept of, 3
definition of, 3
distribution, 299
regional, 298–300
dynamic qualities, 5, 6
establishment on open land, 100
Federal, reserved for ship timbers, 22
(*See also* National forests)
growing stock, 261
as impediment to agriculture, 17
influence on climate, 258, 259
keeping of records, 264
liquidation value, 261, 262
of New York State, situation in, 279, 280
organization, 262, 263
protection, against animals, 86–90
employment in, 356, 361, 362, 394
from insect attack, 78, 282
from unfavorable weather, 90–94
as protection of soil and water, 250–252
public ownership, 334, 335
rainfall interception, 251
school, 320
services of, 6
state (*see* Forestry, state)
trees, life span of, 72
tropical (*see* Tropical forests)
types, number of, 44
use of, 7
for recreation, 237
stages in, 9–12
use of water by, 252, 253
valuation as productive property, 261, 262
virgin, 104–106
definition of, 104
watershed, 320
Form class, definition of, 113, 114
Forsaith, C. C., 155*n.*, 157*n.*, 159*n.*
Foster, Clifford H., 273*n.*
4-H Clubs, 322
Fourdrinier machine, 188–190
France, forests, 290
community, 14
regulation of timber cutting, 11, 15, 337
timber shortage, 11
Frank, Bernard, 249*n.*, 459*n.*
Franklin, Benjamin, 18
Friends of the Land, 322
conservation promotion, 466
Frost action, 90, 91
Fruits and nuts, 207
Fuel wood, 198, 199
Furniture Mart, 51
Future Farmers of America, 322

G

Garrison, Paul M., 58*n.*, 391*n.*
Gaylord Container Corporation, 40
Georgia, 58
early studies in, 427
Germany, forests, 290
state, establishment of, 15
land clearing forbidden in, 14
regulation of timber cutting, 11
timber shortage, 11
Gill, Lake S., 82*n.*
Gill, Tom, 52*n.*, 293*n.*, 313*n.*, 319*n.*–321*n.*
Gillet, Alfred, 202*n.*, 203*n.*
Gillett, Charles A., 321*n.*
Girard, James, 113, 114
Girl Scouts of America, 386
Glesinger, Egon, 304
Glues, 164
Goodman Lumber Company, 152, 269
forest practice, 40
program, 325
Graham, E. H., 397*n.*
Grand fir, 60
Granger, C. M., 313*n.*
Grants-in-aid for fire control, 28, 29
Gravatt, G. F., 86*n.*
Graves, Henry S., 28, 29, 39, 332*n.*, 336*n.*, 417*n.*, 428*n.*
Gray birch occurrence, 5, 47, 104
Grazing, beneficial effects, 87
damage from, 86, 87
Grazing districts, 226
Grazing regulation on national forests, 223
Grazing Service, 223, 226, 227
Great-basin region, 60
Great Britain, 337, 338
forest area, 290
regulation, of forest use, 14
of timber cutting, 11
timber shortage, 11
Greeks and Romans, sacred groves, 13
timber shortage, 11
Greeley, William B., 29–31
Greenberg, David B., 293*n.*

Growth rates, by regions, commercial forest areas, 49
forest formations, central, 49, 54
northern, 48, 49
Pacific Coast, 49, 65
Rocky Mountain, 49
southern, 49, 54–57
Lake States, 49, 51
Middle Atlantic states and New England, 49
tropical, 68
Guaraguao, 68
Guayule, 34
Guise, Cedric H., 346, 417*n.*, 418*n.*
Gulf states forests, 59
Gulick, Luther, 442*n.*
Gullies, 253
Gums, 160, 195
Gutermuth, C. R., 397*n.*
Guthrie, John D., 29, 72*n.*
Gypsy moth, 79, 81

H

Hackberry, 53
Hagenstein, W. D., 64*n.*, 393*n.*
Hail, 93
Hall, William L., 30
Hammatt, R. F., 73*n.*
Hardwoods, cellular structure, 156–158
pulping, 306
Harrar, E. S., 162*n.*, 164*n.*, 169*n.*, 196*n.*
Harrison, William Henry, 25
Hartig, Theodor, 15
Hawaii forests, 68–70
Hawaiian mahogany, 70
Hawley, Ralph C., 3*n.*, 76*n.*, 94*n.*, 106*n.*, 108*n.*, 110*n.*
Hawthorn, occurrence, 5
seed, 96
Heart rots, 84
Heartwood, 155
Hectare, definition of, 111
Heintzleman, B. Frank, 67*n.*, 315*n.*
Hemicelluloses, definition of, 179
Hemlock, growth from seed, 100
occurrence, 5, 44, 46
uses, 48, 195, 202
Henderson, Hiram L., 143*n.*
Hepting, George H., 82*n.*, 84*n.*
Herbert, Paul A., 420*n.*
Hickory, direct seeding, 100
use of, 195
Hirt, Ray R., 83*n.*
Holbrook, Stewart H., 134*n.*
Holland, 337
Holly, 206
Holocellulose, 180
Holst, Monterey L., 39*n.*
Hoover, Herbert, 31
Hoover Committee on Reorganization of Federal Activities, 359
Horseshoe Trail, 460
Hosmer, Ralph S., 30*n.*, 412*n.*
Hostetter, Robert D., 416*n.*
Hough, Franklin B., 24, 25, 428
Hoyle, Raymond J., 168*n.*, 280*n.*
Huberman, M. A., 293*n.*
Hursh, Charles R., 82*n.*, 253*n.*
Hutchins, L. M., 82*n.*

I

Idaho, 62
control of tussock moth, 81
fires, 72
national forests area, 311
regulation of timber cutting, 318
Illick, Joseph S., 20*n.*, 36*n.*
Illinois, 223
furniture manufacturing, 169
Illinois Central Company, 322
Imperial Forestry Institute, 304
Impreg, 275
Incense-cedar, 65
India, earliest organized communities, 8, 9
early tree-planting programs, 13
Indian reservations, 366, 367
forests, 315
range management, 367
in western range, 226
Indian Service, 219, 323
Indiana, furniture manufacturing, 169
shelter-belt planting, 258
Industries, cellulose and plastics, 177, 193, 194
Christmas tree and greenery, 204–206, 385
cooperage, 195, 196
explosives, 177
extractive, 177
forest, Lake States, 51
New England, 49, 50
forest research by, 429
forestry programs, 322, 323
integrated, outlook for, 465, 466
light construction, 146, 147, 152
lumber, forest practice of, 324, 325
outlook for, 464
stabilizing of, 151, 152
naval-stores, 56, 177, 199–202
packaging, 173

Industries, plastics, 177
private, employment of foresters in, 371–378
(*See also* Employment)
pulp and paper, 177, 409, 410
New York, 50, 193, 280
Ohio, 193
outlook, 409, 410, 465
scope, 182
Wisconsin, 51
veneer, outlook, 465
wood-chemical, 177, 203, 409, 410
outlook, 465
wood-distillation, 202, 203
wood-processing, 168–175, 406–409
New York, 168, 169, 406
Pennsylvania, 168
wood-using, New York, 279, 280, 406
(*See also* Employment)
Ineson, Frank A., 377*n.*
Ingas, 68
Insect attack, protection of forest from, 78–82
losses from, 78
(*See also* Forest insects; Wood-destroying insects)
Institute, for International Collaboration in Agriculture and Forestry, 304
of Paper Chemistry, 193, 432
Insurance, forest, 265, 332
Integrated forestry operations, 383, 384
Integrated industries, outlook, 465, 466
International Institute of Agriculture at Rome, 303
International Paper Company, 40
International rule, 112
International Union of Forest Research Organizations, 303
Iowa, 223
Iowa State College of Agriculture and Mechanic Arts, 412, 419
Irrigation water, 254–256
protection of, 255, 256
Israel, 293, 457
Izaak Walton League of America, 322

J

Jack pine, occurrence, 46, 105, 106
use, 48, 51, 281, 282
Jackson, Andrew, 22
Jackson timber harvester, 141
James, Lee M., 268*n.*
Japan, 295
Java man, 8
Jeffrey pine, 65
Jordan, David Starr, 222
Journal of Forestry, 341
Judd, C. S., 70*n.*
Juniper, 104
seed of, 96

K

Kansas, 223
Keen, F. P., 81*n.*
Keep Green program, 322, 466
Kelley, Evan W., 29
Kellogg, Royal S., 30
Kimmey, James W., 82*n.*, 84*n.*
King, R. T., 211
Kirkland, Burt P., 273*n.*
Kittredge, Joseph, 251*n.*, 253*n.*
Kling, John B., 28*n.*, 29*n.*
Koa, 70
Koksaghys, 34, 208
Koroleff, Alexander M., 126*n.*
Korstian, Clarence F., 98*n.*, 100*n.*, 109*n.*

L

Labor, organized, programs, 333, 335
Ladybird beetle, 78
Lake States, forest growth rate, 49, 51
forest industries, 51
forests, 48, 51, 52, 280–282
timber cutting, 281
unrestricted, social dislocations of, 281
Laminated wood, 164–167
Land, agricultural, 36, 308
abandonment of, 36, 37
depletion of, 289
clearing, 17, 18
forbidden, Austria, 14
Germany, 14
economic surveys, 37
forest, definition of, 2
use of, Albania and Asia Minor, 283
Broome County, N.Y., 276–279
democratic approaches to, 286
economic, 460, 461
forestry's place in, 276, 283, 284
Holland, Norway, Denmark, and Sweden, 282
Italy, Greece, and North and South Africa, 283
pattern for adjusting, 284, 285
present, United States, 297–302
public measures for, 285
Switzerland, 282, 283
world-wide examples of, 282, 283
Landscape architecture, 247
employment in, 399, 400
Landslides, 253

Langdale, Harley, Jr., 58*n.*, 391*n.*
Langmuir, Irving, 259*n.*
Larch, growth from seed, 100
  use, 195
Larch sawfly, 79
Laurel, 206
Laws (*see* Forest laws and other legislation)
Leconte sawfly, 79
Legislation needed, 336
Lehigh River avalanches, 253
Leloup, Marcel, 304
Lentz, A. N., 318*n.*
Leopold, Aldo, 220*n.*, 222
Lewison, Wayne C., 50*n.*
Libby, C. Earl, 184*n.*
Light construction, 146, 147
  cycles in, 152
Lignin, 178, 181
Lignumvitae, occurrence, 290
  properties, 68, 160
Lillard, Richard G., 19*n.*, 21*n.*
Limber pine, 60
Lincoln, Abraham, 21
Loblolly pine, 55, 58, 59
Local governments, forestry responsibilities, 462
Locust, 198
Lodgepole pine, occurrence, 60, 65
  use, 195
Log floating, 128
Log prices, 267
Log rules, 112
Log scaling, 112
Logging, 120–132
  accident rates, 402
  cable, 125, 126
  development in America, 120, 121
  entering jobs, 404
  as hazardous occupation, 132
  in Northeast, 127, 128
  place of forester in, 131, 132
  planning and layout, 121–123
  in South, 128, 129
  waste in, 129–131, 307
  in West, 129
Logging equipment, 123–126, 128
  arch, 124
  loaders, mechanical, 126
    power, 128
  power saws, 123, 124
  tractors, ordinary and tomcat, 125
  trucks, 126
Logging operations, 123–126
  bucking, 123, 124
  felling and limbing, 123
  hauling, 126
Logging operations, skidding, 124–126
  swamping, 124
Long-Bell Company, 40
Long Trail, 460
Longleaf pine, occurrence, 55, 56, 59
  properties, 56, 74, 155, 160
  use, 56, 199
Louisiana, 59
  regulation of timber cutting, 318
Louisiana State University, 419
Lowdermilk, Walter C., 11*n.*, 12*n.*
Lumber, concentration yard, 145
  consumption per capita, 136, 137
  exporting, 145
  grading, 143
  measuring, 111, 112
  planing, 143
  retail dealers, 146, 147
  retail handling, 145, 146
  "round-edge," 138
  seasoning, 143, 144
  small-dimension, 149
  wholesale selling, 145
    (*See also* Employment)
Lumber industry, forest practice, 324, 325
  outlook, 464
  stabilizing of, 151, 152
Lumber manufacture, 133–152
  development, 134
  economic importance of, 133, 134
  skilled jobs in, 142–144
    (*See also* Employment)
  waste in, 149, 150
Lumber prices, long-term trends, 267, 268
  wholesale, 136
Lumber production, outlook, 152, 153
  shifts in, 23
  size of mills in relation to, 141, 142
  trend, 134–137
Lumber code, 32, 41
Luther property, 384, 385

M

McAtee, William L., 222
McCulloch, Walter F., 448*n.*
McGee, W. J., 26
McGill University, 193
Magnolia splendens, 68
Mahogany, 68, 290
Maine, cutting cycle, 118, 437
  fire of 1947, 265
  fires, 73
    control of, 77
  forest insect attack, 78, 79

Maine, forest research, 428
lumber production, 134, 135
paper mills, 193
private estate employment, 375
private forestry, 39
river driving, 197, 401
state forestry, 36
virgin forest, 48, 49
wood-processing industries, 168
Malaya, 295
Manzanita, 65
Maple, growth rate, 48
occurrence, 46, 60
seed, 100
Maple-basswood climax, 105
Maple products, 203–205
Martin, J. F., 82*n.*
Maryland, regulation of timber cutting, 37, 38, 318
Schöpf's observations on forests of, 427
Mason, Charles W., 448*n.*
Mason, Earl George, 345*n.*
Masonite process, 186, 187
Massachusetts, 193
regulation of private forestry, 335
of timber cutting, 319
Massachusetts Institute of Technology, 193
Mathewson, J. S., 140*n.*, 143*n.*
Matthews, Donald M., 263*n.*, 269*n.*
Maxwell, George H., 26
May beetle, 79
Medicinal herbs and pharmaceuticals, 207, 208
Mellon Institute, 432
Mencius, 11
Merriam, C. Hart, 222
Mesopotamia, 8
Mesquite, 60, 70
Methyl alcohol, 202
Mexico, 293
Meyer, H. Arthur, 12*n.*
Meyer, Walter H., 113*n.*, 262*n.*
Mice, 88, 215
Michaux, André, 428
Michaux, François-André, 21, 428
Michigan, county and municipal forests area, 319
fires, 72
forest insect investigations, 428, 429
forest receipts, 319
furniture manufacturing, 51, 169
land economic surveys, 285, 286
national forests area, 311
paper products, 193
state forestry, 36, 318, 319
Michigan State College of Agriculture and Applied Science, 412, 419
Middle Atlantic states, 48, 50
forest growth rate, 49
forests and forest industries, 50
Middle East forest destruction, 293
Military government, 369
Miller, Edwin C., 97*n.*, 99*n.*
Miller, John M., 80*n.*
Milpac, 150
Mine timbers, 111, 198
Minnesota, county and municipal forests area, 319, 320
fires, 72
insect outbreaks, 79
land economic surveys, 285, 286
lumber production, 135
national forests area, 311
regulation of timber cutting, 318
state forestry, 36, 318
Minnesota and Ontario Paper Company, 40
Mississippi, 54, 58
lumber production, 135
national forests area, 311
regulation of timber cutting, 318
Missouri national forests area, 311
Mistletoes, 85, 86
use of, 206
Moir, Stuart, 64*n.*, 393*n.*
Montana, 62
fires, 73
national forests area, 311
state forestry, 319
Montana State University, 419
Moore, Barrington, 336*n.*, 338*n.*
Moore, E. B., 318*n.*
More Trees program, 322
Morton, Sterling, 22, 25
Mosquito bomber, 166
Mountain hemlock occurrence, 60, 66
Mountain mahogany, 65
Multiple-use forestry, 6, 384

N

Nancy, French School of Forestry, 15
National Academy of Sciences, 25
National Conference on State Parks, technical jobs with, 247
National Conservation Commission, 28
National Forestry Program Committee, 30
National forests, Allegheny and White Mountain, 29
Chippewa, 281, 282
creation, 41

National forests, in East, 29
expansion, 28
fire control, 312
Pisgah, 29
program, 311–313
purchase units, 29
Saint Jo, 73
Superior, 460
Tongass, 66
western range, 226
National Industrial Recovery Act, 32
National Lumber Manufacturers Association, 334, 429
National monuments, 238
National Park Service, 244, 245, 246*n.* (*See also* Employment)
National parks, 238, 241
administration, 244
area, 315
National Production Authority, 368
National Research Council, 336, 342
National Resources Council of America, 342
National Resources Planning Board, 256
National Security Resources Board, 368
National Wool Growers Association, 226
Nature trails, 241
Naval stores, 199–202
industry, 177
output, 1947–1948, 202
principal species used, 56
Nebraska, 258
Ne-Ha-Sa-Ne Park, 39
Nekoosa-Edwards Company, 40
Nelson, DeWitt, 64*n.*, 393*n.*
Netboy, Anthony, 249*n.*, 459*n.*
Netherlands East Indies, 295
Nevada, 61
regulation of timber cutting, 318
New Brunswick, 79
New England, forest growth rate, 49
forest industries, 49, 50
forests, 48
New England and Texas Timber Salvage Administrations, 369
New Hampshire, 78
forest insect investigations, 428
regulation of timber cutting, 38, 318
state forestry, 36
yield tax, 463
New Jersey, 50, 82, 318
forest insect investigations, 428
regulation of timber cutting, 38
New Mexico, 61
national forests area, 311
New York, county and municipal forests area, 319
New York, forest insect investigations, 428, 429
forest situation, 279, 280
furniture manufacturing, 169
insect damage, 79
land use, 285
lumber production, 134–136
pulp and paper industry, 50, 193, 280
regulation of private forestry, 335, 336
state forestry, 36, 318, 319
stream-pollution control, 465
wood-processing industries, 168, 169, 406
wood-using industries, 279, 280
New York State College of Forestry, establishment at Cornell University, 412
(*See also* State University of New York, College of Forestry)
New Zealand, 296
Nezahualcoyotl, 13
Nitrocellulose, 180
Normal stocking, 117
Norris-Doxey farm-woodland-management program, 424
North Africa forest destruction, 293
North American Conservation Congress, 28
North Carolina, 58
furniture manufacturing, 169
national forests area, 311
state forestry, 36
North Carolina State College of Agriculture and Engineering, 419
North Dakota, 223
North Pacific region, forests, use of, 63
logging methods, 63
Northeast Forestry Foundation, 378
Northeastern Forest Experiment Station, 459
Northeastern states, fire-control pact, 77
Northern forest formations, 44–52
description, 44, 46
growth rate, 48, 49
uses, 46–48
Northern Forestry Union, 304
Northern hardwood, 51
Northern white cedar, occurrence, 44, 46
use, 196, 198
Northern white pine, 44, 108
(*See also* White pine)
Norway, 336, 339
forests, 291
regulation of timber cutting, 14
Nuremberg, early reforestation practice, 14
Nuttall, Thomas, 428

O

Oak, direct seeding, 100
  occurrence, 51
  use, 195, 198
Obet, Viktor, 292*n*.
Office of Price Administration, 369
Ohio, effects of grazing, 87
  forest insect investigation, 428
  forestry commission, 86
  pulp and paper mills, 193
Oklahoma, 54, 59
  shelter-belt planting history, 257, 258
Oliver Iron Mining Company, 325
O'Neil, Frederic W., 184*n*.
Ontario insect outbreak, 79
On-the-job training, Federal service, 356, 357
Oregon, 64
  forests area, county and municipal, 319
    national, 311
    state, 37
  regulation, of private forestry, 335
    of timber cutting, 318
Oregon and California Railroad revested lands, 34, 365, 366
  history of, 314
Oregon State College, 419
  Wood Products Laboratory, 432
Organizations promoting forestry, American Association for the Advancement of Science, 23, 24, 341, 342
  American Bankers Association, 322
  American Forest Products Industries, Inc. (*see* American Forest Products Industries, Inc.)
  American Forestry Association (*see* American Forestry Association)
  American Geophysical Union, 342
  American Institute of Biological Sciences, 342
  American Institute of Park Executives, 347, 400
  American Legion, 322
  American National Cattlemen's Association, 226, 227
  American Paper and Pulp Association, 322
  American Pulpwood Association, 333, 425, 429
  American Society of Mechanical Engineers, 342
  American Turpentine Farmers Association, 322
  Appalachian Hardwood Manufacturers Association, 325
  Boy Scouts of America, 320, 322
Organizations promoting forestry, Caribbean Research Council, 305
  Civil Works Administration, 31
  Civilian Conservation Corps, 31, 335, 369
  Clearwater Forest Protective Association, 41
  Food and Agriculture Organization (*see* United Nations, below)
  Forest Products Research Society, 1, 394, 434, 435
  4-H Clubs, 322
  Friends of the Land (*see* Friends of the Land)
  Future Farmers of America, 322
  Girl Scouts of America, 386
  Imperial Forestry Institute, 304
  Institute for International Collaboration in Agriculture and Forestry, 304
  Institute of Paper Chemistry, 193, 432
  International Institute of Agriculture at Rome, 303
  International Union of Forest Research Organizations, 303
  Izaak Walton League of America, 322
  National Lumber Manufacturers Association, 334, 429
  National Wool Growers Association, 226
  Northeast Forestry Foundation, 378
  Northern Forestry Union, 304
  Pacific Science Association, 304
  Pan-American Union, 304
  Southern Pulpwood Conservation Association, 322
  Technical Association of the Pulp and Paper Industry, 194
  Trees for Tomorrow, Inc. (*see* Trees for Tomorrow, Inc.)
  United Nations, Food and Agriculture Organization, 29, 303, 457
    Division of Forestry and Forest Products, history of, 304, 305
  U.S. Chamber of Commerce, 322
  U.S. Junior Chamber of Commerce, 322
  West Coast Lumbermen's Association, 30, 322
  Western Pine Association, 322
  world, 303–306
Osborn, Fairfield, 289*n*., 298*n*.
Otsego Forest Products Cooperative Association, 377, 378
Ownership rights, 11

P

Pacific area, 295, 296
Pacific Coast forest formations, 62–66
description, 62
growth rate, 49, 65
logging methods, 63, 65
uses, 63–65
Pacific Northwest forests, 64
Pacific Science Association, Forestry Committee, 304
Pack, Charles Lathrop, 30
Pack Demonstration Forest, 97, 117
Packaging industry, 173
Palestine forestry programs, 13
Pan-American Union, 304
Panshin, A. J., 155*n.*, 157*n.*, 159*n.*, 162*n.*, 164*n.*, 169*n.*, 196*n.*
Paper, consumption, 191
production, 191
waste, consumption, 192
Paper birch, occurrence, 46, 47, 104
uses, 48
Paper machine costs, 192
Paper-manufacture technology, 187–190
Paper mills, distribution, 193
water used, 193
Paper-plastics combinations, technology, 190
Papreg, 275
Paraná pine, imports, 457
occurrence, 289, 290, 293, 294
Park administration, 244, 245
Park management, 237
(*See also* Employment)
Park planners, promoting understanding, 240, 241
surveying facilities, 241
Park systems, growth, 237
planning, 238
projecting, 241
Parker, D. E., 86*n.*
Parks, attractions for all, 238
avoiding conflicting use, 239, 240
building, 242, 243
detailed planning, 242
maintenance, 243, 244
preserving natural conditions, 239
Pasto, Jerome K., 277*n.*
Paton, Alan, 283*n.*
Patten Timber Company, 40
Paxson, Frederic L., 20*n.*
Peach boughs, use, 206
Peck, Allen Steele, 29
Peking man, 8
Penn, William, 19
Pennsylvania, 50
Pennsylvania, cord used in, 197
erosion studies, 252
forest insect investigations, 428
furniture manufacturing, 169
paper mills, 193
Schöpf's observations on forests of, 427
state forestry, 36, 318, 319
stream-pollution control, 465
wood-processing industries, 168
Pennsylvania State College, 419
Performance ratings, Federal service, 357
Peru, 295
Philippine Islands, 295
Photosynthesis, 97, 177
Piling, 198
Pinchot, Gifford, 25, 26, 27, 27*n.*, 30, 39
removal from office, 28
Pine, 51
(*See also* specific names)
Pine shoot moth, 78
Pingree, David, 39, 40
Piñon pine occurrence, 60, 104
Plant quarantine, 86
Plastics industry, 177
Plunguian, Mark, 179*n.*
Plywood, 163, 164
Point 4 programs, 457
Poland forests, 290
Pole-sized timber, caring for, 106, 107
Poles, 197
Ponderosa pine, cause of loss in, 85
occurrence, 60, 62, 65
Population, by continents, 291
growth of, 288
U.S.S.R., 291–293
United States of America, 291
of world, 287
Porcupine, 88
Port-Orford-cedar occurrence, 62, 66
Posts, 198
Powder-post beetle, 80, 150
Power saws, 123, 124
Prefabrication, 152
buildings, 147
houses, 306
Price, William H., 64*n.*, 393*n.*
Primitive forest dwellers, 9
Private forestry, 38–41
Private industries, employment of foresters in, 371–378
(*See also* Employment)
Private landowners, aids to, 333
forestry responsibilities, 462, 463
Probation period, Federal service, 356
Proctor, P. B., 162*n.*, 164*n.*, 169*n.*, 196*n.*
Production and Marketing Administration, 316, 333, 362

Programs of forestry (*see* Forestry)
Property rights, development, 10
Public, education of, 331
Public domain, 20–23, 41
Public-private cooperation, 466
Public Works Administration, 31
Puerto Rico forests, 68, 208
  national, 311
Pulp and paper industry, 177
  employment in, 372, 373, 409, 410
  forest practice, 324
  importance of, 191–193
  New York, 50, 193, 280
  outlook, 465
  pulp consumption, 192
  raw material, 192, 193
  Scandinavian, 192
  scope, 182
  Technical Association of, 194
  world trade, 193
Pulping processes, 182–187
  Asplund, 187
  groundwood, 182–184
  mechanical, 182–184
  semichemical, 183, 186
  soda, 183, 185
  sulfate, 183, 185, 186
  sulfite, 183–185
  whole wood fiber, 186, 187
Pulpwood, 196, 197
Purdue University, 419

Q

Quality control, 172, 173
Quebec, 79

R

Rabbits, cottontail, 215
  snowshoe, 88
Railway land grants, 21
Rainfall, 93
  induced, 259
Ranch management, 228–230
Range lands, controversy over control, 226–230
  management, 223–234
    employment in (*see* Employment)
    importance of, 223
    Indian reservations, 367
  manager, requirements of, 230–234
  reseeding, 232
  western, condition, 223
    depletion, 224, 225
    ownership, 226
    use, 226
  types and conditions of, 225
Raushenbush, Stephen, 335*n.*, 336*n.*
Rayon, 180, 181
Rayonier, Inc., 276
Rays, 156
Recreation, conflicting forest use, 239, 240
  use of forests for, 237, 459, 460
  value of, 235, 236
    commercial, 236
  varied needs for, 238
  (*See also* Employment)
Recruitment, Federal service, 356
Red alder, 66
Red fir occurrence, 60, 62, 65
Red gum, 54
Red maple, 5
Red oak, 52
Red pine, occurrence, 46, 105, 106, 281
  use, 48, 282
Red spruce, occurrence, 46
  use, 48, 202
Redbud, 206
Redcedar, occurrence, 5, 104
  use, 206
Redwood, growth, 65
  logging, 65
  occurrence, 62
  properties, 160
  use, 196, 296
Redwood region, 64
Reforestation, farmlands, 37
Regional employment, trends in, 388–393
  (*See also* Employment)
Regulation, of forest use, in Great Britain, 14
  of private forestry, 335, 336, 342, 463
    California, 335
    Finland, 338, 339
    Massachusetts, 335
    New York, 335, 336
    Oregon, 335
    U.S. Forest Service opinion on, 335
    Washington, 335
  of timber cutting, 11, 30, 33, 42
    California, 318
    Colorado, 319
    compulsory, 38
    Connecticut, 38
    Denmark, 15
    European states, 15
    France, 11, 15, 337
    Germany, 11
    Great Britain, 11
    Idaho, 318
    Louisiana, 318
    Maryland, 37, 38, 318
    Massachusetts, 319

Regulation, of timber cutting, Minnesota, 318
Mississippi, 318
Nevada, 318
New Hampshire, 38, 318
New Jersey, 38
New York, 319
Norway, 14
Oregon, 318
Scandinavia, 14
Vermont, 38
Virginia, 319
Washington, 318
Report of the President's Water Resources Policy Commission, 459*n.*
Research, forest (*see* Forest research)
Resins, 160
Retirement, Federal service, 359
Rettie, James C., 377*n.*
Revested lands, Oregon and California Railroad, 365
Rhett, Beverly, 70*n.*
Ribes, 83
Riegel, John W., 447*n.*
Rio Grande, 223
River-basin planning, 256
River birch, 53
River driving of logs, 128, 401
Robinson, R. R., 252*n.*
Rocky Mountain forest formations, 59–62
description, 59, 60
growth rate, 49
use, 60–62
Rocky Mountain juniper, 60
Rohwer, S. A., 80*n.*
Romans and Greeks, sacred groves, 13
timber shortage, 11
Roosevelt, Franklin D., 31
Roosevelt, Theodore, 25, 26, 28
Rosewood, 13, 290
Rosin, 199, 201
Roth, Filibert, 25
Rubber, 208
Ruffed grouse, 88, 215
Russia timber resources, 292, 293
Rutherford, R. M., 397*n.*

S

Sacred groves, 13
Saint Jo National Forest, 73
St. Paul and Tacoma Lumber Company, 40
St. Regis Company, 40
Salaries and promotions, Federal service, 358
Sandalwood, occurrence, 13
use, 70
Sapwood, 155
Sargent, Charles S., 25, 428
Satinwood, 290
Saw kerf, 149
definition of, 112
Sawmills, 137–144
beginning jobs in, 404, 405
(*See also* Employment)
hazard in operations, 404
number, 142
portable, 140, 141
unionized labor, 404
Sawtimber stands, management of, 107
Scale insects, 78
Scandinavia, cooperatives, 378
pulp, 192
regulation of timber cutting, 14
Schenck, Carl A., 39
Schoepf (Schöpf), Johann David, 19*n.*, 427
School forests, 320
Schumacher, Francis X., 113*n.*
Schurz, Carl, 22, 24, 25
Scribner rule, 112
Scrub oaks, 60
use, 202
Seaboard Air Line Company, 322
Sears, Paul B., 289*n.*, 298*n.*
Section of land, definition of, 110
Seed, forest tree, 100
production of, 99
of trees, 96
Seedbeds, 101
Seeding, direct, 100
Seedling and sapling stands, caring for, 106
Seedlings, needs of, 97
Seigworth, Kenneth J., 58*n.*, 391*n.*
Selection system, 108, 109
Selective cutting, 109
Selective logging, 109
Self-development, opportunities for, in Federal service, 358, 359
Self-employment, 380–386
consultant, 380–383
integrated operations, 383, 384
landowner and operator, 384, 385
landscape and nursery operations, 386
resort and camp operation, 386
sawmill operation, 405
timber and sawmill operator, 385
wood fabrication, 385, 386
(*See also* Employment)
Semichemical pulping, 306
Sequoia, 65

Seven-year locust, 79
Sewall, Joseph, 49*n.*
Shavings and sawdust use, 199
Shelter-belt program, 33, 396
effect, 254
history, 36, 257, 258
Shelton Sustained Yield Forestry Unit, 312
Shingles, 195, 196
Ship timbers, Federal reserves for sustained yield, 22
Shirley, Hardy L., 39*n.*, 40*n.*, 293*n.*, 325*n.*
Shortleaf pine occurrence, 52, 53, 55
Show, S. B., 304
Shriner, Ralph, 202*n.*
Siam, 295
Sick and annual leave, Federal service, 358
Sierra forests, 65, 66
Sihlwald, 14
Silcox, Ferdinand, 32
Silk cottonwood, 68
Silting, 253, 254
Silver maple, 52
Silversides, C. R., 126*n.*
Silvics, definition of, 109
Silviculture, 110, 262
Simeone, J. B., 80*n.*
Simmons, Fred C., 126*n.*, 129*n.*, 269*n.*
Simpson Logging Co., 276, 312
Sitka spruce occurrence, 62, 66
Slash pine, occurrence, 55, 58
use, 56, 199
Smith, Arthur D., 223*n.*
Smith, Howard, 448*n.*
Smoke chaser, 75
Smoke jumper, 77
Snow and ice damage, 93, 253
Society, of American Foresters, 1, 341–344
code of ethics, 342–344, 382
colleges of forestry, accrediting of, 418, 419
consulting foresters, fields of specialization, 380–382
list of, 380, 381
employment survey of members, 355
forest landowners in, 326
forest-policy study, 336
junior-member requirements, 415, 416
regulation of private forestry, 335
research cooperation, 434, 435
salary survey of members, 349, 350
training of foresters, 420
trends in employment, 389
Society, of Range Management, 1
of Wildlife Management, 1
committee on employment outlook, 397
Softwoods, world distribution, 289, 290
Soil, 2, 3
protecting, importance of, 249–252
Soil Conservation Service, cooperation, 219, 230, 428
position on aids to private owners, 333
program, 316
public-relations work with, 425
range-management controversy, approach to problem, 228
shelter-belt program, 33
Soil Erosion Service, transfer from, 34
(*See also* Employment)
Soil Erosion Service, 34
South, employment, 390, 391
working conditions, 391
(*See also* Employment)
South Atlantic states forests, 58
South Carolina, 58
South and Central America, 293–295
South Dakota, Black Hills, 62
shelter-belt planting, 258
South and East Asia, 295
Southeastern states forests, 58, 59
Southern forest formations, 54–59
description, 54, 55
growth rate, 49, 54–57
use, 56, 58, 59
Southern pine forest, 55–58
properties, 160
use, 195
Southern Pulpwood Conservation Association, 322
Sowder, Arthur M., 204*n.*, 385*n.*
Spanish cedar occurrence, 68, 290
Sparhawk, William N., 22*n.*, 24*n.*, 25*n.*, 29
Spaulding, Perley, 82*n.*
Spero, Sterling D., 359*n.*
Spoehr, H. A., 437*n.*
Spruce, growth rate, 48
occurrence, 44, 281
use, 383
Spruce budworm, 79, 81
Standard cord, definition of, 111
Stapack, 275
Starr, Reverend Frederick, 22
State conservation departments, 219, 220
wildlife posts, 222
(*See also* Employment)
State forestry (*see* Forestry)
State forests (*see* Forestry)

State University of New York, College of Forestry, 193, 419
employment of graduates, 410, 416, 417, 440
research program, 432
(*See also* New York State College of Forestry)
States, forestry responsibilities of, 461
Steer, Henry B., 137*n.*, 145*n.*, 167*n.*
Stephenson, J. Newell, 182*n.*
Stickel, Paul W., 76*n.*, 94*n.*
Stillinger, John R., 168*n.*, 280*n.*
Stoddart, Laurence A., 223*n.*
Stoeckeler, Joseph H., 254*n.*, 257*n.*
Stuart, Robert Y., 30, 32
Stumpage prices, 266
Sugar maple occurrence, 5, 44, 52
Sugar pine occurrence, 62, 65
Sunscald, 91
Superior National Forest, 460
Sustained-yield management, forcing of, 333
Sustained-yield units, Federal, 312
Shelton, 312
Swamp oak, 54
Swanson, Gustav A., 397
Sweden, forests, 291, 336
growth rate, 302
imports from, 457
regulation of forest practice, 338, 339
Swift, Lloyd W., 397*n.*
Swingle, R. U., 82*n.*
Switzerland, community forests, 14
forestry program, 337
Sycamore, 52, 53

T

Tabanucos, 68
Taft, William Howard, 28
Tamarack, occurrence, 46, 315
use, 198
Tannins, 202, 208
Tarzwell, Clarence M., 397*n.*
Taylor grazing districts, 398
Teacher, attributes for success, 421
Tead, Ordway, 448*n.*
Teak occurrence, 13, 68, 290
Technical Association of the Pulp and Paper Industry, 194
Tennessee, 58, 82
Tennessee Valley Authority, 391
cooperation, 428
river-basin planning, 256
work of, 315, 316
(*See also* Employment)
Termites, 80
Texas, 54, 59
forest insect investigations, 428, 429
shelter-belt planting, 258
Tharandt forestry college, 15
Thelen, Rolf, 140*n.*, 143*n.*
Thinning, 106
Third World Forestry Congress, 305
Thompson, Warren S., 287*n.*, 288*n.*
Tillamook fire, 64
Timber, buying and selling, units for, 110–112
cutting regulation (*see* Regulation)
drain on, 300, 301
estimating, 112–117
accuracy of, 116
on intermediate-sized tracts, 114, 115
on large tracts, 116
on small tracts, 114
statistical control, 116
use of aerial photographs, 115, 116
famine, 23
growing of, basic factors, 261–264
business of, 260, 261
acumen in, 265, 266
and business cycles, 266–268
financial planning, 269
growth of, 116, 117, 300, 301
determination, 117
effect of forest management on, 117, 118
(*See also* Growth rates, by regions)
losses due to natural causes, 300
management plan, 269
operations, planning of, 269
requirements for, 300–302
shortage of, 11
standing, decline in, 299, 300
use, in America, 20
volume, average, 105
decline of, 298
world outlook for, 302
Timber Engineering Company, 321*n.*
Research Laboratory, 429, 432, 434
Timber line, 63
Timber Production War Project, 34
Tongass National Forest, 66
Topography, 2
Toumey, James W., 98*n.*, 100*n.*, 109*n.*
Township, definition of, 111
Tracheids, 156
Transpiration, 252, 253
Tree, definition of, 3
Tree diseases, 82–86

Tree-farm, movement, origin, 64
program, 207, 321, 466
Tree grades, 116
Tree injuries, 90–95
Tree pests introduced, 86
(*See also* Forest insects; Wood-destroying insects)
Tree plantations, protection of, 102, 103
Tree seed, 100, 101, 206, 207
Tree seedlings, field planting of, 101
Tree stem structure, 155, 156
Trees, climatic requirements for, 98, 99
death, 99
growth, 97, 154, 155
requirements for, 97–99
(*See also* Timber, growing, growth of)
hazard to seedling survival, 98
kinds, 3
maturity, 99
protection of planted, 102, 103
shade tolerance, 98
Trees for Tomorrow, Inc., 466
program, 322
Trippensee, Reuben Edwin, 217*n.*, 220*n.*
Tropical forests, 67–70
experiment station, 438
need for, 305
products of, 208
(*See also* Employment)
Truman, Harry S, 459
Tupelo, 54
Turner, David B., 397*n.*
Turpentine recovery, 199, 201
Tussock moth, 79, 81

U

Unasylva, 304
Union membership, 403
United Nations Food and Agriculture Organization, 29, 70*n.*, 293*n.*, 303, 457
Division of Forestry and Forest Products, 289*n.*, 290*n.*
history of 304, 305
(*See also* Employment)
United States, forests, 297–302
growth rate, 302
original area, 17, 297
present land use, 297–302
responsibility for leadership, 456, 457
in world forestry, 302, 303
vegetative types and areas covered, 297
U.S. Chamber of Commerce, 322
U.S. Corps of Engineers, 249
U.S. Department, of Agriculture, 23*n.*, 28*n.*, 33*n.*, 37*n.*, 62*n.*, 73*n.*, 74*n.*, 77*n.*, 225*n.*, 428, 434*n.*
Forest Service, 49*n.*, 96*n.*, 99*n.*, 100*n.*, 130*n.*, 142*n.*, 174*n.*, 195*n.*–198*n.*, 207*n.*, 208*n.*, 223*n.*, 225*n.*, 226*n.*, 312*n.*, 323*n.*, 390*n.*, 436*n.*
Forest Products Laboratory, 160*n.*, 168*n.*
Lake States Forest Experiment Station, 428*n.*
of Commerce, 368
Bureau of Census, 133*n.*, 191*n.*
of the Interior, 363
Bureau of Land Management, 20*n.*, 21*n.*
U.S. Forest Products Laboratory, 275
U.S. Forest Service, changed from Bureau of Forestry, 26, 27
employment opportunities in, 222, 356, 357
flood-control surveys, 396
living and working conditions, 360, 361
public-relations foresters, 425
range specialists, 398
forest resources estimate, 297
functions of Chief, 442–447
budgeting, 444, 445
coordinating, 445
directing, 446, 447
organizing, 443, 444
planning, 443
reporting, 445, 446
staffing, 443
grazing regulation, national forests, 223
improved management practices, 230
public ownership of forests, 334
public-private cooperation, 466
range control, controversy over, 227
regulation of private forestry, 335
U.S. Junior Chamber of Commerce, 322
U.S. Senate, 429*n.*, 435*n.*
University, of California, 419, 432
of Florida, 419
of Georgia, 419
of Idaho, 419
of Maine, 193, 419
of Massachusetts, 419
of Michigan, 412, 419
of Minnesota, 412, 419
of Missouri, 419
of Nebraska, 412
of Washington, 419
of Wisconsin, 419, 466
Urania Lumber Company, 40, 152

Utah, 61
  national forests area, 311
Utah State Agricultural College, 419
Utilization, integrated, 393

V

Vanderbilt, George, 39
Veneer, 162, 163
  industry outlook, 465
  waste in, 174, 175
Vermont, regulation of timber cutting, 38
  state forestry, 36
Virgin forest, 104–106
  definition of, 104
Virgin Islands, 68, 69
Virginia, 58
  furniture manufacturing, 169
  national forests area, 311
  regulation of timber cutting, 319
  Schöpf's observations on forests of, 427, 428
Virginia pine, 52, 53
Viscose, 180
Vogt, William, 289*n.*
Volume tables, 113, 114
Von Platen-Fox Company, 40
von Steuben, Friedrich Wilhelm, 24

W

Wackerman, A. E., 126*n.*, 129*n.*, 269*n.*
Walnuts, direct seeding, 100
War Assets Administration, 369
War Production Board, 34, 368
Ward, Jay, 202*n.*
Washington, national forests area, 311
  private timber companies, 64
  regulation, of private forestry, 335
    of timber cutting, 318
  state forestry, 319
Wasps, 80
Waste (*see* Wood, waste of)
Water, detention of, 251
  retention of, 251
  use of, by forests, 252, 253, 459
Watershed forests, 320
  municipal, management, 256
  protection, importance of, 249–252
Watts, Lyle F., 34, 334*n.*, 335*n.*
Weather Bureau, 428
Webb, Seward, 39
Weevils, wood-boring, 80
West, Clarence J., 178*n.*, 202*n.*
West Coast Lumbermen's Association, 30, 322
West Virginia, 50
West Virginia University, 419
Western hemlock, 66
Western larch, 60
Western live oak, 60
Western Michigan College, 193
Western Pine Association, 322
"Western Range, The," 33, 224
Western redcedar, occurrence, 62, 66
  use, 196, 198
Western white pine, occurrence, 60, 62, 65
  (*See also* White pine)
Weyerhaeuser Timber Company, 40, 145
  integrated operations, 1, 152
Wheeler, W. M., 222
White ash, growth rate, 48
  occurrence, 52
  use, 47
White birch, 315
White cedar, growth from seed, 100
White fir occurrence, 60, 62, 65
White House Conference of Governors, 28
White oak, occurrence, 52
  use, 196
White pine, 104, 105
  growth rate, 48
  occurrence, 46, 52, 281
  properties, 160
  use, 19, 282
White pine blister rust, 83
White pine weevil, 78
White spruce, growth from seed, 100
  occurrence, 46, 67, 315
  use, 48, 206
Whitebark pine occurrence, 60, 65
Whitney preserve, 39
Whitten, R. R., 82*n.*
Wilde, S. A., 104*n.*
Wildlife, attitudes toward, 213
  behavior, 215
  damage by, control of, 89, 90
  definition of, 210
  factors restricting, 216, 217
  forests for, 459, 460
  interspersion of growth requirements for, 217
  overpopulation of, 219
  predators, destruction of, 214
  protection against, 87–89
  protection of, 214, 215
  range for, 216, 217
  refuges, 214, 219, 315
  values of, 211–213
    negative, 213

Wildlife management, developing program for, 217, 218
employment in, 220, 397, 398
(*See also* Employment)
policies and programs, 218
responsibility for, 220
for sustained yield, 218
Wildlife manager, job of, 211, 221, 222
requirements for, 222
working conditions, 221, 222
Wilkins, Austin H., 73*n.*
Williams, J. Randall, 30
Williams, Ross A., 254*n.*, 257*n.*
Willows, occurrence, 54, 60, **315**
seed, 100
use, 224
Wilson, James, 27, 28
Wind, 92, 93
erosion by, 254
losses from, 78
Winter burning, 91
Winters, Robert K., 29*n.*, 40*n.*, 41*n.*
Wisconsin, 61
Conservation Department, 466
county and municipal forests area, 319, 320
Department of Public Instruction, 466
fires, 72
land economic surveys, 285, 286
national forests area, 311
pulp and paper industry, 51
shelter-belt planting, 258
Trees for Tomorrow program, 322
zoning law, 285
Wise, Louis E., 178*n.*–180*n.*, 202*n.*
Wolf, Robert E., 36*n.*
Wood, ash content, 178
chemical components, 178
"compregnated," 435
conditioning for manufacture, 169, 170
decay, 151
control of, 84
fastening of, 170, 171
fiber saturation point, **161**
fibers, 158
fire-retardant, 168
gluing of, 171, 172
identification, 158
impregnated, 435
integrated use, 64
laminated, 164–167
machining of, 170
preserving, 167
properties of, 154, 160, **161**
physical, 158–161
Wood, sanding and finishing, 172
secondary products of, 195–208
shortages, colonial America, 19
shrinkage, 161
treating of, 167, 168
turning, 174
uses of, 161–167
waste of, in use, 151
construction industry, 150
logging, 129–131, 307
lumber yards, 150
secondary products, 175, 208
veneer, 174, 175
wood turning, 174
Wood-based plastics industry outlook, 465
Wood-chemical industry, 177, 203, 409, 410
Wood chemistry, 177
industry outlook, 465
Wood-destroying insects, 80, 150
(*See also* Forest insects)
Wood-distillation industry, 203
products, 202, 203
Wood extractives and derivatives, 199–202
Wood gas, 202
Wood processing, 168–175, 406–409
general administration, 173
plant layout and design, 173
production and sales management, 173
waste, 174, 175
(*See also* Wood, waste of)
Wood pulp, 181–187
Asplund, 187
groundwood, 182–184
manufacture, 181
mechanical, 182–184
production, 183
semichemical, 183, 186
soda, 183, 185
sulfate, 183, 185, 186, 201
sulfite, 183–185
whole wood fibre, 186, 187
Wood pyrolysis, 203
Wood tar, 202
Wood-using industries in New York, 279, 280, 406
Woodbury, Levi, 22
Wooden structures, design of, 161, 162
Woods, John B., 204*n.*
"Woody-plant Seed Manual," 436
Working conditions in forestry, 350–352
(*See also* Employment)
Works Progress Administration, 31, 369
World forestry organizations, 303–306

World forestry responsibilities of United States, 302, 303
World population, 287
Wyoming, 61, 62
  national forests area, 311

Y

Yale University, School of Forestry, accredited, 419
  establishment of, 412
  graduate study, 417, 418
Yellow birch occurrence, 5, 44
Yellow pines, uses, 196
Yellow-poplar, occurrence, 52
  use, 53, 196
Yellowstone National Park, 246
Yield tables, 117

Z

Zivnuska, John A., 267*n.*, 464*n.*
Zon, Raphael, 29, 258*n.*, 428